Biological Management and Conservation

Biological Management and Conservation

ECOLOGICAL THEORY, APPLICATION AND PLANNING

M. B. USHER

Lecturer in Biology,
University of York

LONDON
Chapman and Hall

First published 1973
by Chapman and Hall Ltd
11 New Fetter Lane, London EC4P 4EE

Filmset in Photon Times 12 on 13½ pt by
Richard Clay (The Chaucer Press), Ltd, Bungay, Suffolk
and printed in Great Britain by
Fletcher & Son, Ltd, Norwich

SBN 412 11330 9

Distributed in the U.S.A.
by Halsted Press, a Division
of John Wiley & Sons, Inc.
New York

To G.B.U.

AND

to the memory of G.P.P.

Contents

Preface *page* xi

Introduction 1
The development of man 1
Population and technology 2
The processes of change 6
Climatic development and changes 6
Environmental changes and the influence of man 10
Legislative action 13

PART ONE. ECOLOGICAL THEORY

1. ***The Distribution of Organisms in Space*** 19
Introduction 19
Large scale distribution 20
Biogeography 20
Genetical implications 31
The spread of alien species 33
The Oxford ragwort, *Senecio squalidus* 33
The New Zealand willow-herb, *Epilobium nerterioides* 34
The grey squirrel, *Sciurus carolinensis* 38
Small scale distribution 41
Detection of pattern 41
Analytic techniques 45
Ecological implications 48
Simulation of pattern analyses 53

2. ***The Distribution of Organisms in Time*** 60
The analysis of time series 60
Diurnal rhythms 67
Environmental modification of rhythms 67
Photoperiodic responses 70

Seasonal rhythms *page* 73
Long-term cycles 78
Ecological succession 83
Descriptive methods 83
Analytic methods 88
Random fluctuations 92

3. ***The Concept of the Ecosystem*** 93
Definition 93
The flow of energy 94
Input of energy 95
Movement of energy in the ecosystem 97
An example 101
The cycling of nutrients 109
Ecosystem modelling 116

4. ***Classification of Ecosystems*** 127
Introduction 127
European forestry classifications 128
The search for objective criteria 131
A subjective approach 132
A numerical approach 133
The search for ecological significance 142

5. ***The Response of Ecosystems to Exploitation*** 147
Introduction 147
Introduction to population mathematics 148
The responses of populations to harvesting: laboratory studies 158
The characteristics of balanced harvesting 169
The characteristics of over-exploitation 176
The characteristics of under-exploitation 183
The exploitation of plant populations 192

PART TWO. APPLICATION

6. ***Conservation and Biological Management*** *page* 201
Definition 201
Definitions and uses of the word 'conservation' 201
Application of conservation principles 205

7. ***Conservation and Preservation*** 213
Introduction 213
Research and management in the field 214
Conservation management for plants 214
Conservation management for animals 219
Supplementation of field populations 224
Planning the management of a whole environment 231
Research and management away from the field 244
Botanical gardens 244
Zoological gardens 246
Amateur successes 250
Monkey orchid in England 250
Ospreys in Scotland 254

8. ***Conservation and Education*** 256
Introduction 256
Adults and conservation education 258
Nature trails 258
Organisations 262
Schools and the teaching of conservation 263
The demand for facilities 265
Planning for school use 268
School projects 270
Conclusion 273

9. ***Conservation and Recreation*** 274
Introduction 274
The effects of recreation pressure on the ecosystem 276
Wildfowling 277
Trampling 279
Sand dunes 281

The assessment of demand *page* 284
The design of questionnaires 285
Economic analysis of survey data 289
Recreational planning 296
Studland Heath National Nature Reserve 297

PART THREE. PLANNING

10. ***The Management Plan*** 307
Introduction 307
A *pro forma* for a management plan 309
The Aberlady Bay Local Nature Reserve management plan 314

APPENDICES

Appendix I. Matrix Operations 365
Definition and names 365
Addition and subtraction 366
Multiplication 366
Inversion 367
Latent roots and vectors 367

Appendix II. Some Commonly Used Statistics 370

References 372

Index 386

Preface

Whilst I have been writing this book two developments have been occurring which have influenced ecological thinking, and which undoubtedly will have a great impact on ecologists in the future.

One of these developments concerns the relation between the ecologist and the public. On the public's side there has been an increasing awareness of ecological processes, and more emphasis on subjects such as the environment and pollution in newspapers and magazines. Maybe it was European Conservation Year 1970 (ECY 1970) that succeeded in stimulating this interest. On the ecologist's side there has been a search for the relevance of his research in the world of today. The concern for relevance has been clearly reflected in the 'Comments' that have been written for the first few parts of the British Ecological Society's members' bulletin. The word 'conservation' has been widely used in the context of this relation between the public and the ecologist; indeed it might well be said that the word has been over-used, being applied to any form of protectionist operation.

The second of the developments concerns the quantification of ecological processes. Statistical analysis of experimental data has been applied for several decades, but the recent general availability of computers has meant that mathematical analysis and computer modelling are tools that the ecologist can now use. Indeed, the very nature of ecology, dealing with complex relations between organisms, between organisms and the abiotic environment, and the quantity of data from research projects means that computer analysis is becoming essential. The papers and discussions of the British Ecological Society's symposium on mathematical models in ecology, held early in 1971, showed the range of techniques now available to the ecologist, and also showed that new mathematical theory may be needed to analyse ecological situations (Jeffers, 1972).

In writing this book I have endeavoured to retain elements of both of these developments in ecological thinking. The title *Biological Man-*

agement and Conservation was chosen since it reflects the essence of the relation between the public and the ecologist whilst at the same time underlining the management aspect of any conservation operation. The treatment of the subject has been kept relatively analytic and mathematical since I am sure that management for biological conservation must be based on accurate data and on good predictions.

The subject of conservation is indeed broad, including as it does the application of ecological principles together with the legal, economic and social responsibilities and constraints. In looking at this book one of my own greatest criticisms is that a single book of limited size can hardly do justice to such a broad subject. I have therefore concentrated upon wildlife and conservation in relation to nature reserves, largely excluding from discussion the relations between conservation and forestry, agriculture, fisheries and water supply. However, despite the fact that the book does not cover the whole subject, it is my hope that the breadth of biological conservation has been preserved, whilst at the same time looking at a selection of aspects in depth. I have endeavoured to choose examples, mainly drawn from conservation activities in Britain, that relate to the analytic or mathematical approach to the subject. In many cases I have preferred to use little known examples (ones that have not been used frequently in text-books), though at times the better known examples have been used when they illustrate a general point very well.

It is impossible to write such a book without the help and support of one's family, and also from people working on various models and conservation problems. I am very grateful for the support that I have received. I should particularly like to thank Mr. D. H. Adams and Professor M. H. Williamson, both of whom have read through drafts of most of the chapters, making many suggestions for improving the clarity of the text and for remedying mistakes. The errors that remain are entirely my own. I should also like to thank Mrs. Ann Fisher who has drawn many of the illustrations, and Mrs. Pat Hatfield and Miss Margaret Eslick who undertook the final typing. I should like to express my gratitude to Dr. Eric Duffey and The Nature Conservancy, The Forestry Commission and Mr. A. D. Vizoso for permission to include their photographs (Plates 6, 3 and 4 respectively). The other plates are from my own photographs, and I am very grateful to Mr. R. Hunter for advice and for preparing the prints.

For permission to make quotations I should like to thank Professor

J. N. Black and The University of Edinburgh Press; Mr. S. T. Broad and The British Association for The Advancement of Science; The Clarendon Press, Oxford; Professor J. B. Cragg; Dr. J. L. Fisher; The Longman Group Limited; Professor A. Macfadyen; Professor R. Margalef and The University of Chicago Press; The Nature Conservancy; The Nuffield Foundation and UNESCO.

I should like to thank successive classes of students reading biology in York University, who, in receiving lectures on applied ecology and in discussing conservation in tutorials, have helped me to clarify my own ideas on the subject. Many people have made suggestions or have allowed me to quote their own work. For this help I should like to thank: Mr. R. M. Bere; Dr. R. G. H. Bunce; The Countryside Commission; Miss M. E. Dennis; The East Lothian County Council (for permission to include the majority of the Aberlady Bay Local Nature Reserve management plan in Chapter 10); Miss Noelle Hamilton (who collated the data and prepared preliminary maps on which Figs 1.9 and 1.10 are based); Mr. J. R. W. Harris; Mr. D. H. Kent; Dr. M. C. Lewis; Mr. B. C. Longstaff; Miss M. McCallum Webster; Mr P. J. Murphy; The Nature Conservancy (for permission to include a section on the Studland Heath National Nature Reserve in Chapter 9); The Office of Ecology, United States Department of the Interior; Mrs. J. Paton; Dr. F. H. Perring (for allowing access to data in The Biological Records Centre); Miss C. M. Rob; Dr. A. M. Tittensor; Mrs. M. Vizoso; Mr. H. Wilks; and The Yorkshire Naturalists' Trust (on whose nature reserves I have been able to try out some of my ideas of conservation management).

Dunnington, York M. B. U.
July 1971

Introduction

The development of man

Biological conservation and management are concerned with the relations between one species, Man, and the remaining million or so species of the biosphere.

Before the study of conservation can progress it is important to see why man has come to have such a dominant position in the biosphere. In reviewing the evolution of the present situation one can begin some 900,000 years ago at a time when the hominid development was well established, and represented by the genus *Australopithecus*, animals that had ape-like qualities but many man-like features. The species *A.robustus* and *A.boisei* had large molars and premolars and small canines and incisors, and were also certainly herbivorous. The third species of the genus, *A.africanus*, had smaller cheek teeth and larger front teeth and would have been an omnivore or carnivore.

Uncertainty as to generic placing surrounds the remains excavated at Olduvai but these can provisionally be referred to as belonging to *Homo habilis*. It is, however, certain both that *H.habilis* was man-like in posture and that he used stone tools. There is no site where *Australopithecus* remains alone have been found that has yielded stone implements, but every site in which *H.habilis* remains have been found has contained implements. *H.habilis* lived about 900,000 years ago, and it is important to realize that it is at about this time that implements started to be fashioned and used.

About 500,000 years ago there was a species, or group of species, referred to as *Homo erectus*. The remains show a species with less facial prognathism than *Australopithecus*, a relatively smaller face, and a developed chin. It is likely that these men were able to flake boulders in two directions to form simple tools for cutting, chopping and scraping. There are more fossil remains dating from between 500,000 and 250,000 years ago, but it is beyond the scope of a brief review to document the various types. However, by 250,000 years ago, hand axes and effective cutting tools are found amongst the stone implements. After a further 200,000 years the introduction of tools of bone, horn and ivory can be determined from the fossil records, and it is evident that the species was capable of precision flaking. From about this time onwards the species of man can be referred to as *Homo sapiens*.

Modern man, in the form that we know today, first appeared about 10,000 years ago. He had developed from the *H.erectus* stock, with three major directions of anatomical change. There had been an enlargement of the cranial cavity with a subsequent change in the formation of the head. There had been the development of bipedalism with the associated skeletal modification, and there had been the increased manipulative ability of the hands, resulting in the ability to construct and use far more complicated tools than those of *H.erectus*.

Fuller accounts of the ancestral development of man are given in anthropological textbooks, but an account which particularly concentrates on the relation of man with his environment is given by Arthur (1969).

The production of tools has enabled man to have a greater impact upon his environment. From early in the development of *H.sapiens* man has also been able to use fire. These developments of fire and tools were used at a slow pace, but in the last thousand years or so the impact slowly became more widespread and extended over larger areas, as, for example, the Viking invasion of Scotland which resulted in extensive areas being destroyed by fire. These early forms of modern man lived within the 'natural' environment, but their population size was sufficiently small and their technology so poorly developed that their waste products were dealt with adequately by the decomposer organisms. However, the ability of man to live in harmony with the natural environment is very closely related to the development of a larger population size and a more advanced technology.

Population and technology

During the last 200 years or so two parallel developments, those of technological advance and of human population, have resulted in man's greater impact on his environment. Looking firstly at human populations we can see the phenomenal increase in population size particularly during the last century. Fig. I.1 shows the population size given by Arthur (1969) (data before 1900 are based on estimates). A criterion for judging population increase is the length of time taken for an additional 1000 million to be accumulated (Arvill, 1967). From the data in Table I.1 it

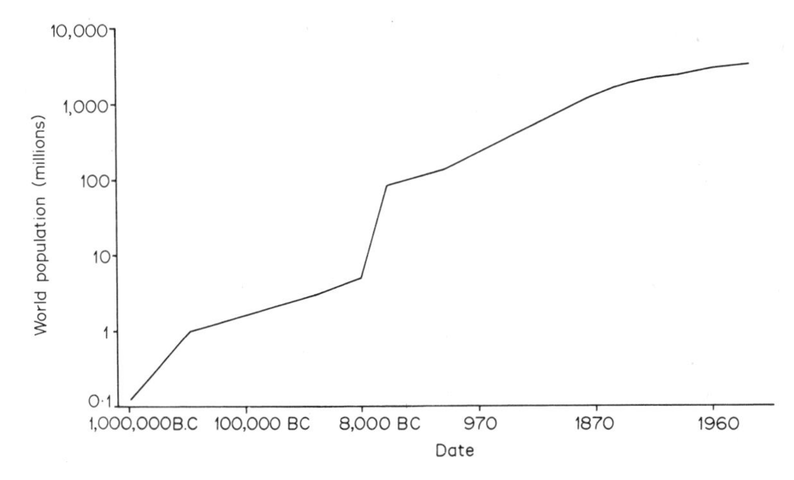

Fig. I.1 The population growth of man. (Note that both axes are logarithmic). (Drawn from data given by Arthur, 1969)

can be seen that the length of time required to increase the world population by 1000 million is decreasing, but this alone would be expected if the rate of increase were constant (see Chapter 5). Thus, a better criterion is to measure the length of time required for the world population to double (Arthur, 1969). These data are shown in Table I.2. It is particularly

TABLE I.1 The growth of the world's population (based on Arvill, 1967)

Population size in millions	*Approximate year*	*Years taken to increase the population by 1000 million*
1000	1830	900,000
2000	1930	100
3000	1960	30
4000	1975 (estimated)	15

clear from this table that the rate of increase is not constant, but is in fact increasing. Some estimates suggest that the world's population will reach 8000 million soon after the turn of the century, which implies that the doubling of the population from 4000 million will occur in 25–30 years. It will be seen in Chapter 5 that this correlation between population size and rate of increase is an unusual feature of animal populations.

TABLE I.2 The growth of the world's population (based on Arthur, 1969)

Population size in millions	*Approximate year*	*Years taken to double the population size*
500	1650	–
1000	1830	180
2000	1930	100
4000	1975 (estimated)	45

The figures given above refer to estimates of the total world population, but it is of greater importance to the conservationist to study the local, regional or national population changes. Standard texts on demography break down the world population into continental or regional components (see, for example, the discussion by Arthur, 1969). But before we can leave the question of population increase we must examine the causes. There are, on a world scale, only three factors which can influence the population size, namely births, deaths and age of commencement of breeding activity in the females of the population. In any unit smaller than the world, population movements in the form of emigration and immigration will also need to be considered.

In Great Britain the birth-rates per year, over a period of 250 years are given in Table I.3. Since about 1870, when there was a birth-rate of approximately 35 per 1000, the rate steadily decreased to about 15 per 1000 in the late 1930s. Immediately post-war the rate increased but it has not exceeded 20 per 1000 since the war. A similar pattern of birth-rate

TABLE I.3 Birth- and death-rates in Great Britain. The figures are approximate means for 50 year periods, and are expressed as birth or death per 1000 of the population per year

			DATES		
	1700–1750	*1750–1800*	*1800–1850*	*1850–1900*	*1900–1950*
Birth-rate	33	37	36	31	20
Death-rate	31	29	23	20	14

can be seen in other Western European countries and in the United States of America. In the developing countries of Asia and Africa the birth-rate remains more or less uniformly at a high level as indeed it did in Britain over the period from 1750 to about 1870. The birth-rate is not changing appreciably, although its global average is probably decreasing slightly. We must therefore look to other factors for the increasing rate of population increase that is demonstrated in Table I.2.

The death-rate in Britain is given in Table I.3, where it can be seen that during the first half of the eighteenth century the birth- and death-rates were more or less in balance. From about 1740 the death-rate has irregularly decreased from about 34 per 1000 to about 11 to 12 per 1000 in the middle of the twentieth century. Reference to the Table clearly shows that during the nineteenth century there was an excess of births over deaths averaging about 12 per 1000 per year. In tropical areas the immediate post-war application of insecticides to control the Anopheline mosquitoes has drastically reduced the impact of malaria, previously the most important mortality factor in the tropics. In India, for example, before the war malaria accounted for about half of the deaths.

Thus the main factor accelerating the rate of population increase has been the dramatic decrease in the death-rate, which has been accompanied by only a slowly falling or stable birth-rate. The death-rate has been reduced by the scientific and technological advances, the second of the main areas of impact of man on his environment, that has been mentioned previously.

The technological impact manifests itself not only in such medical and scientific applications of reducing death-rate and improving nutrition, but also in the raw materials extracted and in the construction that is required for the manufacture and implementation of the technological products. In order to examine these latter impacts let us use, as an example, the motor car (see Table I.4). Here it can be seen that the number of private cars is increasing rapidly (four-fold over the period 1950–1970), and similarly that the rate of car ownership is increasing. These figures do not, however, show the ancillary requirements of raw materials for building the vehicles, of oil for propelling them, of open spaces covered with roads, or of the waste products of internal combustion engines. Technological advance, then, can lead to increased demands on the environment for producing raw materials, for space in the form of roads, runways, factories, etc., and for absorbing the waste products, popularly termed 'pollution'. We must,

TABLE I.4 Private cars in Britain, based on data of the Countryside Commission (1969a)

	YEARS				
	1950	*1960*	*1968*	*1980*	*2000*
Private cars in Britain	2,257,873	5,525,828	10,816,100	20,100,000	27,800,000
Private cars per 1000 persons in England and Wales	87	112	197	350*	420*

* These figures are the author's estimates from other sources.

however, investigate how the other components of the biosphere are developing and how they are affected by man's population and technology.

The processes of change

We have just considered the Man end of the 'Man–Biosphere' relation, analysing the twin developments of man in achieving a large population size and in developing an advanced technology. We are, however, here concerned with a review of the factors that can affect the population size, extent or survival of the other species of the biosphere, distinguished from man by their inability or relative inability to control their environment. These factors can, broadly, be separated into four categories, though the boundaries between categories are seldom well-defined. These categories are:

(1) changes in the climate;
(2) ecological changes in the environment;
(3) changes resulting from the direct influence of man; and
(4) unexplained or random changes.

It is impossible by its very nature to discuss the fourth of these categories, though some of the random events that could be considered are freak storms, abnormal tidal surges or volcanic activity. The first three of these categories will be discussed in more detail.

Climatic development and changes

Climate can be considered on various time scales. To the conservationist two of these are important. The climatic changes since the last glaciation, which was coming to a close about 20,000 years ago, are important in understanding the development of the present plant and animal communities. On the shorter time scale the conservationist might wish to know how the climate is changing at the present time.

Investigations of climatic and vegetational changes since the last ice-age are often based on the analysis of pollen in various deposits such as

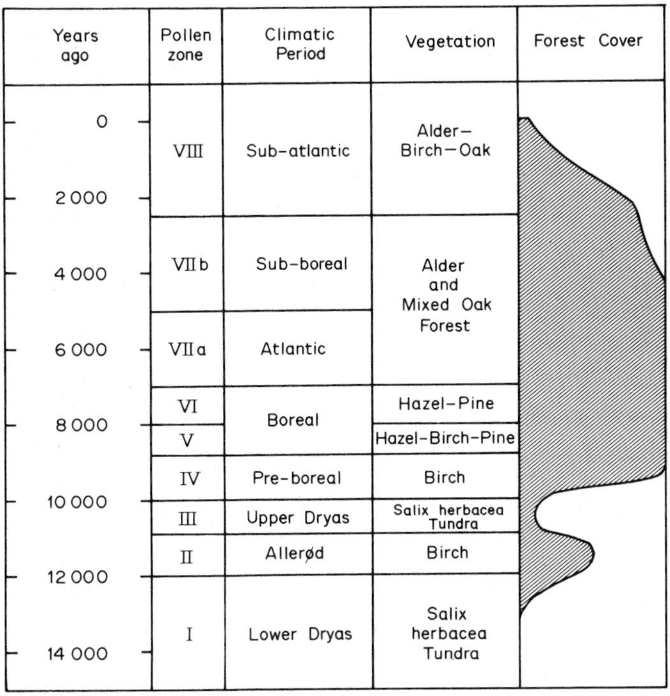

Years ago	Pollen zone	Climatic Period	Vegetation	Forest Cover
0 2 000	VIII	Sub-atlantic	Alder–Birch–Oak	
4 000	VII b	Sub-boreal	Alder and Mixed Oak Forest	
6 000	VII a	Atlantic		
8 000	VI	Boreal	Hazel–Pine	
	V		Hazel-Birch-Pine	
	IV	Pre-boreal	Birch	
10 000	III	Upper Dryas	Salix herbacea Tundra	
	II	Allerød	Birch	
12 000 14 000	I	Lower Dryas	Salix herbacea Tundra	

Fig. I.2 Climatological periods and pollen zones since the last glaciation. (Redrawn from Godwin, 1956)

bogs, mosses and lacustrine sediments (see, for example, Godwin, 1956). A schematic representation of the changes since the last ice-age is shown in Fig. I.2, where the first three pollen zones are referred to as late-glacial, and the subsequent zones as post-glacial. The late-glacial period, ending about 10,000 years ago, can be divided into three pollen zones which correlate closely with climate. During pollen zone I the climate gradually warmed, and the pollen record shows that the vegetation was of a tundra type. The zone has been named after the species *Dryas octopetala* which is frequently represented in the pollen record. During the Allerød period, zone II, the climate continued to warm and then to cool. This zone is characterized by a recession of non-tree pollen, by the frequency of pine pollen and by the absence of dwarf birch, *Betula nana*, pollen. The climate then went through a cold phase, zone III, when there was a rise in non-tree pollen (particularly Gramineae) and birch pollen became more frequent than pine pollen. Also the pollen of *B.nana* and sea buckthorn, *Hippophaë rhamnoides*, returned and willow pollen became more frequent.

After the late-glacial period the climate gradually warmed, and zone IV is characterized by an increased proportion of tree pollen. The division of the Pre-Boreal and Boreal times is characterized by the rapid increase in hazel pollen. During zone V, in southern Britain pine pollen is more frequent than birch pollen, but the reverse is true in northern Britain. The second zone VI of Boreal time is characterized by the preponderance of pine over birch pollen, and by the presence and increase of pollen of the mixed-oak-forest trees, especially oak and elm. Hazel pollen continues to be abundant. During the subsequent Atlantic time a climatic optimum was reached, and a characteristic of this zone VIIa is the rapid increase in the amount of alder pollen, and the simultaneous decrease in pine pollen. With the passing of the climatic optimum Sub-Boreal time is entered. The structure of the vegetation remains similar to that of the Atlantic time, but one characteristic of the change is the decrease in elm pollen. Towards the end of zone VIIb the neolithic and bronze-age cultures in Britain started clearing the forest.

About 500 B.C. there is evidence of increased oceanicity in the climate. This period, Sub-Atlantic or zone VIII, is characterized by an increase in birch pollen and a decrease in pollen of some of the species associated with mixed-oak-forests. However, the clearing of forests by man continued at a greater pace, and hence pollen of Gramineae, Cyperaceae and Ericaceae becomes more abundant. The pollen of plants associated with disturbed or cultivated ground also increases, as for example the *Plantago* species. These changes in the pollen record are illustrated in Fig. I.3 for Tarn Moss at Malham in Yorkshire. The figure only illustrates pollen of the characteristic species previously mentioned, but the full pollen diagrams are given by Pigott and Pigott (1959).

Quaternary ecology is concerned with these long-term climatic changes since the last glaciation. However it may be questioned whether the climate has remained stable throughout the last 2500 years of the Sub-

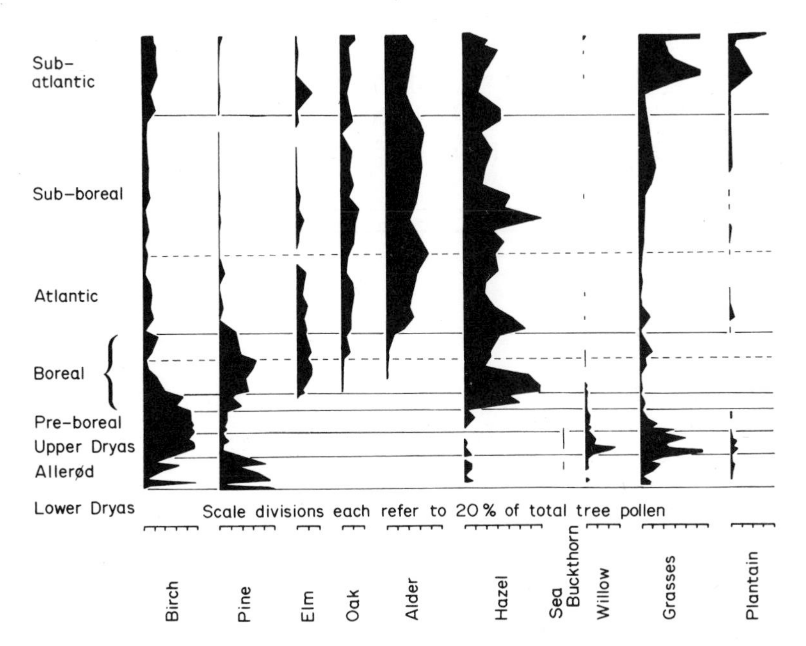

Fig. I.3 The pollen of selected species from Tarn Moss, Malham, Yorkshire. The pollen zones are indicated. The scale divisions refer to the pollen of a species as a percentage of the total tree pollen. (Redrawn from Pigott and Pigott, 1959)

Atlantic period. A paper by Lamb (1969) analyses recent trends in many component features of the British climate, but we will examine just two of these components, namely temperature and wind. The temperature data are shown in Fig. I.4, where it will be seen that it was particularly cold in the 1690s, warm in the 1730s, and warm for a longer period in the first half of the twentieth century. During the middle of the twentieth century the average temperature has been falling. Although the amount fallen is only of the order of 0·5°C, Lamb shows that this has reduced the average length of the growing season by about two weeks and it has doubled the frequency of snow lying in inland districts from 5–7 to 10–15 days. A historical analysis leads Lamb to postulate a temperature cycle of approximately 200 years duration, with the coldest recorded temperatures in the seventeenth century.

The frequency of westerly weather situations over Britain is also shown in Fig. I.4. The westerly winds, coming from the Atlantic Ocean, maintain a prevalence of mild weather. The recent change has meant that there are more winds from, particularly, the northern points, and that there is a tendency towards slow-moving weather systems. Using both direct and indirect evidence Lamb shows that there was a great predominance of south-westerly winds in the early 1300s, 1500s, 1700s and 1900s, with sharp falls in these winds during the 1560s and 1760s similar to that

Fig. I.4 Changes in the temperature and westerly winds in Britain. The lower three graphs refer to the summer, mean and winter temperatures, each expressed as an average for the decade. The upper graph shows the average number of days per annum with westerly winds for periods of five years. (Redrawn from Lamb, 1969)

occurring during the 1960s. Lamb notes that both the westerly weather conditions and the average temperature have a constant phase relationship with the ^{14}C variation in the earth's atmosphere.

Environmental changes and the influence of man

Looking now at the second and third groups of factors causing population change it can be seen that environmental changes might be closely linked with these short-term climatic changes, or result from indirect action of man, or be the result of ecological changes elsewhere in the biosphere. Thus a pathogen acting to reduce drastically the population of herbivores will influence many other species since the vegetation may become taller and predators will either be faced with starvation or with changing their attention to other possible prey species. Studies such as that of Lockie (1966) in the Carron Valley in Perthshire have demonstrated the complex relationships existing between change in land use (afforestation), the population density of small mammals and the response and territorality of the predators (stoats and weasels). The size of the territory defended by males of the predators is not determined solely by food availability, i.e. by the population density of the small mammals, but also by the number of predators competing to set up territories. Since the latter factor depends upon the previous breeding success of the predator species, indirectly a function of the food available at that time, it can be seen that both the past and the present density of the small mammal population influence the present response of the predator.

The occurrence of the large blue butterfly, *Maculinea arion*, in Britain demonstrates both the direct effects of man and the effects of changes in the environment. This species is perhaps particularly susceptible to environmental changes since three sorts of environment are required for its development. Frohawk (1924) described the life history and shows that the flowers of the wild thyme, *Thymus drucei*, are the only known food plant of the young larvae. Larvae of the first three instars, lasting on average 20 days, feed on the flowers, which they resemble in both colour and pubescence. Thus, the first essential feature of the environment is the occurrence of the wild thyme, a species of the closely grazed sward. The effects of myxomatosis, eliminating or drastically reducing the rabbit population, have resulted in a reduction in the number of localities in which the large blue occurs due to the taller growth of the vegetation and the elimination of wild thyme.

After the third moult the larva gains access to an ant's nest (generally *Myrmica scabrinoides* or *M.laevonoides*) (Ford, 1945) where it becomes a predator on the young ant larvae. It feeds thus for about six weeks in the autumn, hibernates in the ant's nest, and feeds again during the following spring. The larva pupates in the ant's nest, and the butterfly, on emerging from the pupa-case, makes its way above ground and expands its wings on any suitable stem. Thus a second component of the large blue's environment is the ant, and it is thought that the two species of ant previously mentioned are the only suitable hosts. Ford (1945) mentions that unusually small specimens of the large blue, seen in most localities, may result from development in nests of ants other than *Myrmica* species. The exact conditions which favour *Myrmica* ants in the large blue localities are unknown. Frohawk (1924) describes and illustrates the behaviour of the large blue larva when it is first approached by an ant, and it is possible that there are behavioural mechanisms that confine the predation to ants of the genus *Myrmica*.

After the adult butterfly has expanded its wings, a third type of habitat is required. Both Ford and Frohawk comment on the short flight period, generally in the morning and just before the evening. The Cornish Naturalists' Trust has found that the species only flies in the sunshine and that imagos do not fly far from the site where they emerged. Marked specimens have flown just under a mile, so the species is unable to travel far to colonize other suitable areas. Frohawk states that the imagos feed on the nectar of the wild thyme, and he describes the behaviour of butterflies seeking shelter in gorse bushes either when disturbed or during cold and windy weather. The butterfly has a preference for spending the night on spikes of rushes. The Cornish Naturalists' Trust has found that a mosaic of these three vegetation types – a close grazed grass sward with the wild thyme, shelter with stands of gorse or heather bushes and small wet or marshy areas – is essential to the well-being of a colony of the large blue.

Thus the species demonstrates clearly its dependence upon its environment, and it can be seen that changing patterns of land-use can easily change any of the habitats upon which the species depends. But environmental change has not been the only factor contributing to the population decline. Frohawk gives a short history of the collecting of this species, and it is certain that the collection of large numbers of specimens has resulted in the elimination of this species from many of its former inland haunts. The species, in Britain, is now more or less confined to coastal habitats in the South West of England, and even here the populations have been reduced by over-collecting. Populations, as discussed in Chapter 5, can withstand a degree of exploitation, but it is likely that the demand of collectors has resulted in over-exploitation of this species. Conservation efforts have initially concentrated on reducing the rate of exploitation by the wardening of known localities and by secrecy to prevent too many collectors arriving at little-known localities. However, emphasis is also placed upon the provision of the correct environment. Periodic burning of gorse and scrub prevents the development of these plants to the exclusion of all others. Intensive sheep grazing has been attempted, as also have cutting and clearing, so as to encourage the development of short grassland with wild thyme.

Thus, the conservation of the large blue demonstrates two important principles. Firstly, the level of exploitation of the population, in the form of adults for collections, has to be reduced to, or below, a level that can be sustained by the population. The fact that only 85 imagos were seen in all localities in 1963, and 170 in 1964 (Stamp, 1969), suggests that the number of specimens that could be taken each year without impairing the success of the species in Britain must be very small. Secondly, the life history of the species must be well documented so that the environment correct to each stage can be managed. The fact that the large blue has the most complicated life history of any British butterfly underlines the dependence of this one species on so many other species that occur in the same ecosystem.

Legislative action

Action on the balance between man and the biosphere can come from many directions. We have seen how man's impact comes both from his population size and from his technology, and action could control either or both of these factors. In the long term they are the only meaningful balancing mechanisms, but controls here will require a much deeper knowledge of our society and a much greater willingness for society to accept controls. The relation can hardly be balanced from the biosphere end, unless there are pathogens whose existence is unsuspected and whose effect is both lethal and quick. During the latter years of the last century and during this century man has attempted to safeguard some species and some tracts of countryside, not by the drastic action of controlling himself, but by the smaller step in this direction of limited legislative action. In Britain at the present time there are many societies whose aim is either environmental conservation or preservation of some group of wild life. Membership of these societies is increasing exponentially, forming pressure groups, which I believe reflects the increasingly popular interest in the 'Man–Biosphere' relation.

During the eighteenth and nineteenth centuries the emphasis was on discovery and collection. Raven and Walters (1956) recount the collecting visit to Snowdon, North Wales, by Thomas Johnson in 1639, when in one day he found a long list of mountain flowers. The botanist today would be unlikely to see so much in so short a time, since Snowdon, being a very accessible mountain, has suffered from collecting. It is very difficult to prove that a species has decreased in abundance or become extinct in a habitat as a direct result of over-collecting, but alpine flora has suffered particularly since collectors both for herbaria and for gardens have removed the plants. Evidence of the number of people searching for alpine flora can be gathered from any of the well-known botanical sites on the Scottish mountains during the summer, when paths are made and picked flowers can be found occasionally beside the paths. This urge to collect also affects many animal species. Brown and Waterston (1962), discussing the history of the osprey in Scotland, document the destruction of the species by collectors. They recount the collecting expeditions into Sutherland during 1847–50 when young and eggs were removed from all nests found and the adults were shot whenever possible. In Inverness-shire from 1846 onwards the fate of 24 nests is recorded. On 13 occasions the eggs were taken by collectors, and in only two instances did the young reach the flying stage. During this period the demands of collectors for specimens of wildlife was tremendous.

An awareness of the need for some control became apparent. In the United States of America the Yellowstone National Park was established in 1872. In Britain the Society for the Protection of Birds was formed in 1889 and the National Trust in 1895. British legislative action centred initially on game and then birds. Many of the rarer species were given a measure of protection by an Act of Parliament during the first half of the twentieth century, but the problem of enforcement was that a protected species could not be identified until it had been shot. The Protection of Birds Act 1954 changed the emphasis since it gave protection to all birds, and their eggs, making as exceptions a relatively short list of common birds which were regarded as pests or harmful. This Act still remains, in Britain, the most comprehensive legislation for any group of wildlife.

Other developments were occurring within the legislative field. The emphasis for protection became less evident as the human population became more mobile and more aware of the countryside. The National Parks and Access to the Countryside Act 1949 pioneered several important developments. Firstly, the Nature Conservancy was created, charged with the provision of scientific advice on conservation, with the establishment and management of nature reserves, and with the organization and development of scientific services in this field. The Act also created the National Parks Commission, which was charged with designating National Parks, areas of countryside where there was strict control of development. At about the same time another Act allowed for the designation of important scientific areas as Sites of Special Scientific Interest (S.S.S.I.). These sites are given a degree of statutory protection by being scheduled by the Local Planning Authority. The provision of long-distance footpaths, of country parks with facilities for open-air recreation and of picnic sites is made by two recent Acts, the Countryside (Scotland) Act 1967, and the Countryside Act 1968. Under these Acts the Countryside Commissions (one for Scotland and one for England and Wales) were formed, and Local Authorities and the Forestry Commission were given wider powers for providing access to the amenities in the countryside. The development of the conservation movement in Britain is reviewed by Stamp (1969).

The growing awareness of conservation of wildlife can also be seen on the African Continent, where the Organization of African Unity (O.A.U.) has drawn up 'The African Convention for the Conservation of Nature and Natural Resources'. Burhenne (1970) discusses this Convention and the development of tropical conservation policy since the London Convention of 1933. The African Convention sets out a policy of recognizing the increasing importance of the soil, water, floral and faunal resources from economic, nutritional, scientific, educational, cultural and aesthetic points of view, and of accepting that management of these resources should aim at preserving these irreplaceable assets whilst also satisfying the needs of mankind. The Convention contains two schedules of protected species, one for species given total protection throughout the Continent, and the other listing species that should be given a measure of protection. Using the primates as an example we see that the totally protected species include all the Malagasy lemuroids, Barbary ape, Gelada baboon, Tana River mangabey, Diana monkey, Zanzibar red colobus, Tana River red colobus, Uhehe red colobus, green colobus, chimpanzee, pygmy chimpanzee and gorilla. Those on the second schedule include all prosimians of the family Lorisidae and all monkeys except the common baboon. Thus nearly all the primates of the African Continent would be protected by the Convention. Perhaps the most important feature, however, of the Convention is that it provides for the international planning of a conservation effort.

In the United States of America the concern for the environment during the past decade has perhaps developed further than elsewhere in the world. Arvill (1967) reproduces an extract of President Johnson's message to Congress of February 1966 when he put forward a creed to preserve the natural heritage. President Johnson affirmed the rights of the population to clean water, to clean air, to surroundings free from man-made ugliness, to recreational use of the countryside, and to enjoy wildlife in its natural habitat. However, with each right he also mentioned the duty of the population not to pollute water, land or air, not to foul the countryside, and not to remove wildlife from its habitat. This message to Congress will, undoubtedly, be looked upon as an important creed for conservation, but future generations may seek for its implementation. President Nixon announced a legislative programme in February 1970. In this he puts forward 23 major legislative proposals and 14 new measures within the areas of water pollution control, air pollution control, solid waste management, parklands and public recreation, and organizing for action. These proposals add to President Johnson's creed, showing the way in which government can attempt to legislate to maintain the balance between man and the biosphere.

Action will be seen to involve three stages. First, one must understand the processes that are at work: the subject of 'pure' ecology. Secondly, a plan for management of the resource must be formulated: the subject of 'applied' ecology. Thirdly, and only following the first two phases, a campaign can be undertaken either by the implementation of a management plan or by the preparation of legislative proposals. These three stages – investigation, formulation and implementation – set the pattern for the remainder of the book.

PART ONE
Ecological Theory

1. The Distribution of Organisms in Space

Introduction

For a long time ecologists have studied the spatial distribution of species. On a very large scale these studies have been biogeographical, often concerned with assessing evolutionary significance or adaptive radiation. On a smaller scale there has been the collection of 'faunistic' data, which are essentially lists of species occurring within some predetermined habitat. The biogeographical studies are important in understanding some of the ecological influences acting upon populations, and must form a part of conservation management. Management techniques will, for example, be different for a species on the edge of its geographical range and for the same species clearly within its range. The geographical study of distribution has been the subject of some theoretical treatment, especially in relation to island biogeography, and this will be discussed later in this chapter.

On the smaller scale the study of the local distribution is necessary for an understanding of the regulation of population numbers. This small-scale distribution will be referred to as pattern, which is not a static feature of a population, but essentially dynamic. Although the pattern is measured at a moment in time, if it is repeatedly measured it will be found to vary, demonstrating the existence of ecological progression or of cyclic processes. A branch of mathematical ecology has been concerned with the detection and quantification of pattern, the analysis depending on sophisticated statistical techniques. The changes in the geographical distribution of species and the establishment, detection and quantification of the patterns of distribution will be discussed in this chapter.

Field data can therefore be used for the detection of large and small-scale distribution of species, and as will be emphasized later in this chapter this detection can be based on mathematical and statistical analyses. However, these analyses do not always yield results that are immediately obvious. Sometimes it is difficult to predict how an analysis will behave when applied to unusual field conditions. It is often useful to create artificially simple cases in which there is a known pattern and to experiment with the analysis. This process can be repeated by building more and more complex patterns into the artificial data until the results of the analysis resemble the results of the analysis of field data. These methods are known as simulation, and provide the ecologist with a powerful analytic method, although care must be exercised in equating the simulated pattern with the field pattern. Questions that are difficult to answer theoretically may be answered approximately by repeated simulation of a postulated situation. What, for example, is the effect of quadrat size upon the parameters of a population estimated by sampling, or how do the results of an analysis depend upon the starting position in a pattern? With the availability of computers to speed the process, such questions can be answered by simulation methods. In the final section of this chapter such a computer simulation of pattern analysis will be investigated, demonstrating the practical application of simple simulation methods.

Large scale distribution

Biogeography

Biogeography, as a science, has been concerned with the distribution of plant and animal species over the earth's surface, and with investigating the evolutionary significance of the distributional ranges of species, of genera, of families and so forth. In the study of conservation perhaps the most important aspects of biogeography are in relation to the theory of islands, and closely linked with such theories we must also consider the genetics of the populations themselves. This 'island' nature of populations may not have been of importance in times when natural or semi-natural communities formed a continuum, but with agricultural reforms areas of monoculture (crops of one species) have been created, leaving the wildlife to inhabit small pockets of land, the 'islands'. It is within these islands that the conservationist will be concerned with such ecological processes as invasion and extinction, dispersal and competition, and the genetical processes of gene flow between isolated units and the maintenance of heterozygosity. The work of Curtis (1956) clearly shows that the fragmentation of natural and semi-natural communities results in the formation of ecological 'islands' (Fig. 1.1). He shows that a continuous woodland area in the Cadiz Township of Wisconsin of 21,548 acres in 1831 was reduced to 786 acres in 55 blocks by 1950. The destruction of woodland resulted in a reduction of 35 per cent in the permanently flowing streams by 1935, reflecting a decrease in the subsoil water storage. Similar work in Britain (Moore *et al*, 1967) shows that the mechanization of farming operations

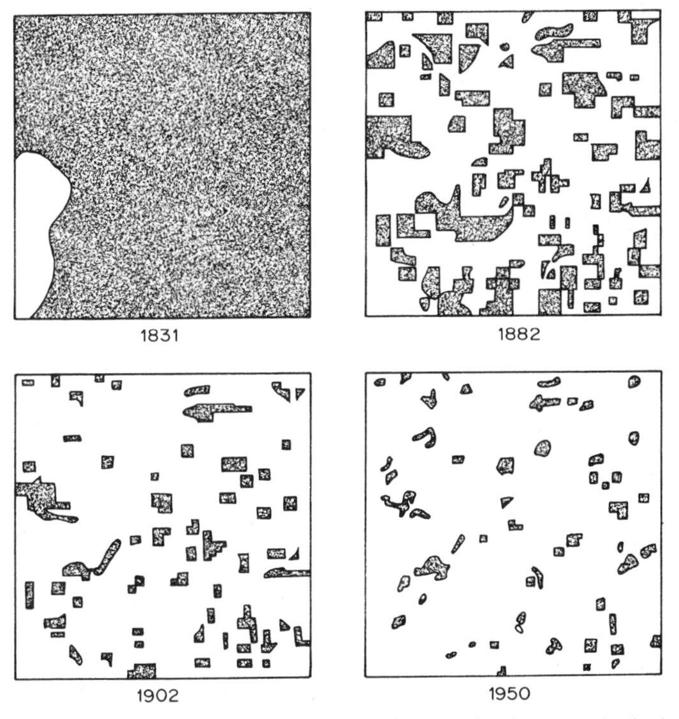

Fig. 1.1 The fragmentation of woodland in Wisconsin into ecological 'islands'. (Redrawn from Curtis, 1956)

has led to the loss of hedgerows and the formation of 'islands' of this type of habitat (Fig. 1.2). The destruction of these hedges has been particularly fast during the last decade as is shown by the graph of hedgerow length in Fig. 1.2.

The theory of island biogeography has been developed by MacArthur and Wilson (1967), where they show that the most striking of all relations is that between the area of an island and the number of species. Such 'species–area' curves (Fig. 1.3) show that there is a linear relation between the logarithm of the number of species of a particular taxon on an island and the logarithm of the island's area, which can be expressed by the equation

$$S = CA^z$$

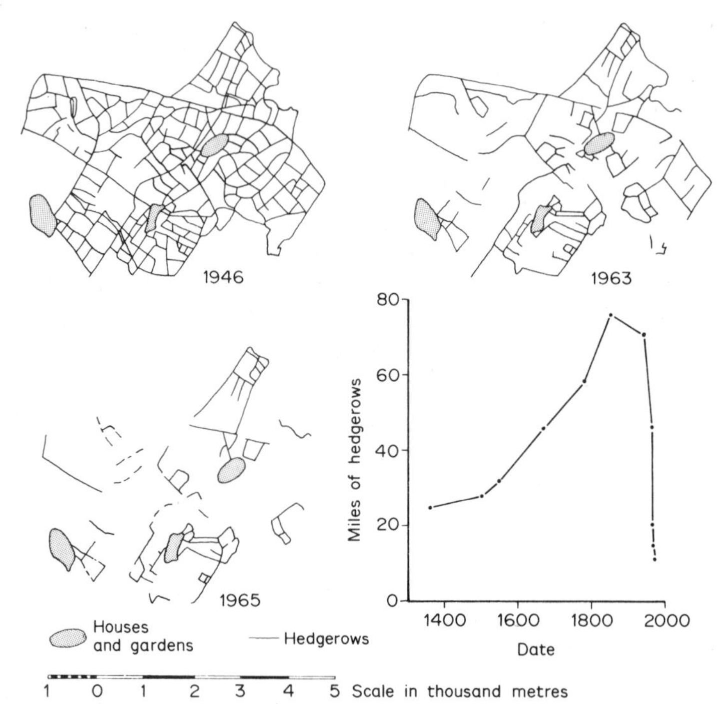

Fig. 1.2 The fragmentation of a hedgerow mosaic in Huntingdonshire during a 20-year period. The graph shows the estimated mileage of hedgerows on 4000 ac in Huntingdonshire during the period 1364–1965. (Redrawn from Moore, Hooper and Davis, 1967)

where S is the number of species of the taxon found on the island, A is the area of the island, and C and z are parameters. C depends on the taxon and biogeographic region, and z appears to be remarkably constant, varying between 0·20 and 0·35 in the majority of studies. Such 'species–area' curves are also characteristic of nature reserves (Fig. 1.4), since they show the relation between species diversity and the size of the reserve. Using the flora occurring in the 12 nature reserves in Yorkshire for which data are available the 'species–area' curve is described by the equation

$$S = 69{\cdot}02A^{0.2072}$$

Although a linear relation is approximately followed the correlation is not so marked as in the case of islands since nature reserves have been selected in areas of biological diversity. Nevertheless, an important first result in biogeographic theory can also be applied, at least approximately, in the occurrence of species on nature reserves.

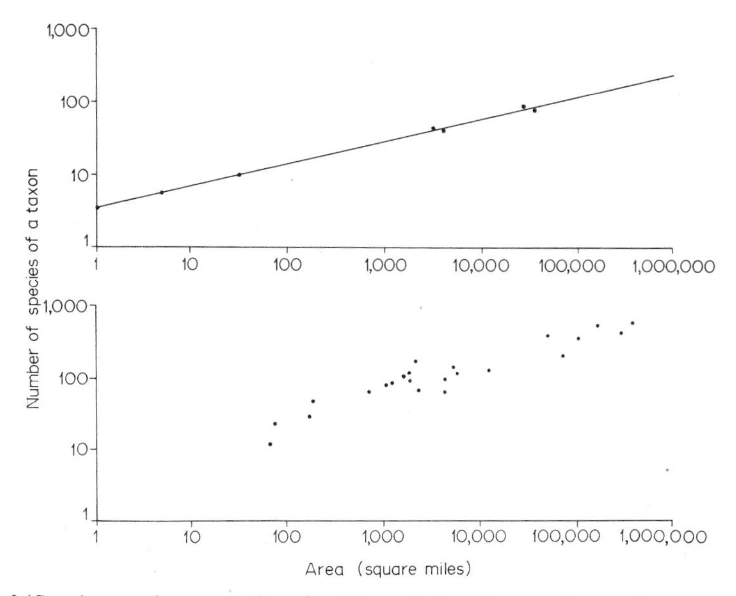

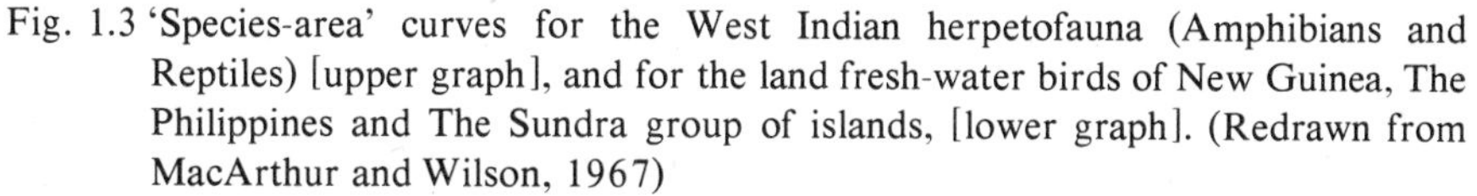
Fig. 1.3 'Species-area' curves for the West Indian herpetofauna (Amphibians and Reptiles) [upper graph], and for the land fresh-water birds of New Guinea, The Philippines and The Sundra group of islands, [lower graph]. (Redrawn from MacArthur and Wilson, 1967)

Wilson and Simberloff (1969) and Simberloff and Wilson (1969) report the results of controlled experiments in biogeography, and these throw considerable light on the dynamic situation that faces the conservationist. The experimental work was carried out in the Florida Keys using

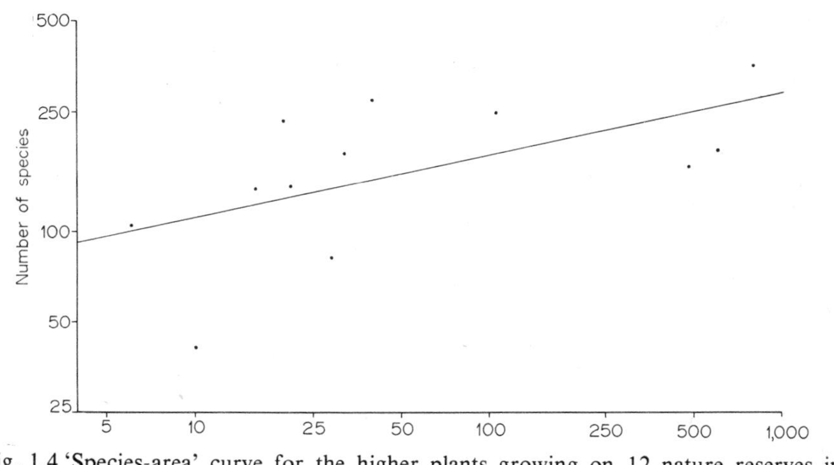

Fig. 1.4 'Species-area' curve for the higher plants growing on 12 nature reserves in Yorkshire

nine small islands that were ecologically simple, consisting only of red mangrove trees (*Rhizophora mangle*) with no ground above the high tide level. The islands, varying in size between 11 and 18 m diameter, were situated at different distances and directions from the immigrant sources elsewhere in the Keys. The islands were defaunated by enclosing them in a tent and fumigating with methyl bromide at a concentration lethal to arthropods but not to plants. Recording consisted of monitoring the island faunas approximately every 18 days and in surveying the fauna of the Florida Keys in order to estimate the size and composition of the species pool from which immigration had occurred.

To understand the processes of colonization that operate in the Florida Keys islands we must first discover how immigrants reach the islands. Aerial transport must be considered to be the most important with the active flight of butterflies, moths, beetles and wasps, and with the passive transport by wind of such groups as psocids, thrips and spiders, mostly the smaller organisms. Other colonists arrived by phoresy, or by parasitic relations with birds and snakes, or by association with nesting material brought by birds. Considerably less colonization was achieved by water transport, either free or on rafts. The fish in the sea around the keys are numerous and fed upon the arthropods in the water, whether the arthropods were actively swimming or passively drifting. Rafting was of rare occurrence due to the scarcity of floating debris and to the lack of ground on which the rafts could land. Thus we are essentially concerned with aerial transport aided either by wind or by birds.

The nature of ecological succession on the islands must also be understood. Since the islands contained but one plant species a progression of discrete and stable communities could not be expected to occur. The islands are situated in an area which has very small seasonal fluctuations in climate, and no seasonal component in the colonization was observed. There was, however, a regular pattern of invasion of species, but this was not accompanied by the wholesale extinction of distinct animal associations. To discuss the pattern of these invasions we shall consider just a few of the islands in the study. These are shown in the map in Fig. 1.5, where it will be seen that island E2 is very close to a Key, islands E3 and ST2 have intermediate and similar distances, and island E1 is relatively isolated. Island E6 is a control island that was not defaunated.

The first species to colonize the islands were often the smaller arthropods passively transported to the island by the wind. The time of arrival

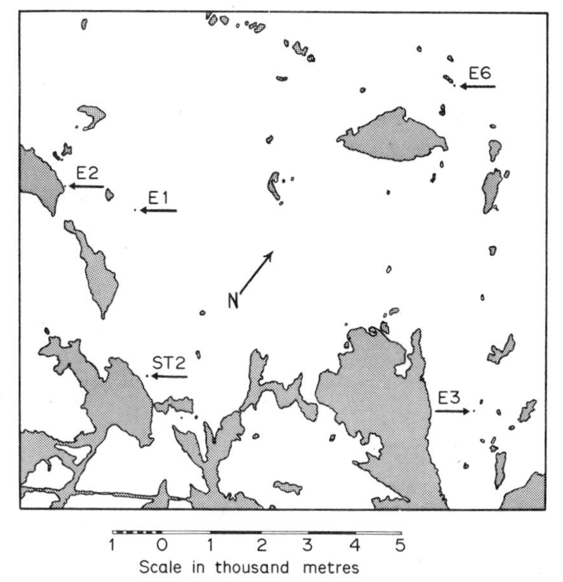

Fig. 1.5 Map showing the location and numbering of experimental islands. (See the text for a discussion). (Redrawn from Wilson and Simberloff, 1969)

and species composition of the psocids (Corrodentia) was unpredictable, but from the second month after defaunation there were usually one to four species on an island at any census period, though a total of 24 species were recorded during the period of the experiment. As a group they invaded and quickly built up a large population, though they became extinct almost as readily. Many of the species persisted on an island for less than a month (see Fig. 1.6). The spiders (Araneae) as a group were less variable in their pattern of colonization. Many of the species behaved as the psocids, immigrating readily, becoming extinct quickly and occurring on only one or two islands. A few, however, which included the species found on the islands before defaunation, colonized most of the islands and persisted for a longer period of time (Fig. 1.6).

The ants displayed the most orderly pattern of colonization. Surveys before defaunation showed two trends in the distribution of ants. First, on islands of approximately equal size, *Crematogaster ashmeadi* was the only species to occur on the most distant islands. Coming closer to the source area *C.ashmeadi* occurred with *Pseudomyrmex elongatus*, and coming still closer these two species occurred with *Paracryptocerus varians*, *Tapinoma littorale* and the *Camponotus* sp. (see list in Table 1.1).

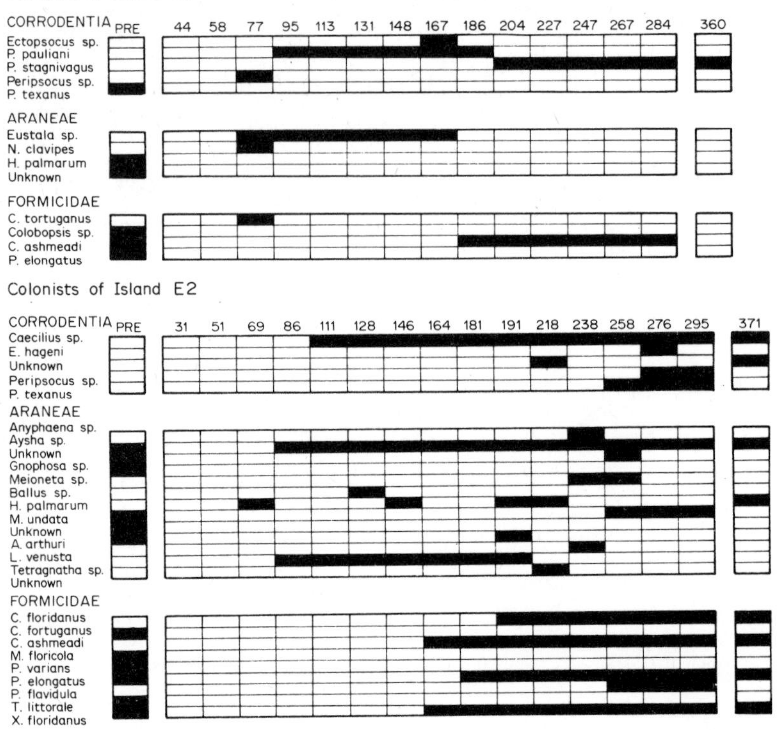

Fig. 1.6 Selected groups of colonists on two of the defaunated islands. Data are given for the pre-defaunation survey and for surveys following defaunation (indicated by numbers of days). Solid entries in the tables imply that a species was seen or inferred to be present. (Drawn from data given by Simberloff and Wilson, 1969)

TABLE 1.1 The colonization of experimental islands by ants. (Source: Simberloff and Wilson, 1969)

Island	*Species before defaunation*	*Species in order of colonization*
E1	3, 4, 9	4
E2	2, 6, 7, 9, 10, 11	(4, 11), 9, 1
E3	3, 4, 7, 9, 10	4, 9, 11, (6, 8), 7
ST2	3, 4, 7, 9, 10, 11	(4, 5, 9), 11, 1, 10, 7

Species codes

1 *Camponotus floridanus*
2 *Camponotus tortuganus*
3 *Camponotus sp.*
4 *Crematogaster ashmeadi*
5 *Crematogaster atkinsoni*
6 *Monomorium floricola*
7 *Paracryptocerus varians*
8 *Parathechina bourbonica*
9 *Pseudomyrmex elongatus*
10 *Tapinoma littorale*
11 *Xenomyrmex floridanus*

On the islands very close to the source area most or all of the above species occurred together with one or more species such as *Camponotus*, *Pseudomyrmex 'flavidula'*, *Monomorium floricola* or *Xenomyrmex floridanus*. Secondly, at a fixed and relatively short distance from the source area, the only species to occur on the smallest islands (small bushes about 1 m in height) is *C.ashmeadi*. On slightly larger islands this species and/or *P.elongatus* occur, whilst on larger islands still these two species occur together with about two more species drawn from *Colobopsis*, *Monomorium*, *Paracryptocerus*, *Tapinoma* and *Xenomyrmex*. On islands the size of the experimental islands the full complement of species for the appropriate distance from shore could be expected. Thus, the ability to colonize small islands is closely related to the ability to colonize distant islands.

On the defaunated islands *C.ashmeadi* was nearly always the first colonist, and on all islands except for the most distant (E1) *P.elongatus* was also an early colonist. Other colonists follow closely the predictions that have been discussed above. The sequence of colonization is given in Table 1.1, and is shown for two of the islands in Fig. 1.6.

Selected groups of species show these patterns of colonization, but if all species, irrespective of their taxon, are included, colonization curves (Fig. 1.7) can be produced, showing the relation between the number of species

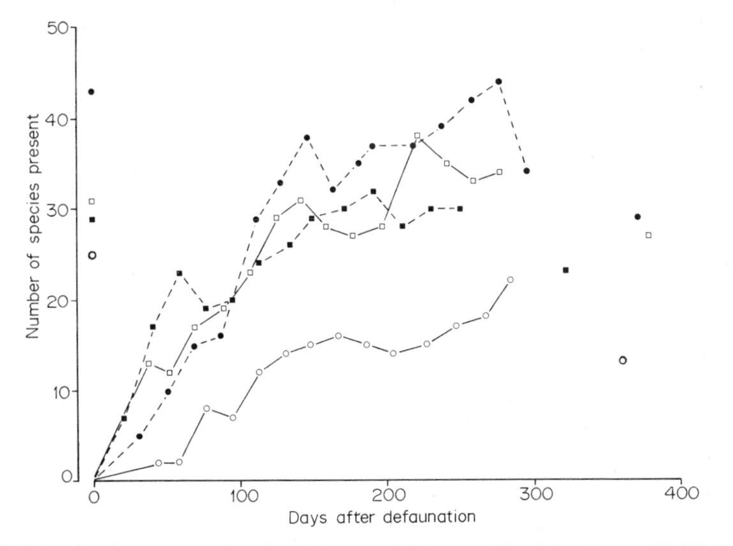

Fig. 1.7 Colonization curves for the defaunated islands. The islands are El (○), E2 (●), E3 (□) and ST2 (■). (Redrawn from Simberloff and Wilson, 1969)

present on an island and the time since defaunation. If we denote the number of species at a particular time, t, by S_t, then the curves in Fig. 1.7 demonstrate that S_t approaches an equilibrium value $\check{S}$. The fact that $\check{S}$ does exist can be derived from three series of facts. Firstly, on the control island (island E6) the number of species present decreased from 30 to 28 during the enumeration period, not a significant change, although the species composition changed to a greater extent. Secondly, islands of a similar size and a similar distance from a source area have similar numbers of species (compare islands E3 and ST2 in Fig. 1.7). Thirdly, the curves shown in Fig. 1.7 increase to approximately the value before defaunation, and then they oscillate about this value. It can be seen from Fig. 1.7 that the value of $\check{S}$ is dependent upon the distance of the island from a source area (compare islands E1 and E2), and it will also depend upon the size of the island, though the experimental work that used similarly sized islands cannot demonstrate this relationship. The colonization curve can be summarized by the equation

$$S_t = \sum_{r=1}^{p} \{ i_r(1 - e^{-(i_r + e_r)t})/(i_r + e_r) \}$$

where p is the number of species in the pool, i_r is the invasion rate of the rth species, and e_r is the probability of extinction of the rth species, though this would presumably be related to the species population density. As a population approaches its equilibrium value (i.e. letting $t \rightarrow \infty$) we can see that

$$\check{S} = \sum_{r=1}^{p} \{ i_r/(i_r + e_r) \}$$

The precise form of these equations and their stochastic application is given by Simberloff (1969). He shows that while S_t is small compared with $\check{S}$ there is little interaction between the species, but that after S_t has increased to between 75 per cent and 90 per cent of $\check{S}$ then interaction between the species becomes important. $\check{S}$ is thus determined not only by the size of the island and the distance from a source, but also by the interaction between the species occurring on the island.

The dispersal of the animals in the area and the resultant colonization of islands is one feature of island biogeography, but the picture would not be complete without a fuller consideration of the two parameters of the equation above, namely immigration and extinction. In the Florida study the census was made every 18 days, a time period too long to demonstrate all of the immigrations (arrivals of a new species) and extinctions.

However, rough estimates of the rates can be made, and these suggest that the immigration rates during the first 150 days of the experiment were between 0·05 and 0·50 species per day. As the value of S_t rises from zero towards $\check{S}$ the immigration rate exceeds the extinction rate, and as S_t approaches $\check{S}$ the two rates become approximately equal. It is suggested that the invasion rates for all species (i_r) remain more or less constant with time, and that the extinction probability (e_r) for all species is more or less constant whilst there is no interaction between the species. When interaction occurs the probability of extinction increases and is thus dependent upon the value of S_t. Oscillations of the colonization curve about $\check{S}$ are caused by the temporary divergence of the immigration and extinction rates. Simberloff and Wilson (1969) show that these rates can be estimated from equations

$$I_t = \sum_{r=1}^{p} \{ i_r - i_r^2 (1 - e^{-(i_r + e_r)t}) / (i_r + e_r) \}$$

and

$$E_t = \sum_{r=1}^{p} \{ i_r e_r (1 - e^{-(i_r + e_r)t}) / (i_r + e_r) \}$$

where I_t and E_t are the immigration and extinction rates at time t respectively.

Experimental studies such as that on the Florida Keys are rare, but they reflect the dynamic processes involved in biogeography. More traditional approaches, emphasizing the static geographical range of species, also contribute much to ideas of conservation. Raven and Walters (1956), discussing the British alpine flora, describe three main types of geographical distribution. The distribution of *Saxifraga caespitosa* (Fig. 1.8) clearly shows an arctic distribution. Although in Fig. 1.8 the distribution is only shown between 60°W and 60°E, the species is circumpolar, occurring 81°N in Greenland and 38°N in western North America. Throughout its range it usually occurs between 60° and 77°N. The distribution of *Cherleria sedoides* (Fig. 1.8) is an example of an alpine distribution since this species is to be found in the main mountain ranges of Europe. The most northerly occurrence of the species is in Scotland, where it is one of the very few truly alpine elements in the flora. The third group is the most numerous of the mountain plants in Britain, their distribution being described as arctic-alpine. The distribution of *Veronica fruticans* in Fig. 1.8 shows that this species occurs in very similar areas to *C.sedoides* south of 60°N, but that it is also widely distributed in the arctic regions of Norway, Iceland and Greenland.

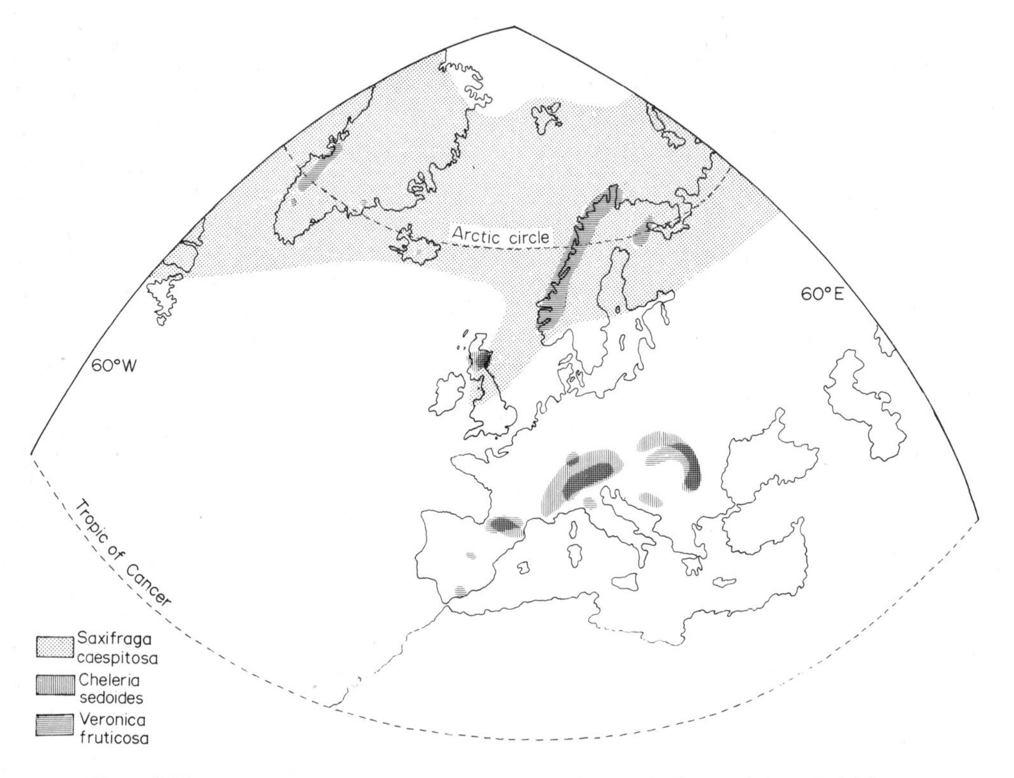

Fig. 1.8 The geographical distribution between 60°E and 60°W of three British mountain plants. (Redrawn from illustrations given by Raven and Walters, 1956)

Although these three forms of geographical distribution can explain many of the species of the British mountain flora, a few species cannot be associated with these categories. A characteristic of this group is a disjunct distribution, since the species occurs in widely separated areas. An example is the mountain sandworts of the genus *Arenaria*. *A.norvegica* is recorded over a wide area of Iceland and Norway, and it is local in Sweden, Finland and Scotland, in the latter country only occurring on the Shetland Islands, on the Island of Rhum and in West Sutherland. *A.gothica* is known only from four widely separated areas in Sweden, Gotland, Switzerland and the mountain of Ingleborough, Yorkshire. The species *A.ciliata* is represented by ssp. *hibernica*, which is only found on the Ben Bulben range in West Ireland, and by ssp. *pseudofrigida* which occurs extensively in the polar regions of Greenland, Spitzbergen and the Soviet Union west of Novaya Zemlya.

Genetical implications

With populations of the same or related species that are separated by such wide distances, questions about the historial distribution immediately arise, as do considerations of the evolution of the separate species and sub-species by geographical separation. These disjunct distributions immediately suggest that genetic change has taken place in the species between the different regions, but it is also important to realize that genetical adaptations take place within communities in each of the regions. Genetical variation that leads to adaptation through selection is a natural feature of populations, and it must be considered by conservationists. Berry (1971), reviewing the conservation aspects of population genetics, reaches four conclusions:

(1) Any environmental change, be it natural or part of 'management', within the normal tolerance of a species will result in genetical adaptation in the affected population; if the changes exceed the species tolerance the population will become extinct (as happened to the large copper butterfly, *Lycaena dispar* in Britain). There is a slight ambiguity in that 'tolerance' is itself a genetical property.

(2) Adaptation is rapid and precise, and is normally based on variation existing within a population (as has been suggested for the spread of the melanic form of the peppered moth, *Biston betularia*). The probability of random, or deleterious, change taking place is very small and can probably be neglected.

(3) The amount of variation, at least in a normally outbred, diploid species, is extremely large and resistant to loss. Berry quotes examples of the variation remaining in populations subjected to rigorous and long-term directional selection as justification. He also quotes the Skokholm house mouse population, isolated since it was founded by a few individuals 70 years before, containing the same amount of variation as British mainland mouse populations. Thus, he concludes that it is extremely unlikely that any natural management procedures could significantly affect the amount of variation in a local population to the extent of making that population unable to respond to environmental change.

(4) Attempts at preservation (as in the 'gene-bank' in a zoological garden) may provide some alleles available for study or possible introduction into a wild stock, but will certainly not preserve the whole genotype of the form in question. Berry quotes the results of attempting to improve

the native turkey stock in Missouri by releasing commercially bred hybrid birds, in which the released birds proved inferior to the wild birds in every respect of viability that was studied – brain, pituitary and adrenal size, and breeding success.

The conclusions reached by Berry (1971) relate especially to the variation within populations. On the wider biogeographical front we can ask what happens when there are several populations. Maynard Smith (1970) analyses the question – 'If a species is divided into a number of local populations, between which some migration takes place, and if neutral mutation maintains a high degree of polymorphism within populations, what will be the degree of resemblance between populations?'. In finding the answer Maynard Smith investigates three cases. Firstly, when migration between the local populations is relatively infrequent, then almost all loci will be heterozygous in hybrids between local populations. Also in different local populations, different loci will be polymorphic, or the same loci will be polymorphic for different alleles. This case, more precisely, is determined by

$$u\,r \gg m$$

where u is the rate of mutation, m is the rate of migration, and r is the number of local populations. Secondly, when migration is frequent, the species is equivalent to a single population with an effective size equalling the sum of all of the local population sizes. This case is defined by

$$u\,r \ll m$$

In large populations very few loci will be homozygous. Thirdly, when the migration rate is intermediate, the terms ur and m are of the same order of magnitude. Either, then, the total population size is small, or almost all the loci will be heterozygous in hybrids between local populations.

From this theoretical treatment it can be seen that the migration between populations is a feature that interests conservationists. With the Florida Key islands the migration rate determined which species arrived, and by knowing such rates for all species in the pool of colonizers we can predict the number of species likely to be present on an island. Maynard Smith (1970) shows that the genetical effects on the population depend, at least in part, on the migration rate. We can see that as m tends to zero, the results of isolation of populations leads to the process of speciation where isolated populations resemble each other less and less closely.

The spread of alien species

One aspect of interest to biogeographers is the spread of species that has occurred in recent times, as these have been documented and may throw some light on processes that have occurred in the past. Some species have increased their distribution naturally, as for example the fulmar, *Fulmarus glacialis*, which has been progressively nesting further south around the British coasts, or the collared dove, *Streptopelia decaocto*, which has extended its range westwards from Jugoslavia in the 1930s to the Atlantic Coast of Europe by 1950, and first nesting in Britain in 1955. Other species have extended their range following introduction by man, as, for example, the house sparrow, *Passer domesticus*, which was introduced into the North American Continent from Europe, or the grey squirrel, *Sciurus carolinensis*, which was introduced into Britain from the eastern United States and Canada. Many plant species have been introduced to places outside their natural geographical range either by the direct or indirect action of man.

The Florida Keys studies demonstrated that the processes of colonization and extinction within small areas are constantly occurring, but all of the species that are spreading are showing that their ability to colonize is superior to the processes acting to cause their extinction. It is therefore important to examine examples of species that are spreading, and to determine the characteristics of such successful species.

The Oxford ragwort, *Senecio squalidus*

One species of plant that has been introduced into Britain is the Oxford ragwort, *Senecio squalidus*, a species native of Sicily and Southern Italy. It has been cultivated in the Oxford Botanical Garden since at least 1690 (Kent, 1956), though it was a century later before it 'escaped' and became established on the walls of Oxford. Kent (1956) records how this local species attracted collectors who took seeds for cultivation, leading to the establishment of colonies in many parts of Britain after the plumed fruits had been wind-disseminated from these private collections.

Kent (1960) records how indirect action by man was also responsible for the spread of this species. *S.squalidus* reached the Great Western Railway system in Oxford about 1879, and the plumed fruits were carried in the vortex of air following trains. The species spread rapidly along the various railways systems, reaching Tilehurst, Berkshire in 1883; Pewsey and Swindon, North Wiltshire in 1888 and 1890 respectively; Reading, Berkshire in 1890; Fenny Compton, Warwickshire in 1891; Bletchley Junction and Slough, Buckinghamshire in 1896 and 1900 respectively. After the seed had been introduced to these new areas, large colonies of the plant developed in the vicinity of the railway and then the species quickly spread outwards, distribution being aided by the wind-borne seed. The spread of the species from Oxford was also aided by the movement of iron ore and ballast by the railways. Thus by 1939 the plant was well established and spreading over large areas of England and Wales, though always in the vicinity of railways and large towns (Plate 1). In rural areas it was absent or rare and sporadic, and it had not yet reached areas in Britain furthest from Oxford, namely South East and North England, Central and West Wales, or Scotland.

Kent, in a series of papers, has described the spread of the species since 1940. The spread of *S.squalidus* into Scotland during the 1950s shows the same characteristic association of this species with railways (Kent, 1966), but it also demonstrates that the species is associated with disturbed ground. The first Scottish records during this century were from a seaside path in Fife, and from a roadside bank, waste ground and a railway junction near Edinburgh. The species spread north, being recorded by the railway in Morayshire, and west being associated with the railway in Glasgow and Ayrshire. Near Edinburgh the species flourished on waste land around the city.

The distribution of *S.squalidus* in Britain is shown in the map (Fig. 1.9). This species, which was geographically isolated from Britain in historical times, was introduced to Oxford and was rapidly distributed by its wind-borne fruit. Ecologically, it is a species of disturbed habitats, depending for its existence upon the lack of competition from other species, a habitat provided by railway banks and cuttings. In the vicinity of large towns the species was able to spread from the railway system to other disturbed areas and to waste ground, also habitats where there is reduced competition from the native British flora.

The New Zealand willow-herb, *Epilobium nerterioides*

The factors causing the spread of the Oxford ragwort were clearly defined, but such cases are rare in plant biogeography. More often the spread of a species cannot be linked with any specific factors. Such a plant is *Epilobium nerterioides*, a native of New Zealand, which has been

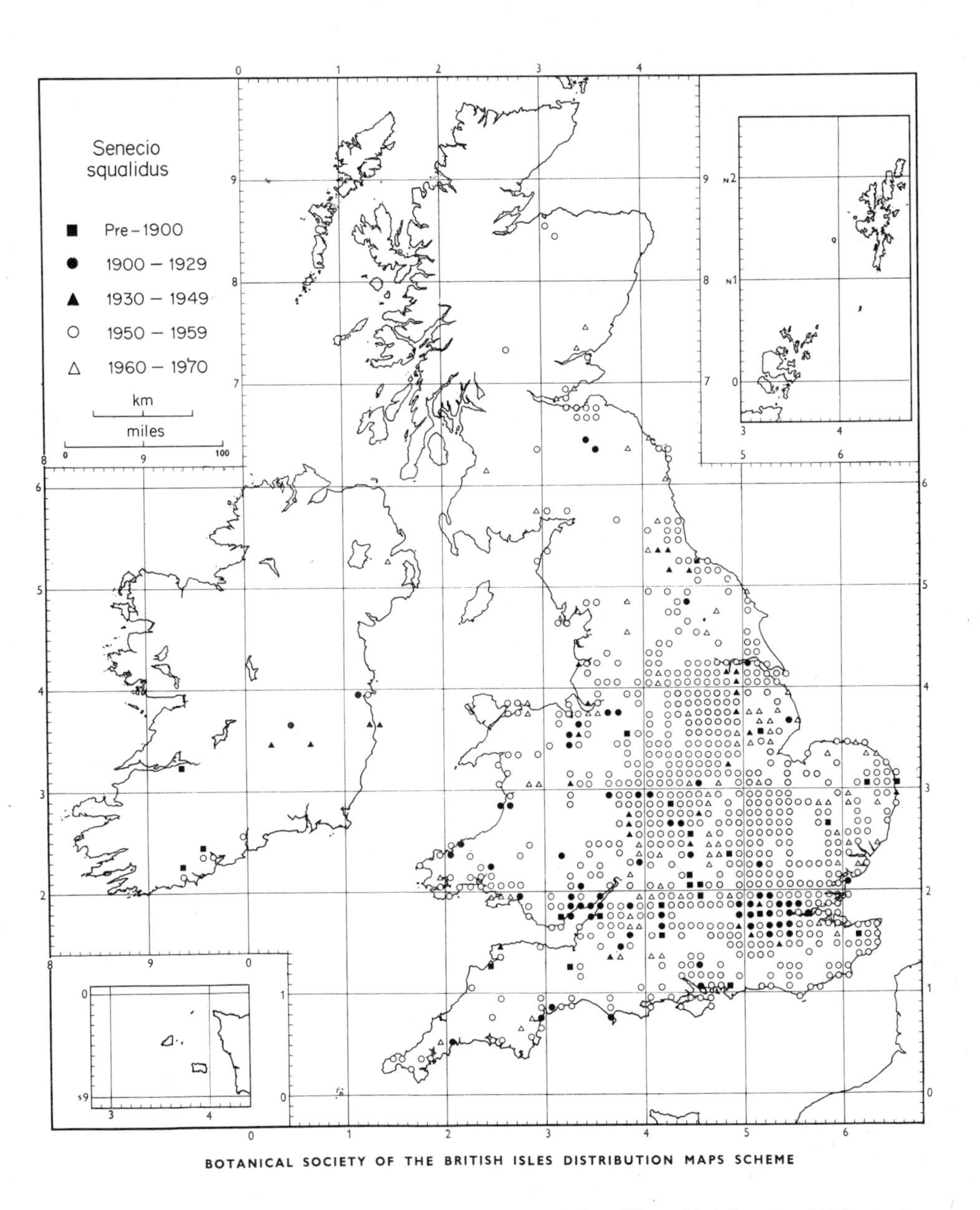

Fig. 1.9 The spread of *Senecio squalidus* in Britain. (Compiled by the Biological Records Centre)

introduced both as a rock-garden plant and accidentally with New Zealand shrubs. The history of the spread has been discussed by Davey (1961). The first record was in 1904 when the species was found as a weed in a garden in Edinburgh, and it was later recorded near Leeds after its introduction with shrubby plants from New Zealand. Its spread in Scotland can be shown by the following records: Ardrishaig, Kintyre in 1911; Stirlingshire in 1920; Lanarkshire in 1926; Renfrewshire and Glasgow in 1932; the islands of Raasay in 1935, Skye and Islay in 1952 and Orkney in 1956. By 1949 it was known to be abundant and spreading throughout the Forth and Clyde valleys, and during the 1950s and 1960s it has extended its range into Perthshire, Inverness-shire and North East Scotland. The spread has been similar down the west side of England and Wales, until by 1960 it was scattered throughout North Wales, the Cardiff area, Dartmoor in Devon and in Cornwall. Altitudinally it occurs from sea level to 3000 ft in England and Wales, and to 2000 ft in Scotland. The distribution is shown in Fig. 1.10.

It seems unlikely that man is aiding the distribution of *E.nerterioides* and hence it is essential to look for locality factors that might be determining the spread. Davey (1961) records that in New Zealand *E.nerterioides* occurs chiefly in open habitats on moist stony soils, and that it is widespread in rocky situations in the alpine and subalpine mountain ranges and in river gorges at lower altitudes. He states that it is among the more important characteristic species of 'stable river bed' formations at both high and lower altitudes, and that it occurs in other 'stony' habitats. In the South Island it is confined to moister situations.

In Britain the species can be found growing in open habitats, on moist and well-drained substrate of stony or gritty texture, and with a wide range of pH (Davey 1961). Its occurrence in three communities in Wales and Scotland will show the similarities between the British colonized habitats and the natural New Zealand habitats. In Cwm Idwal, North Wales, it occurs at an elevation of about 1300 ft on rocks and ledges near small streams, and is associated with such species as *Acrocladium* mosses, *Saxifraga stellaris*, *Thalictrum alpinum*, *Pinguicula vulgaris* and *Festuca ovina*. This is a typical rocky habitat, generally very moist, and not very acidic. Near Campbeltown in Kintyre, Scotland, *E.nerterioides* occurs in the dried-out river beds, near Loch Lussa, following the damming of the rivers for hydro-electric power. At an elevation of about 400 ft *E.nerterioides* forms almost pure stands, growing in cracks in the rocks

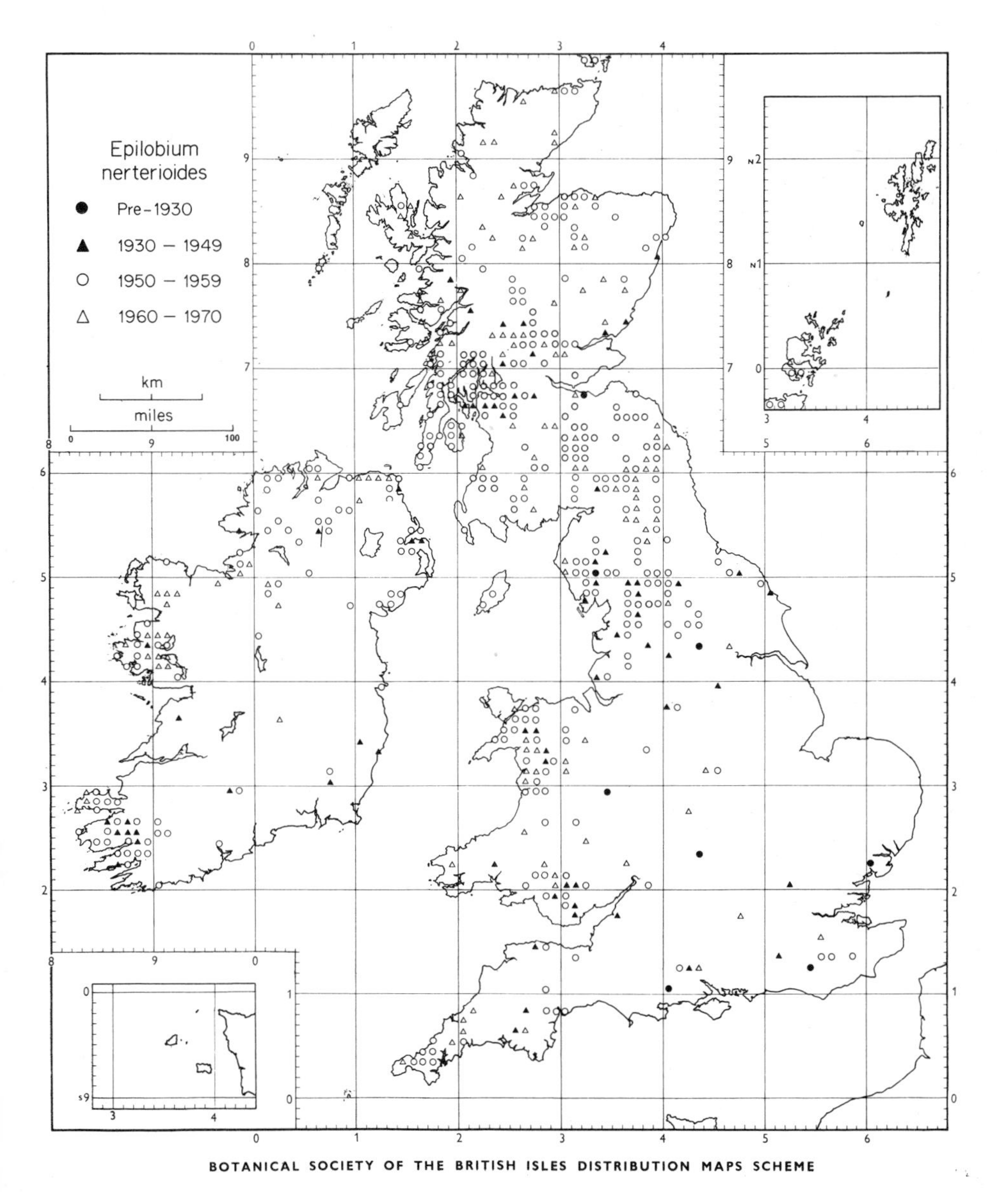

Fig. 1.10 The spread of *Epilobium nerterioides* in Britain. (Compiled by the Biological Records Centre)

and stones that formed the old river bed. The only species associated with it are the pearlworts, *Sagina* spp., though where pockets of soil have developed in the river bed there is growth of *Succisa pratensis*. In a third site, beside a river in Peeblesshire, Scotland, *E.nerterioides* is growing over boulders. The species associated with it in this base-rich habitat are *Sedum villosum*, *Rumex acetosa* and *Agrostis* spp.

E.nerterioides has thus colonized similar habitats in Britain to its native habitats in New Zealand. It is particularly common in habitats where there is not a closed vegetative canopy (Plate 2) and where the micro-climate is relatively damp. The spread of this species from Edinburgh, moving northwards into the Highlands of Scotland, moving westwards and south has coincided with the presence of suitable habitats. The seeds are minute and with long hairs they are wind-dispersed and may be carried for long distances. Davey (1961) records only one parasitic fungus, and no diseases or animal feeders or parasites. Thus, *E.nerterioides* has probably spread so far in Britain because it has effective long-distance seed dispersal, because there are few or no animals or plants harvesting its production or eating its seeds, and because the western side of the British Isles provides the moist rocky habitats that it requires.

The grey squirrel, *Sciurus carolinensis*

The analyses of the plants have indicated that there must be a suitable habitat into which the alien can spread, and that there must be some effective means of seed dispersal. With animals there must similarly be a suitable habitat, although it is possible for the alien to create a habitat either by physically altering an environment or by competitively excluding a species that previously occupied that habitat. The grey squirrel, *Sciurus carolinensis* (Plate 3), provides us with an example of an alien species that competed with a native species, the red squirrel, *Sciurus vulgaris* (Plate 4). Although the grey squirrel was introduced early in the nineteenth century, it was the animals introduced between 1876 and 1930 that have resulted in the dramatic increase in the population size of this species (Forestry Commission, 1960). The main centres of introduction in England and Wales were the Home Counties, including London, Kent, Surrey, Middlesex, Buckinghamshire, Bedfordshire and outliers in Hampshire and Devon, a group in Cheshire and Denbighshire, and another group in Yorkshire. From these regions the grey squirrel consolidated its position (Fig. 1.11), until by 1945 the species was well established in an area bounded by Leicestershire and Kesteven (Lincolnshire) in the north, and Breconshire, Monmouthshire and Herefordshire in the west, and in another large area in the three Ridings of Yorkshire and in the south of County Durham. Between 1945 and 1955 (Shorten, 1953, 1957) the grey squirrel extended its range. In England the population in Yorkshire and that in Nottinghamshire and north Lincolnshire became separated by less than 10 miles, and the species spread into Cornwall and to the border between Norfolk and west Suffolk. During this period the species advanced up to 60 miles westwards in Wales. Lloyd (1962) shows that during the next five years, 1955–1960, there was only a limited spread in the distribution. Although the squirrel appeared in more of the 10 km squares in 1960, none of the new squares, except in Essex, was other than adjacent to a previously colonized square. Thus the squirrels seem to have consolidated their new range rather than continuing to advance on a broad front. By the end of the 1950s the grey squirrel had thus established itself throughout most of England and Wales, but it was still absent from Norfolk, east Suffolk, Anglesey, Westmorland, Cumberland, Northumberland, the Isle of Ely and the Isle of Wight.

As the grey squirrel advanced, so the distribution of the red squirrel, *S.vulgaris*, declined (Fig. 1.11). The red squirrel is common throughout areas that are not occupied by the grey squirrel. Shorten (1957) showed that in 1955 the red squirrel was absent from about 80 per cent of forests lying within the 1945 range of the grey squirrel, and she also showed that the majority of forests that had both species present had been colonized by the grey squirrel since 1945. Lloyd (1962) similarly shows that the red squirrels only occur in areas uncolonized or only recently colonized by the grey squirrel. The extension in the range of the grey squirrel and the contraction in that of the red squirrel are obviously closely correlated, and it would appear that the red squirrel ceases to exist in an area 10–15 years after it has been colonized by the grey squirrel. It is impossible, however, to assign the exact causes of these distributional changes. Southern (1964) discusses the populations, habitats and food of the two species, and although data were not extensive his lists show that the two species have similar requirements. No observational evidence has been gained on aggression between these two species, but the closeness of their habitat requirements does suggest that, where both species occur together, they will be competing for this habitat. The distribution of the

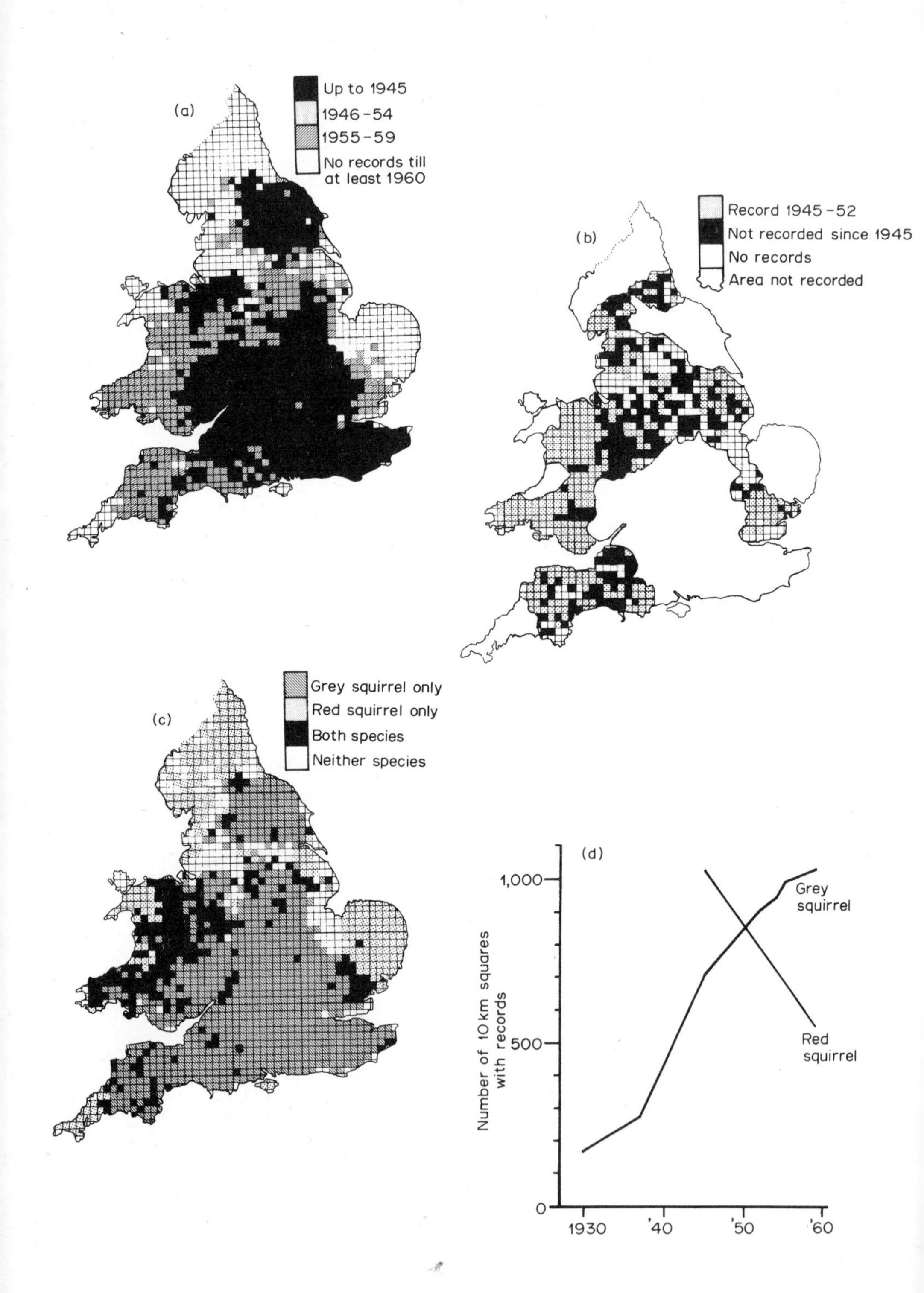

(a)
Up to 1945
1946-54
1955-59
No records till at least 1960
(b)
Record 1945-52
Not recorded since 1945
No records
Area not recorded
(c)
Grey squirrel only
Red squirrel only
Both species
Neither species
(d)
Number of 10 km squares with records
1,000
500
0
1930
'40
'50
'60
Grey squirrel
Red squirrel

squirrels in England and Wales is thus an illustration of the effect of interspecific competition, but it fails to demonstrate the mechanisms of such competition. The case for competition is not, however, proven, since Shorten (personal communication) has indicated that there are examples of the red squirrel becoming extinct in an area prior to colonization of that area by the grey squirrel.

Small scale distribution

In previous sections of this chapter we have considered the distribution of plants and animals on islands and within areas that can be mapped nationally and internationally. However, within the island or the geographical region all species show heterogeneity of distribution, for which the term 'pattern' has been used in plant ecology. In examining this pattern we are faced with three separate questions, each of which must be considered in greater detail. First, how can we detect pattern and what is the yardstick by which we measure the pattern? Secondly, what are the analytic techniques that allow us to quantify the pattern, and thirdly what are the ecological implications or causes of the pattern?

Detection of pattern

To consider the detection of pattern we must start by defining the distribution of organisms in space, initially considering only two dimensional space such as a field or the surface of an agar plate. The six illustrations in Fig. 1.12 show different arrangements of approximately 50 points in a two-dimensional space. In illustrations (*a*) and (*b*) the distribution can be considered as regular, since in Fig. 1.12 (*a*) the points are arranged at the corners of equilateral triangles, such that the distance between any pair of neighbouring points is constant, say D. In Fig. 1.12 (*b*) a certain amount of the regularity has been lost, since now a pair of neighbouring points may be separated by a distance of either D or $\sqrt{2}\,D$. Such regular distributions obviously occur in man's monocultures,

Fig. 1.11 The distribution of the red and grey squirrels in Britain. Each little square on the maps refers to an area 10 km × 10 km. A: Records of the grey squirrel. B: Records of the red squirrel. C: The status of the two species in 1960. D: A graph showing the number of 10 km squares occupied by the grey and red squirrels in England and Wales. (Redrawn from Shorten, 1953 and Lloyd, 1962) such as orchards and young forests, but their occurrence in nature is rare. Fig. 1.12 (*c*) shows a random distribution of the 50 points over the field. The illustration was drawn from a series of random numbers and it illustrates the ease with which the human eye can detect non-existent patterns of groups of points and spaces. Figs. 1.12 (*d*)–(*f*) show three

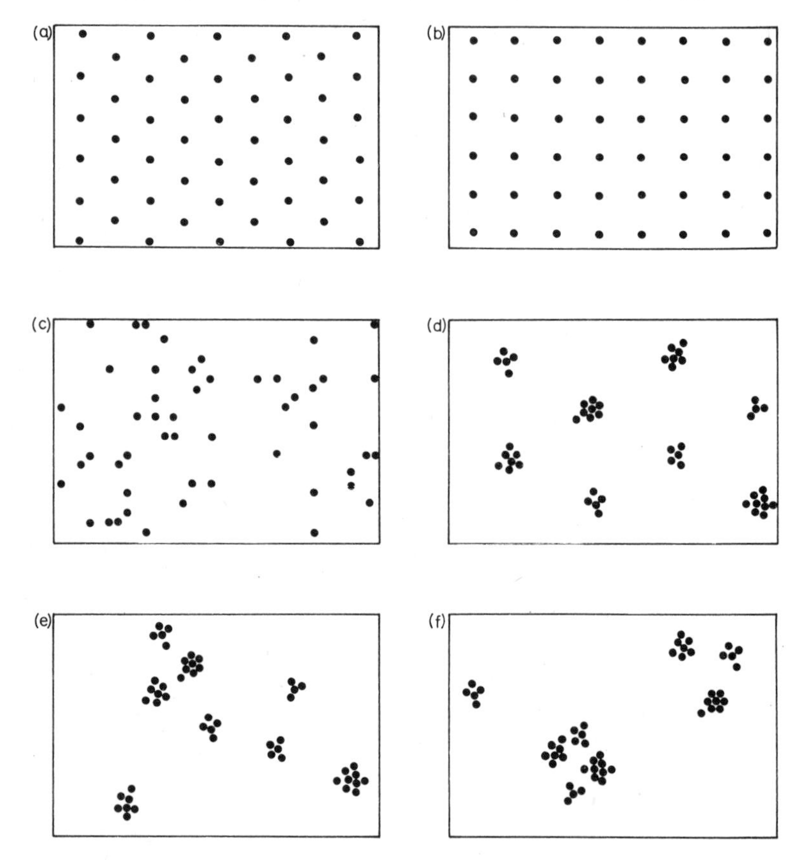

Fig. 1.12 Patterns of approximately 50 points. (*a*) and (*b*) are regularly distributed, (*c*) is random, and (*d*), (*e*) and (*f*) are clumped or aggregated.

types of clumped or aggregated distribution. In all of these illustrations the 50 points have been brought together into eight groups or aggregations, the size of each group varying from between four and nine individual points. In Fig. 1.12 (*d*) the eight groups are arranged at the corners of hexagons, and hence the groups are regularly distributed in space. In Fig. 1.12 (*e*) the eight groups are randomly distributed in space, whilst in

Fig. 1.12 (*f*) the groups are themselves grouped together to form a double hierarchy of aggregation.

From a statistical point of view we are concerned with the features of sampling such situations. Controversy has been created by arguments in favour of random or in favour of systematic sampling. For random sampling a series of samples are located at random positions within the field, and these locations precisely determine the boundaries of the sample. There is thus no relation whatsoever between any pair of samples that are taken. In order to take a systematic sample, we might wish to take a series of transects across the field, and to record our samples at regular intervals along these transects. Thus, the location of one sample is determined by the location of the sample before it. A hybrid method of sampling would be to divide the area in several strata, and then to take random samples within each strata (stratified random sampling). Although there has been controversy over the relative merits of these sampling methods, each method has much to be recommended for a particular purpose. The method by which samples are taken should always be determined by the purpose of sampling. Thus, if we wish to determine if an organism is randomly distributed or aggregated on a field, a series of random samples would be the best statistical approach. If, however, we wish to quantify the type of pattern that the organism is exhibiting, then a systematic sample will allow us to map the presence or absence of the organism, and to reach conclusions about the nature of its spatial distribution.

First, though, we must detect whether an organism is distributed at random in our field. The samples on which the tests are based can be taken either at random or systematically, the latter being the sampling method employed by 'Student' (1907) when the Poisson (random) distribution was formulated for the numbers of yeast cells in a haemocytometer. If the mean number of organisms in a sample is m, then

$$p(x) = \frac{e^{-m} m^x}{x!} \tag{1.1}$$

is the probability of observing exactly x organisms in any one sample. The data in Table 1.2 show the distribution of the Collembola, *Tetracanthella wahlgreni*, in 32 soil cores. Since there are 24 animals the mean is $m = 0{\cdot}75$. Using this value for m, Equation (1.1) generates the values of $p(x)$ shown in Table 1.2. Multiplication of $p(x)$ by 32 gives the

TABLE 1.2 The numbers of *T.wahlgreni* in soil cores

Number of animals in a core (x)	*Observed number of samples*	$p(x)$	*Expected number of samples*
0	17	0·4724	15·116
1	9	0·3543	11·337
2	4	0·1329	4·251
3	1	0·0332	1·063
4	1	0·0062	0·199
More than 4	0	0·0010	0·034
Total	32	1·0000	32·000

expected values, which appear to agree closely with the observed values of the numbers of *T.wahlgreni* in the samples, thus indicating a random distribution at this scale of sampling.

A feature of the Poisson distribution is that the variance (s^2) equals the mean, and hence quick tests for the Poisson distribution can be based on the departure of these two statistics from each other. Green (1966) discusses a number of coefficients which have been used to estimate the divergence of an observed distribution from the Poisson. One of the simplest to apply and to test statistically is

$$\chi^2 = \frac{s^2}{m}(n - 1) \tag{1.2}$$

where the sample size is n, and χ^2 has $(n - 1)$ degrees of freedom. In the example for *T.wahlgreni* shown in Table 1.2 we have values of m, n and s^2 of 0·75, 32 and1·032 respectively. Thus we have $\chi^2 = 42{\cdot}67$, which has a probability slightly greater than 0·05. If we are using the level $p = 0{\cdot}05$ to judge a significant departure of an observed distribution from a Poisson distribution then we can say that there is no evidence to suggest that *T.wahlgreni* is other than randomly distributed at the scale of sampling that was used. Greig-Smith (1964) does caution the uncritical use of this χ^2 criterion of the departure from randomness, since distributions can be constructed that differ markedly from the Poisson distribution, but yet their variance and mean are approximately equal.

The χ^2 criterion does have the advantage that it can be used to test for departure from the random distribution either in the direction of aggregation, or in the direction of regularity. Thus if the probability of χ^2 is greater than, say, 0·95 we can assume that the distribution is tending towards regularity, whilst if the probability is less than, say, 0·05 we can assume that aggregation is present.

Usher (1969a) used the χ^2 criterion to investigate the distribution of a number of species of soil Collembola in a Scots pine forest soil. Out of 234 sets of data analysed the regular distribution was found in only two sets, both for *Folsomia quadrioculata.* The random form of distribution was presumed for 64 sets of data, and the aggregated form of distribution for the remaining 168 sets of data. Most studies on the distribution of plants and soil animals have shown some degree of aggregation.

Usher (1970b) further considered the distribution of the aggregations of predatory mites in the soil, and from these studies it was clear that there was a second level of aggregation. Some species had aggregations that were regularly spaced, other species had aggregations randomly distributed within the three-dimensional environment of the soil, and still other species had aggregations that were themselves aggregated into larger areas of high population density. These are thus similar to the distributions shown in Figs. 1.12 (*d*), (*e*) and (*f*).

The use of statistical tests and indices can thus determine the nature of the distribution of an organism in a two- or three-dimensional universe. These tests cannot, however, demonstrate the actual form of the pattern, and quantify the process. Although the work of Usher (1969a, 1970b) quoted above was able to determine the nature of the distribution of the soil arthropods, it was not possible to extend this to look deeply into the make-up of the pattern. Work in animal ecology has not analysed the causes of pattern in detail, but botanists working with stationary organisms that are distributed on a two-dimensional field have been able to make greater progress.

Analytic techniques

Analysis of pattern requires systematic samples so that the relation between the location of one sample and all the others, at least within a strata, is known. Although the techniques were originally derived for sampling a square grid they have been found to work adequately when applied to a line transect of contiguous samples. It must, however, be mentioned that anomalous results can arise, and results that are more meaningful can be derived from analyses that include the data from several line transects. A sample of artificial data and the subsequent analysis are given in Table 1.3. This data has been very much simplified, but it can be conceived of as

TABLE 1.3 Artificial data for a transect of 32 contiguous quadrats, and the subsequent pattern analysis

Basic data	*11111111000000001111111100000000*
Block size	
2	2 2 2 2 0 0 0 0 2 2 2 2 0 0 0 0
4	4 4 0 0 4 4 0 0
8	8 0 8 0
16	8 8
32	16

Block size	*Degrees of freedom*	*Uncorrected S.S.*	*Corrected S.S.*	*Mean square*
32	–	8·0	–	–
16	1	8·0	0	0
8	2	16·0	8·0	4·0
4	4	16·0	0	0
2	8	16·0	0	0
1	16	16·0	0	0

being the presence (1) or absence (0) of a species, or as a cover of 100 per cent (1) or zero (0), or as a density of one or zero. Data from natural plant communities will consist of quantitative estimates of cover, frequency or density. The term 'block size' is used to represent the number of quadrats that have been grouped together at that stage of the analysis. Thus, block size 2 is derived by adding the data from the first and second quadrats, from the third and fourth quadrats, from the fifth and sixth quadrats, and so on. Block size 4 is derived by adding the data for the first pair of block size 2 (quadrats 1 to 4), the second pair of block size 2 (quadrats 5 to 8), and so forth. Block sizes are thus always powers of 2, 2^n, where n is a non-negative integer. The uncorrected sums of squares are calculated by summing the squared terms for the data of a particular block size, and dividing the total by the block size. Thus for block size 8 the uncorrected sum of squares is

$$(8^2 + 0^2 + 8^2 + 0^2)/8 = 16{\cdot}0$$

The corrected sum of squares is calculated by subtracting from the uncorrected sum of squares the uncorrected sum of squares for the block size larger. Thus, for block size 8, the corrected sum of squares is

$$16{\cdot}0 - 8{\cdot}0 = 8{\cdot}0$$

The mean square is the corrected sum of squares divided by the degrees of freedom. The analysis is analogous to a multiple split-plot analysis of variance, large values of the mean square indicating heterogeneity of distribution at that particular block size. The analysis in Table 1.3 shows only one non-zero value of the mean square, with block size 8, indicating that this is the scale of heterogeneity, as is readily verifiable from the basic data. Further consideration of the meaning of the maximum mean square will be given in the simulation section later in this chapter.

Although such a pattern analysis is simple to carry out, the actual data that is used for the analysis is important. The first applications of pattern analysis relied upon density data. This is, however, of limited use since it is frequent for individuals of a species to be ill-defined. This is particularly true of mat- and clump-forming plants whose pattern may be of more interest than plants that occur as single individuals. Thus, it has become more frequent for the data to consist of cover values. Cover can be estimated subjectively by observing the degree of cover within a quadrat, or it can be measured quantitatively but recording the presence or absence of a species at a number of points within a quadrat. Kershaw (1957) describes a method by which a basic sample unit was divided into five sub-units, points spaced 1 cm apart. His data thus consists of a basic unit which has a dimension of 5 cm, and cover values that range from zero to five inclusive. Errington (1970) questioned the use of basic units and sub-units, and showed that more precise results could be gained from using the data from the points, and thus using only cover values of zero (the species absent from beneath the point) or one (the species present beneath the point). Although either density or cover can be used for plants, density data would be the only suitable data if such analyses were to be applied to animal populations.

This method of pattern analysis leaves the question of significance in the statistical sense unanswered. Greig-Smith (1964) discusses the methods of attempting to fit confidence limits to judge the significance of peaks, and he shows that none of the methods are entirely satisfactory. He gives a table for fitting approximate confidence limits to an expected mean square for each block size.

Other methods of pattern analysis have also been proposed. Yarranton (1969) has described a method based on regression analysis. His method appears to avoid some of the more difficult aspects of the method described above, and to be more easily tested for the significance of the regression coefficient. Goodall (1961, 1963) has criticized Greig-Smith's methods on various statistical grounds. He points out that the method does not take into consideration a trend of abundance along a transect, i.e. a species that becomes more common or less common. It is, however, a relatively easy operation to correct the basic data for any trend by converting the data to deviations about a regression line, as is described in the introductory section of Chapter 2 for time series. Also, one must always consider that pattern analysis is an exploratory technique suggesting further lines of research. The analysis suggests that a measure of heterogeneity exists at a particular dimension, but more detailed field work will be required to show the reasons for the pattern.

Ecological implications

Hutchinson (1953) has discussed the concept of pattern in ecology, in which five kinds of pattern are distinguished, being defined as: 'The distribution of organisms and of their effects on their environment may be determined by external forces, such as light, temperature, humidity or density gradients, changes of state in certain directions, currents, winds, etc. Patterns produced in this way will be termed *vectorial.* The distribution may be determined by genetic continuity, offspring remaining near the parent, giving a *reproductive* pattern. The distribution may be determined by signalling of various kinds, leading either to spacing or aggregation, producing *social* pattern. The distribution may be determined by interaction between species in competition leading to *coactive* pattern. The distribution may depend on random forces producing a *stochastic* pattern.'

The distribution of clover, *Trifolium repens*, in an *Agrostis-Festuca* grassland in North Wales is an example of vectorial pattern (Snaydon, 1962). Both in widespread sampling over 130 sq km and in a localized area of 100 sq m there was a significant correlation between calcium content and the cover of *T.repens*. There were also significant correlations between the cover and phosphate and between phosphate and calcium. Chemical analysis of the soil showed that over a distance of 0·6 m there could be a threefold increase or decrease in the contents of calcium and phosphate. This distance corresponded to the scale of heterogeneity in the distribution of *T.repens*.

Greig-Smith (1961b) discusses the distribution of the marram grass *Ammophila arenaria* at Newborough Warren, Anglesey. Two scales of pattern, at 20–40 cm and at 80–160 cm are exhibited. The latter scale, referred to as the tussock pattern, results from environmental control of seedling establishment, and this pattern is maintained by *A.arenaria* through the later stages of the sand dune succession. Undoubtedly this pattern could be classed as vectorial, though the smaller scale of pattern, referred to as tillering pattern, is reproductive. It results from the production of a number of tillers at the tips of vertical rhizomes, which were more or less unbranched, and it therefore corresponds closely with the morphology of the plants. Social patterns were described by Usher (1969a) for the Collembola, *Isotoma sensibilis*. In this species the young aggregate at the site of their egg cluster, but the adults form aggregations that are unrelated to the location of juvenile clusters. The phenomenon of aggregation at the time of the moult has been widely recorded in the Collembola (Christiansen, 1964). Coactive patterns are more difficult to detect, but the use of the correlation coefficient in pattern analysis (Grieg-Smith, 1961a) may be used to postulate this form of pattern, though the coactive pattern may operate in conjunction with other types of pattern, particularly vectorial. Stochastic patterns are, by their very nature, difficult to discuss and may not even be patterns in the generally accepted sense. Determination of patterns belonging to this category may have resulted from a failure to identify other types of pattern.

TABLE 1.4 Average cover values for three moss species in a Wolds woodland

Species	*Cover (per cent)*
Eurhynchium striatum	43·71
Rhytidiadelphus triquetrus	17·79
Mnium undulatum	4·34

An analysis of pattern of woodland bryophytes in the Wolds, Yorkshire, was carried out by Errington (1970). Three species of moss, detailed in Table 1.4, are used as an example of the methods employed. Cover was measured by the presence or absence of each bryophyte species under points, 2 cm apart, on transects 22·6 m long. The smallest unit of measurement is thus 2 cm. One pattern analysis for each species is shown in Fig. 1.13 (*a*), though the discussion is based on many analyses. *E.striatum* showed three scales of pattern corresponding to block sizes 4, 16–32 and 256. It will be noticed that the values of the mean square for this species are generally greater than those of the other species. In general species with a cover or frequency of 50 per cent have the greatest mean squares whilst species with cover of less than 5 per cent or more than 95 per cent have very small mean squares. *R.triquetrus* has three scales of pattern at block sizes of 2–4, 32 and 256, whilst *M.undulatum* has maximum mean squares at block sizes of 1, 4–8 and 64. In each case the smallest scale of pattern corresponds to the size of the moss species

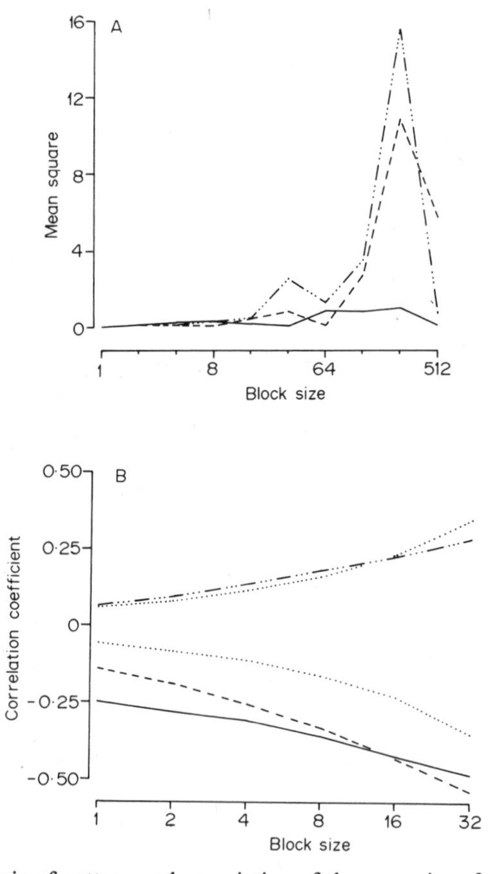

Fig. 1.13 The analysis of pattern and association of three species of moss growing in a Wolds woodland. (a): The species are ———————— *Mnium undulatum*, — — — — *Rhytidiadelphus triquetrus*, —·· ·—·· · *Eurhynchium striatum*. (b): The associations are ———————— *R.triquetrus* and *E.striatum*, — — — — *R.triquetrus* and *M.undulatum*, —·· ·—·· · *M.undulatum* and *E.striatum*. The dotted lines indicate the significance level ($p = 0{\cdot}05$) for the correlation coefficients. (Redrawn from Errington, 1970)

and thus reflects morphological pattern. With *E.striatum* the size of 8 cm is larger than would have been expected, but this is due to the strongly intertwined nature of its growth form. The largest scale of pattern for *E.striatum* and *R.triquetrus*, approximately 5 m, is vectorial, being determined by the ash tree canopy and the gaps in the canopy.

Analysis of the correlation between species is required to understand the nature of the patterns in the community, Fig. 1.13 (*b*). The largest species of the habitat, including *E.striatum* and *R.triquetrus*, are negatively associated with each other, whilst the smallest species, including *M.undulatum*, are positively associated with each other. From a study of pattern and association Errington suggests that the two most frequent (in terms of cover) large species are potentially in competition with each other and hence occur in more or less distinct areas. *R.triquetrus* is one of the least shade-tolerant species and since it grows taller and possibly starts growth earlier in the season it occupies the lighter portions of the forest floor. *E.striatum* is a very shade-tolerant species, and competitive effects of *R.triquetrus* have probably forced it to occupy the more shaded portions of the forest floor. Since the light and shaded portions of the forest floor are determined by the dimensions of the tree canopy the scale of heterogeneity of the distribution of these moss species is similar to that of the tree canopy. Errington considered that the smaller scale of pattern of *M.undulatum*, which has an upright growth form, was probably due both to competition with the other species and to the availability of water.

Such studies tend to give the impression of a stable pattern of distribution since the field collection of data occupies a relatively short period of time. Watt (1960b) presents the results of a study of grass-heath species measured annually over the period 1936–1957. Analysis for pattern can be made on his data for Lakenheath Warren in the Breckland where he recorded the presence or absence of species in $1\frac{1}{4}$ cm square quadrats within an area of 10 cm by 160 cm. Using *Galium saxatile* as an example the frequency of occurrence within the quadrats is shown in Fig. 1.14. It can be seen that there is a periodicity with intervals of nine and 10 years between maxima, which is shown to correlate with the rainfall early in the growing season ($r = 0{\cdot}660$, $p = 0{\cdot}01$). Watt interprets the fall in frequency as a response to spring and summer drought, and a rise in frequency as an expression of the capacity of the plant to increase in a favourable environment, both responses reflecting the plant's Atlantic distribution.

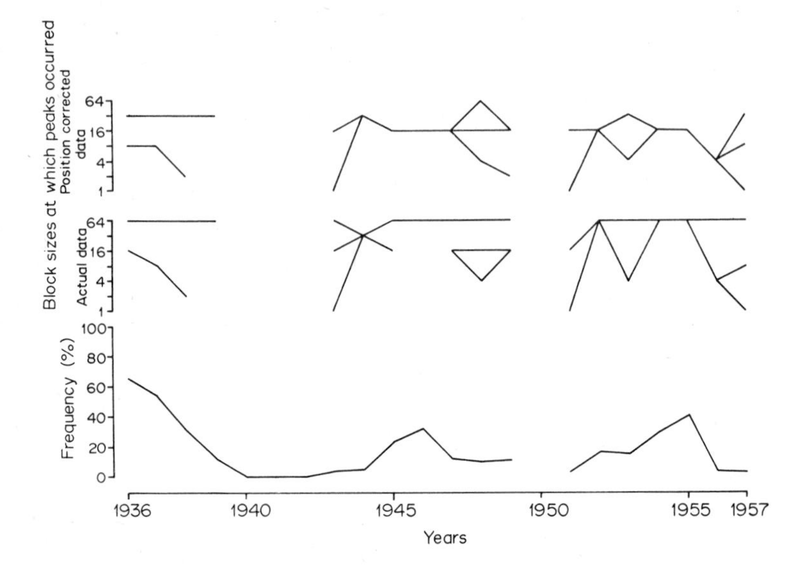

Fig. 1.14 The frequency and pattern of *Galium saxitile* on the same transect during the period 1936–1957. (Based on data given by Watt, 1960)

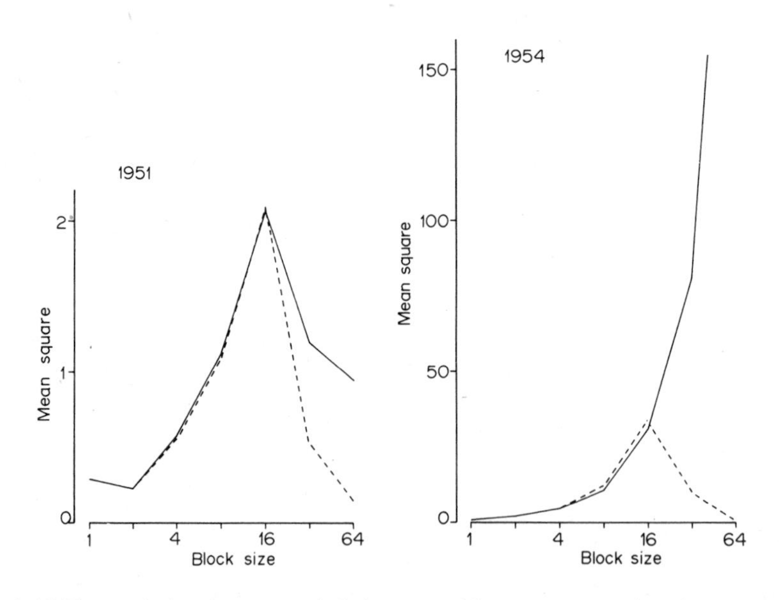

Fig. 1.15 The analysis of pattern of *Galium saxatile* on a transect in 1951 and 1954. The continuous line shows the analysis of the actual data, and the dotted line shows the correction for linear trend in the distribution of *Galium* along the transect. (Calculated from data given by Watt, 1960b)

Carrying out a pattern analysis one is faced with the problem of correcting the data for trend (Greig-Smith, 1961a). The location of peak mean squares is indicated in Fig. 1.14. It can be seen that very frequently the curve of mean square plotted against block size, using the basic data, increases up to the largest block size of 64 (as in the 1954 data in Fig. 1.15), due mainly to the decreasing frequency of *G.saxatile* along the transect. This masks the true scale of heterogeneity in the pattern. Elimination of the linear trend (see Chapter 2 for the elimination of trend in time series analyses) from the data yields more clearly defined peak mean squares. Fig. 1.15 shows analyses for 1951 (frequency 3·4 per cent) and 1954 (frequency 30·0 per cent), and in both of these graphs the correction for trend yields a peak mean square at block size 16. It will be seen that the maximum heterogeneity is often associated with this block size (Fig. 1.14), indicating a scale of pattern of the order of 20 cms. It can also be seen from Fig. 1.14 that at times of low frequency the scale of the pattern alters, and peaks occur at smaller block sizes.

The subject of pattern analysis is given an extensive treatment by Pielou (1969) (Chapters 7–16).

Simulation of pattern analyses

Simulation can be considered from the point of view of a mathematical topic, or it can be used as a relatively simple ecological tool. The latter approach will serve here in order to attempt to answer some questions about the methods of the pattern analysis that have been used in the preceding section. Simulation in this context is thus an attempt to mimic a field situation by mathematical and computational models.

All simulation processes have two features in common. First, there is the basic requirement of a mathematically expressed model of the system that is to be simulated. If one, for example, wishes to simulate the behaviour of an insect in an environmental gradient, the model may consist of dividing the gradient into zones, defining for a period of time how the insect can move in relation to these zones (i.e. a probability function), and formulating the rules under which the insect will respond to the gradient. The model may be deterministic, in which case the result of a simulation will always be the same, or it may be stochastic, embodying the random element, in which case the result should give an expectation together with an estimate of the variance of this expectation.

Secondly, there is the computation of the simulation process. This can be extremely time-consuming if hand machines are to be used, but with the general availability of electronic computers many simulation techniques are extremely fast. In many cases there is no theoretical means, or no mathematically simple theoretical means, of determining the expected outcome and its associated variance. A large number of simulated runs of the model, building in random variation by the use of a random number generator, will give an estimated mean and variance of the outcome. Thus although no theoretical expectation or variance have been determined, working estimates of these parameters can be made.

Before dealing with the problems of pattern analysis it will be useful to give a computer programme for the process. The programme is written in BASIC since this language is almost self-explanatory, and is close enough to both ALGOL and FORTRAN for users of either of these languages to understand it. The programme is:

```
 1 REM PATTERN ANALYSIS PROGRAM. MAXIMUM
 2 REM TRANSECT LENGTH 128 QUADRATS
10 DIM D[128]
11 DIM S[8]
12 REM ARRAY D HOUSES THE TRANSECT DATA
13 REM ARRAY S HOUSES THE UNCORRECTED
14 REM SUMS OF SQUARES
20 PRINT "TRANSECT LENGTH";
21 INPUT N
22 REM N IS THE TRANSECT LENGTH
30 FOR I=1 TO N
31 READ D[I]
32 NEXT I
40 FOR I=1 TO 8
41 LET S[I]=0
42 NEXT I
50 LET M=N
51 LET P=1
59 REM ACCUMULATE SUMS OF SQUARES.
60 REM P=BLOCK SIZE
61 FOR I=1 TO M
62 LET S[P]=S[P]+D[I]*D[I]
63 NEXT I
70 IF M=1 THEN 90
71 LET M=M/2
72 LET P=P+1
80 REM GROUP QUADRATS IN PAIRS
81 FOR I=1 TO M
82 LET D[I]=D[2*I]+D[2*I-1]
```

```
83 NEXT I
84 GOTO 61,55
90 PRINT
91 PRINT
93 PRINT "B.S.                D.F.            ";
94 PRINT "  UNCORRECTED      CORRECTED";
95 PRINT "          M.S."
96 PRINT "                                    ";
97 PRINT "        S.S.            S.S.";
98 PRINT
100 LET S[P]=S[P]/N
101 PRINT N,"                              ",S[P]
109 REM F IS DEGREES OF FREEDOM
110 REM M IS BLOCK SIZE
111 LET P=P-1
112 LET M=N
120 FOR I=P TO 1 STEP-1
121 LET M=M/2
122 LET S[I]=S[I]/M
123 LET F=N/(2*M)
130 LET S1=S[I]-S[I+1]
131 LET S2=S1/F
140 PRINT M,F,S[I],S1,S2
150 NEXT I
160 PRINT
161 PRINT
200 DATA 1,1,1,0,0,0,0,0,0,0,0,1,1,1
201 DATA 0,0,0,0,0,0,0,0,1,1,1
999 END
```

The first question that can be asked about a pattern analysis is 'What does the peak mean square refer to?'. We know that it gives a measure of the heterogeneity of the pattern at that particular block size, but pattern has two components, namely the size of the clump and the size of the space between clumps. Errington (1970) investigated this question by using clump sizes of 1, 2, 3, 5, 8 and 13 units, each in combination with space sizes of 1, 2, 3, 5, 8 and 13 units. The results were identical when the space and clump sizes were interchanged, for example space size 3 and clump size 8 gave the same results on analysis as clump size 3 and space size 8 (see Table 1.5). In this case the largest peak mean square is at block size 4, the peak at block size 16 largely being due to the shortness of the transect (there is no peak when the transect length is 1024 and the same pattern is used). The simulation of the various block and clump sizes, Fig. 1.16, shows that the peak mean squares occur at the block size smaller than the number $\frac{1}{2}$(clump + space size). Thus with clump and space sizes of 3 and 8 respectively (or vice versa) the peak mean square can be predicted as occurring at the block size smaller than $\frac{1}{2}(3 + 8) = 5{\cdot}5$. This can be confirmed from Fig. 1.16 and Table 1.5.

TABLE 1.5 Analysis of a pattern of clump size 3 and a space size 8 in a transect of 128 units

Pattern	11100000000111000000001110000000011100000000111000000000...			
B.S.	*d.f.*	*Uncorrected S.S.*	*Corrected S.S.*	*M.S.*
128	–	10·125	–	–
64	1	10·125	0	0
32	2	10·125	0	0
16	4	11·000	0·875	0·2188
8	8	12·000	1·000	0·1250
4	16	21·000	9·000	0·5625
2	32	30·000	9·000	0·2812
1	64	36·000	6·000	0·0938

If we wish to estimate the mean clump size (c) from a pattern analysis with a peak mean square at block size (b), then we know that

$$b \leqslant \tfrac{1}{2}(c + s) < 2b \tag{1.3}$$

where s is the mean space size. We also know that cover or frequency (f) is given by

$$f = c/(c + s) \tag{1.4}$$

and thus we can put the following limits on the mean clump size,

$$2fb \leqslant c < 4fb \tag{1.5}$$

where the value of f is estimated from the field data and the value of b is given by the analysis.

In these simulation experiments we have had a fixed scale of pattern, and criticism that such patterns never occur in the field is justified. A second question about the analysis might then be 'What is the effect of random fluctuations about some basic pattern?'. Errington (1970) simulated this situation by randomly placing clump sizes of 4 units along a transect of length 1024 units such that clumps did not overlap but could be contiguous. Cover values of 3·5 per cent, 18·6 per cent, 36·6 per cent 57·3 per cent and 69·6 per cent were generated and analysed. Some of the

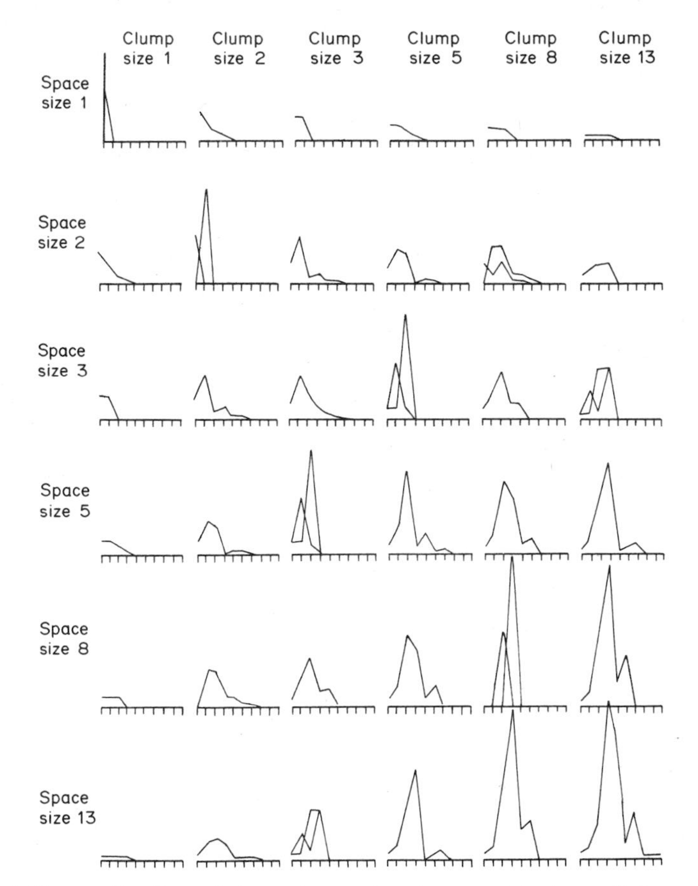

Fig. 1.16 The results of simulating patterns of various combinations of clump and space size on a transect of length 512 units. (Redrawn from Errington, 1970)

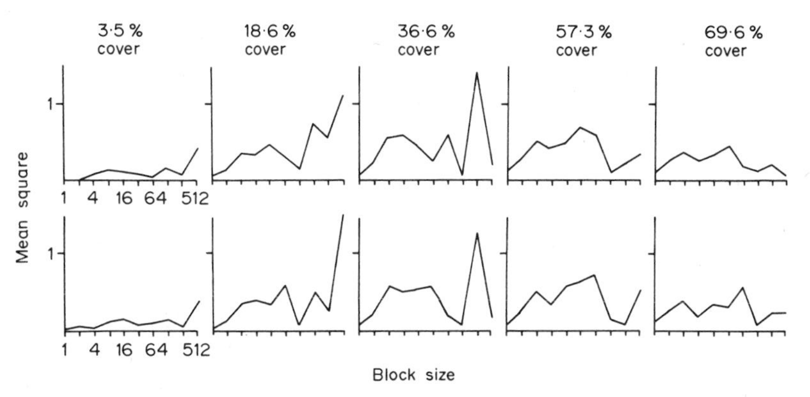

Fig. 1.17 Simulation of patterns of randomly spaced groups of 4 units with various cover values on a transect of length 512 units. (Redrawn from Errington, 1970)

results are shown in Fig. 1.17. Knowing that the clump size is 4 units, Equation (1.4) can be used to estimate the mean space size, and from this we can predict peak mean squares at block sizes of 32, 8, 4, 4 and 2 for the five percentages respectively. The results show fairly good agreement with expectation when the cover value is intermediate, but there are discrepancies at either end of the range of cover values. In this simulation as the cover became greater the clumps coalesced to form groups of 8, 12 or more units. This coalescing makes the assumption of a clump size of 4 and a space size of 1·75 unreal. If both these figures were multiplied by 1·5, as the data suggest would be appropriate, the predicted peak mean square would be at block size 4, which is in accordance with the results of the analysis. When cover values are very low, mean square peaks occur at block sizes 2 or 4 and 16 or 8, and in all cases at 128 as well. It appears that at low cover values the analysis picks out approximately the scale of the clump size since all analyses had a peak at block size 2, 4 or 8. The analysis also indicates the magnitude of the space size (110 units on average) by the peak at the larger block size.

A third question that we may ask is 'What is the effect of the starting position in a pattern on the results of the analysis?'. Usher (1969b) simulated this question with several models of increasing complexity, and showed that the results of the analysis depended upon the starting position. Using clump and space sizes of 8 units, the analysis gave a peak mean square at block size 8 provided that the patterns

1 1 1 1 1 1 1 1 0 0 0 0 0 0 0 0 1 1 1 . . . . .
1 1 1 1 1 1 1 0 0 0 0 0 0 0 0 1 1 1 1 . . . . .
1 1 1 1 1 1 0 0 0 0 0 0 0 0 1 1 1 1 1 . . . . .
1 1 0 0 0 0 0 0 0 0 1 1 1 1 1 1 1 1 0 . . . . .
1 0 0 0 0 0 0 0 0 1 1 1 1 1 1 1 1 0 0 . . . . .,

or the patterns as above with the zeros and ones reversed, were used. The other patterns

1 1 1 1 1 0 0 0 0 0 0 0 0 1 1 1 1 1 1 . . . . .
1 1 1 1 0 0 0 0 0 0 0 0 1 1 1 1 1 1 1 . . . . .
1 1 1 0 0 0 0 0 0 0 0 1 1 1 1 1 1 1 1 . . . . .,

and the similar patterns with zeros and ones reversed, gave a peak mean square at block size 4. This led to a suggestion that in field work a longer transect should be taken, and separate analyses carried out on quadrats

$1 - 2^n$, $2 - 2^n + 1$, $3 - 2^n + 2$, and so on, and the peak mean square to the right should be used for further study.

Errington (1970) carried out similar simulations on many patterns. He shows that the effect of this leftward shift in the peak mean square is most pronounced when the clump and space sizes are approximately equal to each other and to a power of two. Fig. 1.16 shows the various alternative results for his simulation experiments, and it can be seen that a leftward shift occurs whenever the sum of clump plus space size is a power of two. The effect of starting position is minimal when the sum of the clump plus space size is of the form $\frac{1}{2}(2^n + 2^{n+1})$, where n is any integer.

Simulation experiments have thus shown how to interpret the results of field data so as to estimate the average clump and space sizes in patterns. They have also shown that under certain conditions, those of very low cover or frequency, the analysis yields approximately the dimensions of the clump size, though other simulation trials showed that the results were very dependent upon the starting position in the sequence. The simulation experiments have also indicated an improved method of field sampling so that the effect of the starting location can be assessed. Such simulation techniques can be applied to any analysis of a complex field situation, where the components thought to be of importance can be specified and several computer simulation trials undertaken to ascertain the response of the analysis to changes in these components.

2. The Distribution of Organisms in Time

The analysis of time series

The processes with which biological conservation is concerned extend over many time scales, from the diurnal variations in the position or response of a population to the long-term changes in which a species may enter and leave a community during the successional process. A statistical approach to these situations is by the techniques of time series analysis, time series being defined as sets of observations taken at specified times, usually at equal intervals. The time series represented in Table 2.1 is artificial, but it can be taken to be the monthly catches of a moth species in a light trap during a 20-year period. The two immediate observations that can be made

TABLE 2.1 Artificial data for the monthly totals of a moth species caught in a light trap

Year	*Months*											
	J	*F*	*M*	*A*	*M*	*J*	*J*	*A*	*S*	*O*	*N*	*D*
1951	0	1	0	5	17	12	4	27	13	3	0	0
1952	0	0	0	1	23	10	2	19	24	6	1	0
1953	1	0	1	7	14	12	9	19	20	4	0	0
1954	0	0	1	6	45	9	1	3	14	2	0	0
1955	0	1	1	3	17	11	0	5	26	7	1	0
1956	0	1	3	12	13	2	5	24	10	0	0	0
1957	0	0	0	8	9	3	4	26	13	7	3	1
1958	0	0	1	4	12	7	3	13	23	10	0	0
1959	0	1	2	5	18	5	15	26	6	1	0	0
1960	0	0	0	5	21	5	6	22	15	0	0	0
1961	0	0	1	7	12	3	0	6	33	7	0	0
1962	1	0	0	3	23	3	2	17	14	5	1	0
1963	0	0	2	12	11	0	6	32	10	0	0	0
1964	0	0	0	1	16	2	3	27	12	6	3	1
1965	0	0	2	6	18	0	4	15	27	5	1	0
1966	1	1	2	8	14	5	6	12	23	4	1	0
1967	0	0	0	0	16	13	0	22	18	5	0	0
1968	0	0	1	6	9	3	8	12	18	6	0	0
1969	0	0	1	3	15	7	5	18	11	3	1	0
1970	0	0	2	9	7	0	11	21	16	1	0	0

from the data are first that the moth occurs seasonally, and secondly that it has two generations per year which are on the wing in May and during August and September. This is particularly clearly seen when the data are plotted as a graph of monthly catch against time, see Fig. 2.1 (*a*).

In general, all time series can be broken down into four components, defined as:

(1) *Long-term trends* which indicate the general direction in which the time series graph is moving. Fig. 2.2 (*a*) shows a graph with an upward trend, indicating that the species is becoming more abundant during the period of time. Although this graph shows a straight line relationship, a long-term trend might well demand the fitting of a curvilinear line (Kendall, 1951).

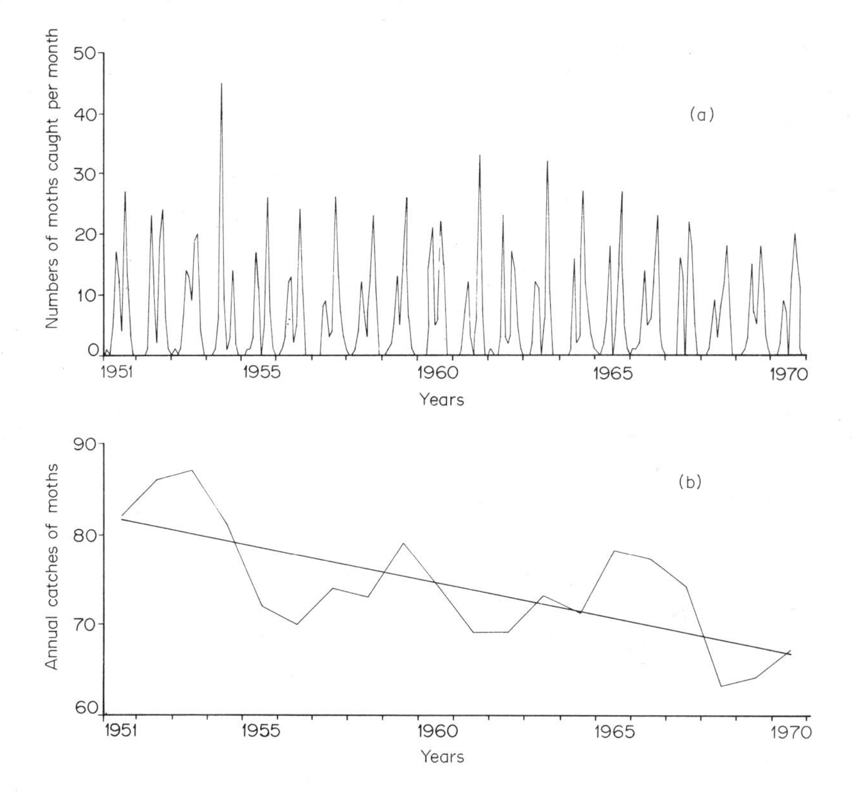

Fig. 2.1 Artificial data for the monthly catches of a moth species in a light trap. (a): The actual monthly data. (b): The annual catches of the moth shown together with the regression equation relating the annual catch to the time, Equation (2.2)

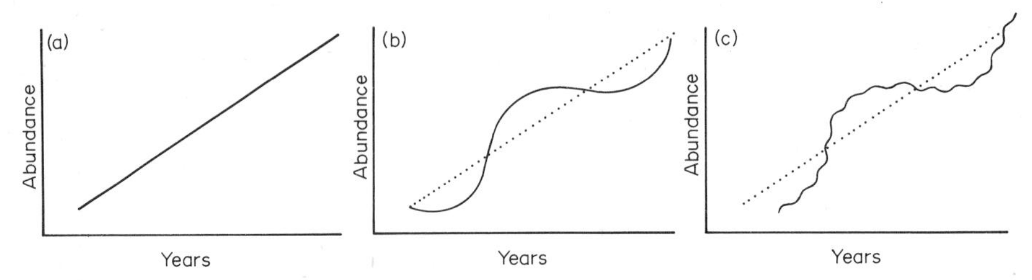

Fig. 2.2 Various components of a time series. (a): Trend. (b): Cyclical movements about a trend line. (c): Seasonal and cyclical movements about a trend line

(2) *Cyclical movements* refer to the long-term oscillations about the trend line. In order not to confuse this component with the third in the list, these oscillations are defined as having a periodicity of greater than one year, though the periodicity may or may not be regular. Such movements are shown in Fig. 2.2 (*b*).

(3) *Seasonal movements* refer to oscillations that a time series follows during the same period in successive years, or during the same time in successive days. The combination of seasonal movements with cyclical movements and a long-term trend is shown in Fig. 2.2 (*c*).

(4) *Random movements* refer to sporadic oscillations that are due to chance events. Biological data is usually very variable, and random movements can reflect the differing genetic qualities of the population, or the influence of random ecological factors such as weather.

To analyse the data in Table 2.1 a regression line was fitted through all the points. This gave the equation

$$m = 6{\cdot}7448 - 0{\cdot}0047\,t \tag{2.1}$$

where m is the monthly catch predicted from the trend line, and t is the number of months after the start of the experiment (January 1951 = month 1). The regression coefficient is, however, not significant ($F = 0{\cdot}39$ with $n_1 = 1$ and $n_2 = 238$ d.f.). The intense seasonal variation has masked the effect of any trend, if such a trend exists. The seasonal differences can be eliminated by taking the yearly catches. Using the total annual catch (Table 2.2), the regression is

$$M = 82{\cdot}3526 - 0{\cdot}7812\,T \tag{2.2}$$

where M is the predicted total annual catch, and T represents the number of years after the start of the experiment (1951 = year 1). The predicted annual catches using this equation are shown in Table 2.2, and the trend

TABLE 2.2 The total annual catches of moths in the light trap, and the predicted catches derived from the trend line, Equation (2.2)

Year	*1951*	*1952*	*1953*	*1954*	*1955*	*1956*	*1957*	*1958*	*1959*	*1960*
Total	82	86	87	81	72	70	74	73	79	74
Trend	81·6	80·8	80·0	79·2	78·4	77·7	76·9	76·1	75·3	74·5
Year	*1961*	*1962*	*1963*	*1964*	*1965*	*1966*	*1967*	*1968*	*1969*	*1970*
Total	69	69	73	71	78	77	74	63	64	67
Trend	73·8	73·0	72·2	71·4	70·6	69·9	69·1	68·3	67·5	66·7

line is shown in Fig. 2.1 (*b*). The regression coefficient is significant ($F = 17{\cdot}4$, $p < 0{\cdot}001$), and the negative sign of b indicates that the moth is becoming less abundant during the 20-year period. For a full discussion of the elimination of trend see Kendall (1951).

In order to estimate the seasonal movement the average percentage method, which expresses the data for each month as a percentage of the monthly average for that year, has been used. Thus in 1951 a total of 82 moths were caught, giving an average of 6·83 per month. (We have made an assumption that we can treat all months as being of equal duration. We could of course take an average, weighted according to the number of days in each month, but the arithmetic would become unnecessarily involved). Since four moths were caught in July 1951 this represents a catch of 4/6·83 or 58·6 per cent of the average monthly catch. Similarly the August catch of 27 moths represents 395·2 per cent. The percentages for all the months are shown in Table 2.3. The seasonal index is represented by the mean or median of each of the months. In this example the mean has been used since the months January, February, March, November and December have so many zero counts. Whenever the mean is used a certain amount of care is necessary as an abnormally high or low value in one year can distort the seasonal index. For this reason many statisticians advocate rejecting extreme values. For a discussion of the procedure see Spiegel (1961).

The cyclical movements can be investigated by first eliminating the trend from the data and then either by eliminating the seasonal movements or by calculating serial correlation coefficients. Elimination of components of the time series is carried out by dividing the table entry by the appropriate index. Thus in July and August 1951 the catches of moths

TABLE 2.3 Moths caught per month, expressed as a percentage of the average monthly catch for each year. The 'mean' row gives the seasonal index for the occurrence of this moth, expressed as a percentage

Year	*Months*											
	J	*F*	*M*	*A*	*M*	*J*	*J*	*A*	*S*	*O*	*N*	*D*
1951	0	14·6	0	73·2	248·8	175·6	58·6	395·2	190·1	43·9	0	0
1952	0	0	0	13·9	320·9	139·6	28·0	265·1	334·8	83·8	13·9	0
1953	13·8	0	13·8	96·6	193·1	165·5	124·1	262·1	275·8	55·2	0	0
1954	0	0	14·8	88·9	666·8	133·3	14·8	44·4	207·4	29·6	0	0
1955	0	16·7	16·7	50·0	283·3	183·4	0	83·3	433·3	116·6	16·7	0
1956	0	17·2	51·5	205·7	222·8	34·3	85·7	411·3	171·5	0	0	0
1957	0	0	0	129·7	145·9	48·6	64·9	421·8	210·8	113·5	48·6	16·2
1958	0	0	16·4	65·8	197·3	115·1	49·3	213·7	378·0	164·4	0	0
1959	0	15·2	30·4	76·0	273·4	76·0	227·9	394·8	91·1	15·2	0	0
1960	0	0	0	81·1	340·4	81·8	97·3	356·9	243·2	0	0	0
1961	0	0	17·4	121·7	208·7	52·2	0	104·4	573·9	121·7	0	0
1962	17·4	0	0	52·2	400·0	52·2	34·8	295·5	243·5	87·0	17·4	0
1963	0	0	32·9	197·3	180·8	0	98·6	526·0	164·4	0	0	0
1964	0	0	0	16·9	270·4	33·8	50·8	456·2	202·8	101·4	50·8	16·9
1965	0	0	30·7	92·3	277·0	0	61·6	230·8	415·3	76·9	15·4	0
1966	15·6	15·6	31·2	124·7	218·2	77·9	93·4	187·0	358·5	62·3	15·6	0
1967	0	0	0	0	259·4	210·8	0	356·8	291·9	81·8	0	0
1968	0	0	19·1	114·2	171·5	57·1	152·4	228·6	342·9	114·2	0	0
1969	0	0	18·7	56·3	281·3	131·3	93·7	337·4	206·3	56·3	18·7	0
1970	0	0	35·9	161·2	125·4	0	197·0	376·1	286·5	17·9	0	0
Total	46·8	79·3	329·5	1817·7	5285·4	1767·8	1532·9	5947·4	5622·0	1341·0	197·1	33·1
Mean	2·3	4·0	16·5	90·0	264·3	88·4	76·6	297·4	281·1	67·0	9·9	1·7

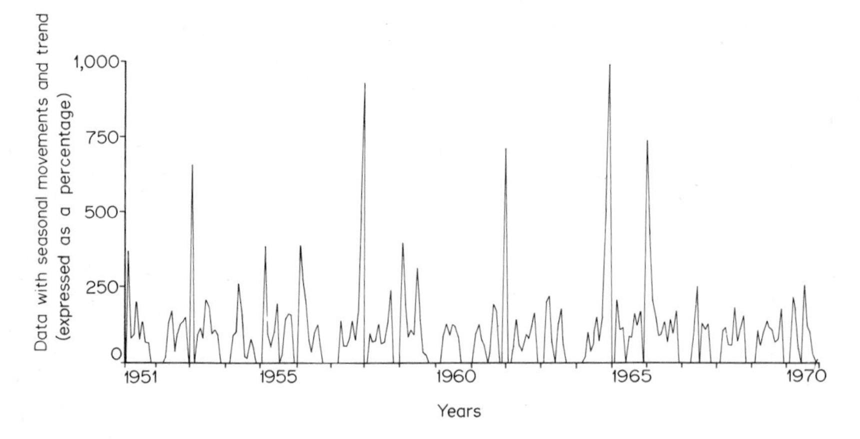

Fig. 2.3 Data for the monthly catches of a moth in a light trap after the elimination of trend and seasonal influences

were 4 and 27 respectively. Division by 0·766 and 2·974 (since the seasonal index is expressed as a percentage in Table 2.3) yields the seasonally corrected values of 5·222 and 9·079 for these two months. Using the monthly data predicted by the trend equation, Equation (2.1), the predicted catches during these months were 6·71195 and 6·70725 respectively. Division of the seasonally corrected values by the predicted value from the trend equation yields a result that reflects only the cyclical and random movements. Thus, for July and August 1951, the seasonally and trend-corrected data are 5·222/6·71195 = 77·80 per cent and 9·079/6·70725 = 135·36 per cent. The data for the whole of the 20-year period are shown in Fig. 2.3, where some evidence of a periodic cyclical movement can be seen.

The alternative method, that of serial correlation, relies upon calculating the correlation coefficient between data collected on the ith occasion and on the i–kth occasion, k being defined as the lag. Thus, with a lag of one year the pairs of variables that are correlated are the totals, corrected for trend, for 1951 and 1952, for 1952 and 1953, and so on until the totals for 1969 and 1970. For this particular series there are 19 pairs of variables to be correlated. The general formula for the serial correlation coefficient (Moran, 1952) is

$$\begin{aligned} r_k &= \frac{n}{n-k}\,\frac{\sum\limits_{i=1}^{n-k}(x_i - \bar{x}_i)(x_{i+k} - \bar{x}_{i+k})}{\sum\limits_{i=1}^{n}(x_i - \bar{x}_i)^2} \\ &= \frac{n}{n-k}\,\frac{\sum\limits_{i=1}^{n-k} x_i x_{i+k} - (\sum\limits_{i=1}^{n-k} x_i)(\sum\limits_{i=1}^{n-k} x_{i+k})/(n-k)}{\sum\limits_{i=1}^{n} x_i^2 - (\sum\limits_{i=1}^{n} x_i)^2/n} \end{aligned} \tag{2.3}$$

where n is the total length of the series. There are several different formulae for r_k, similar to the above, but Equation (2.3) is the simplest to calculate since the denominator summations are independent of k. With data for 20 years the serial correlation coefficients were calculated for lags of one to 15 years inclusive, and were plotted to form the correlogram in Fig. 2.4, from which it will be seen that there are (significant) positive correlations with lags of one, six, seven and 13 years, which indicate a cyclical movement with a periodicity of six or seven years.

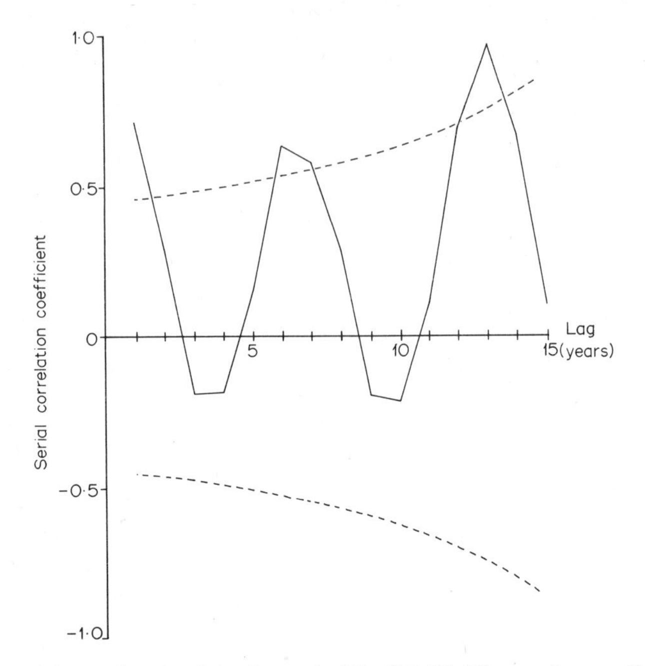

Fig. 2.4 The correlogram for the data shown in Fig. 2.1 (*b*). The continuous line joins the serial correlation coefficients, and the dotted lines give very approximate limits for significance

Negative correlations are, however, more useful in determining a periodicity. The question of testing for significance will be discussed later in this chapter.

The moth trap data thus exhibits all four components of a time series. There is a long-term trend, indicated by the significant regression line, which shows that the moth is tending to become scarcer over the 20-year period. The seasonal movements are obvious in that the species does not fly in winter, and that it has two generations which are to be found in the spring and summer. Elimination of these two components demonstrates that a cyclical movement of period six to seven years exists. The moth, though decreasing, has a maximum or minimum population size approximately every sixth year. The fact that none of the serial correlation coefficients equals $+1$ or -1, and the fact that the graph in Fig. 2.3 does not follow an exactly regular cycle, indicate that random movements are also present in our time series.

Having examined some artificial data which demonstrate the temporal processes that occur in communities, the remainder of this chapter will look at actual instances of these diurnal, seasonal and longer-term oscillations, as well as at the trends that occur within communities.

Diurnal rhythms

Although the term 'diurnal' has frequently been used to denote rhythms lasting in the order of 24 hours, the word 'circadian' is often favoured since it does not carry the implication of day as opposed to night. Danilevsky, Goryshin and Tyshchenko (1970) list seven types of circadian rhythms that are to be observed in terrestrial arthropods, including:

(1) *locomotor*, the general mobility of the animal, its flight activity and taxes of various kinds;
(2) *reproductive*, including the acts of copulation and oviposition;
(3) *ontogenetic*, the emergence of the larva from the egg, the emergence of the imago, and moults;
(4) *metabolic*, in relation to feeding, respiration and excretion;
(5) *morphological*, which is concerned with the volume of cells, the deposition of chemical substance layers in the cuticle, and the state of chromatophores;
(6) *biochemical*, the enzymatic and secretory activity of cells;
(7) *biophysical*, the electrical activity of neurons.

We must, however, ask ourselves which of these rhythms have ecological significance, and it is also pertinent to ask how environmental factors might modify such circadian rhythms.

Environmental modification of rhythms

Species that are demonstrating circadian rhythms will be susceptible to the normal environmental influences, and it is important to see to what extent the rhythms can be modified. Modification can be caused by seasonal cycles, and these will be discussed further in the section dealing with photoperiodism. Modification can be caused by climatic changes and other species in the environment, factors that may not have a regular periodicity. The amount of disturbance in the regularity of the circadian rhythm caused by a climatic change could be regarded as a measure of the strength of that rhythm.

An example of the relation between winter temperature, weather conditions and diurnal activity in the Collembola, *Dicyrtomina rufescens*, is given by Uchida and Fujita (1968). On 13 days during the winter of 1966–67 they recorded the temperature at hourly intervals, and counted the number of insects that were active. For the analysis of their results they distinguished four different weather types, based on the mean temperature and on the daily change in temperature. These four types are:

(1) 'calm and cold' days which had fair weather, no snowfall, and relatively low air temperatures (maximum 0°C and minimum −7°C);

(2) 'calm and warm' which were also fair with no snowfall, but the air temperature was relatively higher and there was a thaw (maximum 5°C, minimum −1°C);

(3) 'decreasing' type, which was characterized by the day starting with 'calm and warm' weather, but during the afternoon the temperature dropping and snow starting to fall;

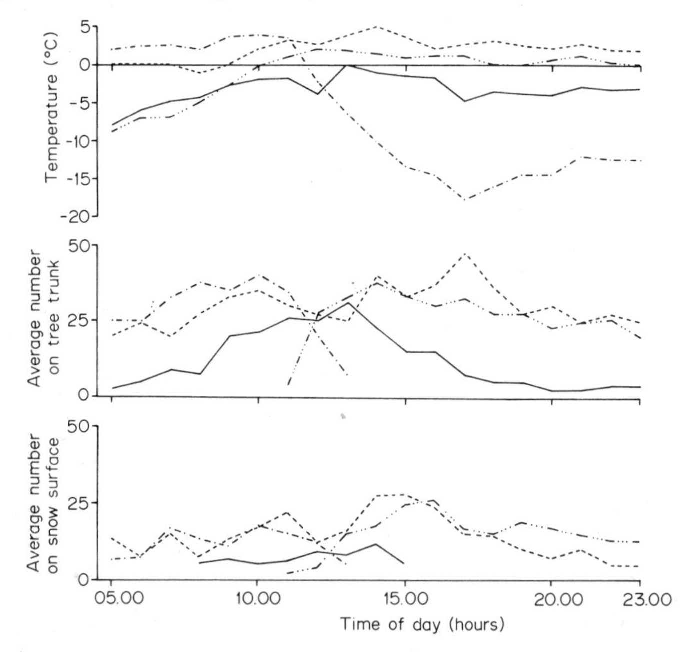

Fig. 2.5 Data for the diurnal activity of *Dicyrtomina rufescens*. The upper graph shows the diurnal temperature movements, whilst the two lower graphs show the activity of the Collembola on the tree trunks and on the snow surface. Lines representing the weather types,

———————— Calm and cold — · — · — Decreasing

— — — — — — Calm and warm —· · ·—· · · Recovery

are described in the text. (Redrawn from Uchida and Fujita, 1968)

(4) 'recovery' type in which the snow ceased to fall in the morning and the afternoon was 'calm and warm'. Temperature recordings taken during examples of each of these weather types are shown in Fig. 2.5 (*a*).

During really adverse weather *D.rufescens* congregate in small spaces beneath the snow. This is usually possible in the vicinity of trees where snow-holes have developed from water running down the tree trunk, and from which Collembola can emerge either onto the surface of the snow or onto the tree trunk where they feed on blue-green algae. During the 'calm and cold' weather type *D.rufescens* move about sluggishly on the snow surface and on the tree trunks, see Figs 2.5 (*b*) and (*c*), and during these conditions they are only to be seen for a few hours during the period of maximum temperature. However, during the 'calm and warm' weather type *D.rufescens* is active and jumps when stimulated. The number of animals on the tree trunks and on the snow surface remains large throughout the day and night, though the activity on the snow surface reflects the increased temperature during the daytime. During the 'decreasing' and 'recovery' weather types the behaviour of *D.rufescens* can be predicted from a knowledge of its behaviour under the stable weather types. As the temperature falls below $-10°C$ activity both on the snow surface and on the tree trunks ceases, and thus this temperature would appear to be a threshold below which all activity stops. After a period of bad weather it is noticeable that the Collembola do not become active again until a higher temperature is reached than that at which they would become active when there are more stable temperature conditions.

These observations demonstrate two factors about the circadian rhythm of this species. First, they do not prove the existence of locomotor and metabolic rhythms. The Collembola, whilst weather conditions are optimal, are active during the whole of the 24 hour period. Since observations were carried out on the population and not on individuals, this statement applies, therefore, only to the average activity of the population. Secondly, the observations demonstrate that a circadian rhythm is imposed on the species by the weather conditions. If the temperature is other than optimal the species is only active during that period of the day when the temperature is highest. As this period will usually be at more or less the same time on successive days it will produce an apparent circadian feeding and locomotor rhythm.

Activity in insects is not, however, only controlled by temperature.

Certain Lepidoptera are strongly attracted to emergent trees and small hills in tropical regions, gathering on and around the summits of such objects. Although the behavioural implications of these flights have never been fully worked out, it is known that most species occur in these positions for only a limited period of time, and that the time of day is specific to a species. Indeed some species that regularly collect in such places for a short period of time are very rarely seen during the rest of the day. When entomological survey work is to be undertaken it is important to realize that there is a diurnal succession of species on these summits. It seems likely that such diurnal behavioural responses are concerned with mating, and act as a mechanism whereby many individuals of the same species congregate in a small area at the same time.

Photoperiodic responses

Circadian rhythms in plants have been the subject of prolonged study by plant physiologists. Perhaps the most important diurnal response is that of photoperiodic reactions, though many other responses have been studied (see, for example, Street and Opik, 1970). The ecological importance of light has been demonstrated by Bainbridge, Evans and Rackham (1966).

The initial work on photoperiodic response was carried out on tobacco and showed that plants of a particular variety grew to a height of 10–15 feet and did not usually flower outdoors in the vicinity of Washington, D.C. However, when the plants were grown in a greenhouse during the winter they only grew a few feet in height, flowered freely and set seed. These observations led to the suggestion that this variety of tobacco was a 'short-day' type, in that the development of the flowers was initiated by the short-day conditions (an account of this work is given by Meyer and Anderson, 1952). Further work has shown that there are four types of photoperiodic response, being shown by:

(1) '*Short-day*' plants, which flower only when they experience a relatively short photoperiod. Flowering may occur, though more slowly or less profusely, within a range of photoperiods longer than the optimal ones, but as the photoperiod is further increased these plants will not flower and remain in the vegetative state indefinitely. Examples of short-day species are strawberry and violet, and most of the early spring and autumnal flowering species of the temperate zones.

(2) '*Long-day*' plants, which flower readily under a range of long

photoperiods. Under shorter photoperiods than the optimal ones flowering will be slower and less profuse, and under still shorter photoperiods flowering will not occur and the plants remain indefinitely in the vegetative state. Examples of long-day species are lettuce, and most late spring and early summer flowering plants of temperate zones.

(3) '*Intermediate*' plants, which flower only in a restricted range of

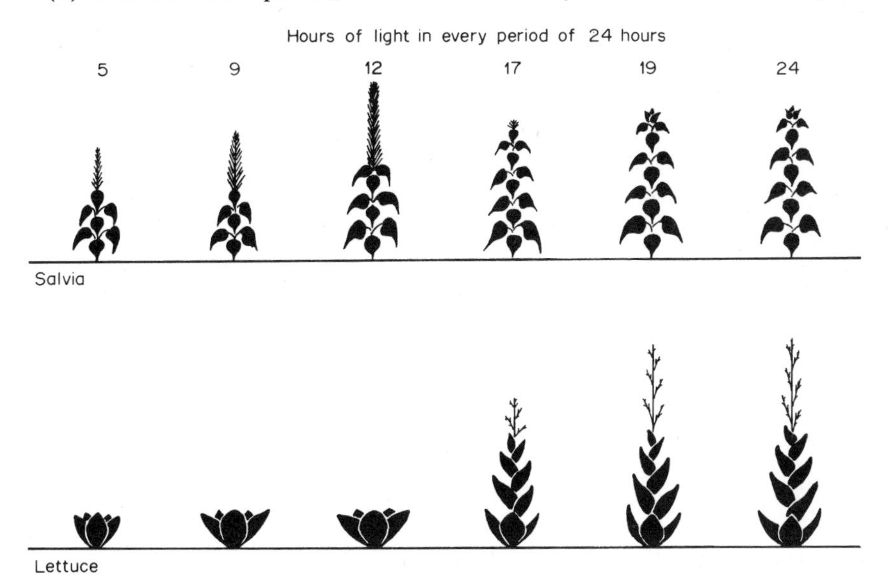

Fig. 2.6 A diagrammatic representation of short-day and long-day species. (Prepared from the photographs of *Salvia* and lettuce in Meyer and Anderson, 1952)

photoperiods. When the photoperiod is longer or shorter the amount of flowering will be reduced, or the plant will remain in the vegetative state.

(4) '*Indeterminate*' plants flower over a very wide range of day lengths. Examples of such species include weeds such as the chickweed and dandelion, as well as tomatoes and most varieties of tobacco.

Fig. 2.6 gives a diagrammatic representation of the responses of salvia (short-day species) and lettuce (long-day species) to various photoperiods between five hours per day and continuous illumination.

Photoperiodic responses are known in animals as well as in plants. Danilevsky, Goryshin and Tyshchenko (1970) illustrate the effects of photoperiod on the diapause of various insect species. Thus the butterfly *Pieris brassicae*, cabbage white, is a long-day type. Specimens from

Leningrad were all in diapause when the day-length was less than 14 hours, whilst none were in diapause when the day-length exceeded 16 hours. The same species from Sukhumi (on the Black Sea 17° south of Leningrad) showed all individuals active with day-lengths above 12 hours, and all in diapause with day-lengths of nine hours or less. The sawfly, *Neodiprion sertifer*, is a short-day species since it enters diapause when day-lengths exceed 15 hours and are shorter than four hours. Whilst the photoperiod is intermediate between these values either none or only a small proportion are in diapause. Intermediate types are known as, for example, the Lepidopteran *Euproctis similis* which enters diapause when the day-length is less than 15 hours or greater than 20 hours. Of most ecological significance are the threshold region of the photoperiodic reaction and the critical photoperiod (that which causes diapause in 50 per cent of individuals) since these determine the phenological time for the diapause. The day-length which causes 100 per cent of individuals to enter diapause is also of importance since it prohibits any further development of the species.

Photoperiodic responses are thus the result of the circadian rhythms of light and dark. Their ecological significance is manifest not only in the species' geographical distribution and phenology, but most importantly in the role of the species in the ecosystem. In plants the flowering season will be related to the photoperiodic response, since long-day plants can only flower during the summer in temperate regions and can never flower in the tropics. Short-day plants, inhibited from flowering during the temperate summer, will flower either during the spring or autumn. The response to photoperiod is rapid and threshold photoperiods can be determined. Thus Meyer and Anderson state that the threshold of cocklebur is 15·5 hours, and any photoperiod shorter than this will stimulate flower initiation. The fact that there is an immediate response to day-length can be demonstrated for the cocklebur. One short day followed by one long night has been found to be sufficient to initiate flower development in plants that have otherwise been kept under long-day conditions. This rapid response obviously has implications to laboratory and growth-room experiments where the supply of electricity for the lighting may be interrupted. This indeed happened during the strike of power workers in Britain in 1970 when short-day plants, being grown vegetatively under long-day conditions, developed flower buds following electricity cuts.

Discussion of the results of the photoperiodic response, often expressed in the phenology of the species, will form the topic for the next section of this chapter.

Seasonal rhythms

The term 'phenology' has been widely used for the study of seasonal variability of the life cycle phases, their activities, and their temporal relationships one to another. The term is, however, best kept for the descriptive or observational study of these occurrences, whilst 'phenometry' is used for analytic and quantitative studies where measurements of such parameters as photosynthetic activity or hormonal changes have been made.

Descriptive phenology has led to the production of diagrams designed to convey all the essential information. Fig. 2.7 shows diagrams for two sea birds, the fulmar, *Fulmarus glacialis*, and the gannet, *Sula bassana*, based on illustrations in Fisher (1954). The fulmar's eggs are found between early May and mid-July, and young are present from the end of June till mid-September. Birds that have lost their eggs or young soon disperse to the North Atlantic and North Sea areas, whilst birds that have bred successfully and their young disperse late in August and during the first half of September. All birds are widely dispersed at sea during October, but between November and the following April the birds gradually return to their breeding sites. Thus, the illustration has selected two features of the birds' life cycle, the reproductive processes and the spatial distribution, and has presented these in a diagrammatic form. One criticism is obviously that the data are shown as mean values, and approximate dates, without any indication of the variability about these mean values.

The illustration for the gannet in Fig. 2.7 is more complicated since it incorporates an attempt to show the variability in the reproductive cycle, as well as incorporating different spatial distributions for two age groups of birds. Reference to the illustration shows that the young migrate to North West Africa along the European coasts. These birds arrive back at the breeding sites later than the adult birds that have spent the winter months in home waters. Phenological studies in animals are thus concerned with a description of the reproductive cycle and with a statement of the distribution of the species, be it of a migratory nature or some form of diapause (for example winter hibernation or summer aestivation).

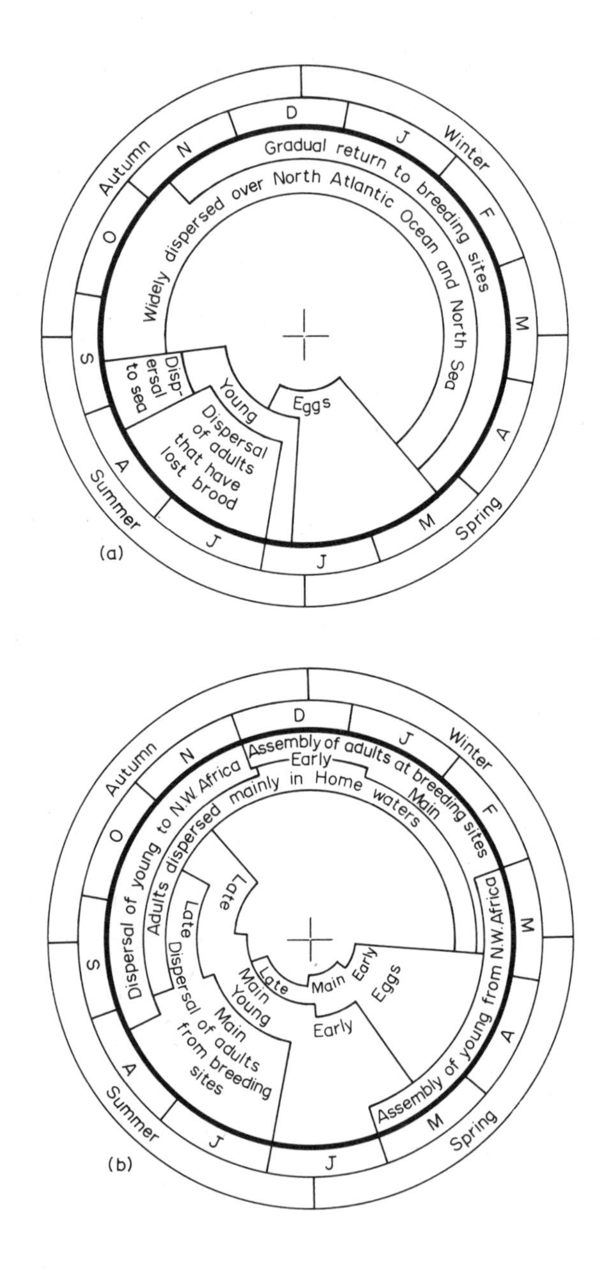

Fig. 2.7 Diagrams showing the life cycles of two species of sea birds. (a): Fulmar. (b): Gannet. (Redrawn from Fisher, 1954)

Phenological studies in plants are similarly concerned with the reproductive cycle, but they are also concerned with the various stages in the development of the plant. Each distinguishable phase within the life cycle of a species is known as a 'phenophase' – typical phenophases would be germination, bud bursting, vegetative growth, flower buds, flowers, unripe seeds or fruits, ripe seeds or fruits, litter fall, death or die back, resting period. Interest is focussed upon the beginning and end of each phenophase since these two dates delimit the duration of that particular phase. When all of the phenophases of a species are compiled into an annual sequence this is referred to as the 'phenodynamic'. The phenodynamics of several species of grass are shown in Fig. 2.8, where it can be seen that *Anthoxanthum odoratum* flowers during the first half of June, and seed dispersal has been completed before the middle of July. In contrast *Phleum pratense* comes into flower a month later, but the flowering and seed ripening and dispersal phases are all of short duration, and reproductive activity is completed by early August. In these phenodynamics the degree of slope of the lines denoting the start and end of a phenophase indicates the abruptness of that transition. Thus, for *Festuca rubra* flowering activity finishes relatively abruptly as indicated in Fig. 2.8 by the near-vertical line. However, the yellowing of the leaves is a very gradual process and hence it is indicated by a line with only a few degrees slope.

If the phenodynamics of all the species forming a community are tabulated the collection is referred to as a 'phenological spectrum'. Fig. 2.8 is not such a spectrum since it contains data for only seven species, whilst in the community there was a total of 34 species. Lieth (1970), however, illustrates the phenological spectrum of the false oatgrass (*Arrhenatherum elatius*) association from which Fig. 2.8 is compiled. 'Phenograms' for communities are constructed by recording either the number or percentage of either individuals or species that are entering a particular phenophase. In the false oatgrass association previously mentioned Lieth's phenological spectrum shows that, on 2nd June, 13 of the 34 species were in flower, whilst only one species was in the phenophase of unripe seeds or fruit. On 2nd July the number of species in flower had increased to 23 and the numbers with unripe and ripe seeds were 14 and 8 respectively. These data are plotted in Fig. 2.9 to show phenograms for this community for the four phenophases of flower buds, flowers, unripe seeds and ripe seeds. It is clear that there is one major flowering season, during late June and early July, when over 50 per cent of the species are

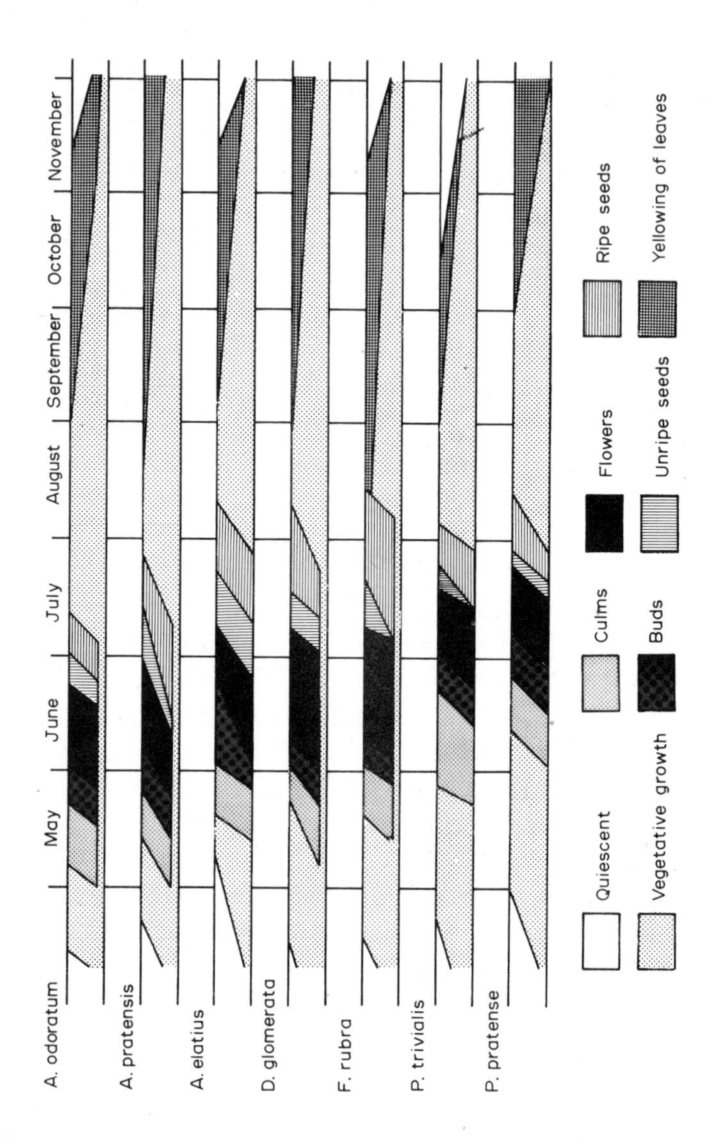

Fig. 2.8 The phenology of some grass species in a Polish *Arrhenatherum elatius* association. (Redrawn from the phenological spectrum in Lieth, 1970)

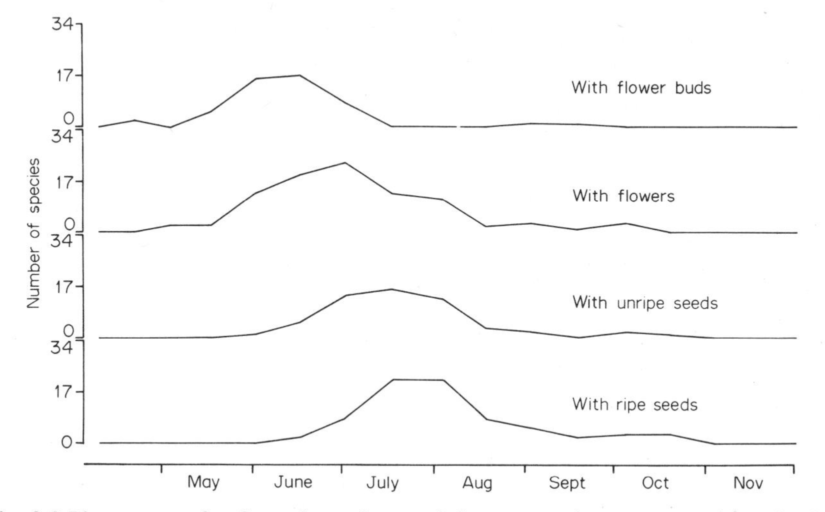

Fig. 2.9 Phenograms for four phenophases of the community represented in Fig. 2.8. The upper line is for buds, the second for flowers, and the third and fourth for unripe and ripe seeds and fruits respectively. (Drawn from data given by Lieth, 1970)

in flower. The series of phenograms in Fig. 2.9 shows the approximate time lag between flower buds and seeding. For the main peaks in these graphs this lag can be estimated as being approximately 50 days. However, in the autumn four of the species flower for a second time, and this is shown in Fig. 2.9 by the slight peak at the right hand end of the four graphs. It can also be seen that there is a time lag of approximately 35 days between the peak for flower buds and that for ripe seeds. Thus phenograms can indicate the relative speed of the processes during the season.

The main use, however, of the phenogram is for comparisons between ecosystems. It has been found in the south eastern United States that in dry areas the phenogram for the flowering phase is bimodal with spring and autumn peaks, whilst in wet land communities there is a pronounced summer peak of flowering.

Perhaps the main interest of phenology in conservation is the descriptive or observational approach that we have just examined, since preliminary questions will be concerned with what to expect and when to expect it. However, such an approach does not quantify the contribution of a species to the community to which it belongs. For a complete understanding of a community the contributions of the individual components are a prerequisite, as we shall see in Chapter 3. Lieth's 1970 paper is very much a plea that ecosystem research should not just be concerned with annual totals of energy accumulation and utilization, but rather that the seasonal influences should be understood and incorporated in ecosystem modelling. Lieth shows that in stands of wild garlic, *Allium ursinum*, in Germany the above ground dry matter is very closely related to the chlorophyll content of the leaves during the period mid-March to mid-June. These are also fairly closely related to the photosynthetic surface, but show no relation to the below ground dry matter. Similarly, discussing the data for the false oatgrass association, depicted in Fig. 2.8, Lieth shows that the maximum dry matter production of both the grass *Dactylis glomerata* and the herb *Rumex acetosa* coincides with the time of flowering. Phenometric studies thus not only quantify the contribution of the species to the ecosystem, but they can also be used to predict this contribution from the morphological or physiological features of the species.

Long-term cycles

In examining the pattern of *Galium saxatile* in Chapter 1 we saw that the species increased and decreased in frequency with a cycle of approximately 9–10 years. Watt (1960) attributed these fluctuations to the climatic effects of eastern England acting on a species that has an Atlantic form of distribution. Earlier, Watt (1947) had drawn attention to the dynamic nature of plant communities and to the interrelationships of temporal and spatial distribution.

It is, however, amongst some animal species that we see the most pronounced cyclical fluctuations, though Williamson (1971) points out that such cycles are of rare occurrence amongst the many species of the animal kingdom. The behaviour of the lemming in migrating from its mountain habitats is well known, and documented by Elton (1942). Using his data for South Norway (areas to the south of South Trøndelag), migration years occurred in 1862–3, 1866, 1868–9, 1871–2, 1875–6, 1879–80, 1883–4, 1887–8, 1890–1, 1894–5, 1897, 1902–3, 1906, 1909–10, 1918, 1920, 1922–3, 1926–7, 1930, 1933–4 and 1938. There are thus 21 migrations during the period 1862–1938, and it is obvious from the dates that there is a tendency towards recurrence of migrations every third or fourth year.

When cycles exist it is important to try to discover the main causes.

For the lemming Elton lists five possible causes, though he concedes that the controlling factors are still far from completely known. Predators tend to have similar cycles to their prey, and it is known that tundra predators, such as the arctic fox, will follow lemming down into the lowlands. Disease, particularly Tularaemia, affects the lemmings and is thought to be transmitted by other parasites or by blood-sucking flies. Emigration removes part of the population, and it has been documented that the first waves of a migration contain adults from the previous year. There is also a removal of the replacing generation since later waves of a migration consist almost entirely of young lemmings born in that same year. Climatic effects are also thought to be contributory, since it is suggested that a warm autumn followed by a mild spring will tend to stimulate a lemming migration. It is also suggested that the autumn of a migration year tends to be colder than an average autumn. Of these five factors listed by Elton three, predators, disease and climate, act upon all animal populations, and the remaining two factors, emigration and lack of replacement, are perhaps more manifestations of the cycle than causal factors. It seems, therefore, that other factors, possibly of a nutritional nature, are acting to maintain the cyclical nature of the lemming populations.

When cycles exist one must have analytic methods of investigating causal factors, of carrying out tests of significance, and of being able to predict with a fair degree of certainty what is likely to happen. The regular periodicity of 10 years in the lynx cycle in Canada, as estimated by the Hudson Bay Company (Elton and Nicholson, 1942), has led to speculation that the sunspot cycle is the causal factor for the lynx cycle (see Fig. 2.10). Using the series of the logarithms (to base 10) of the numbers of lynx trapped and the series of sunspot numbers, the correlation coefficient between the series,

$$r = \frac{\sum_{i=1}^{n}(x_i - \bar{x})(y_i - \bar{y})}{\sqrt{\left(\sum_{i=1}^{n}(x_i - \bar{x})^2 \sum_{i=1}^{n}(y_i - \bar{y})^2\right)}} \tag{2.4}$$

over the period 1821 to 1934, was shown by Moran (1949) to be 0·1329. Williamson (1971) explains the basis for using logarithms of all population counts in analytic procedures. Since there are 114 pairs of observations the usual test of significance of r would be with 112 degrees of freedom, but since each of the variables x_i and y_i is serially correlated this test of significance is inappropriate. However, even with 112 degrees of freedom the value of r is not significant, and hence a more exact test of significance is not required to show that the lynx and sunspot cycles are not related. Reference to Fig. 2.10 (*a*) shows that the two cycles are nearly in phase from 1820 to 1850, that they are nearly in opposite phase from 1850 to 1890, and that they are nearly in phase again from 1890 to 1930.

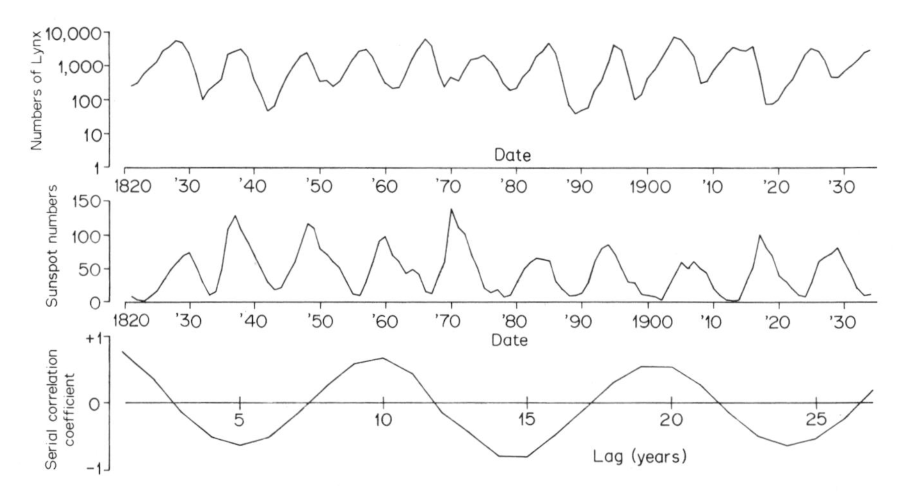

Fig. 2.10 Data for the lynx cycle in North America and for the sunspot cycle. The upper graph show the lynx cycle over the period 1821–1934, plotted as the logarithm (to base 10) of the number trapped. (Drawn from data in the Appendix to Moran's (1953) paper). The centre graph shows the number of sunspots. (Redrawn from Moran, 1949). The lower graph is a correlogram for the lynx series. (Drawn from the values of r_k given by Moran, 1953)

In order to investigate tests for the relationships between such series more fully let us use as an example the four species of game-bird, grouse, ptarmigan, capercailzie and blackgame, shot on Scottish estates during the period 1866 to 1938. The data are given by Mackenzie (1952), and the methods of analysis are discussed by Moran (1952). Correlation coefficients were calculated using Equation (2.4) in order to determine if the cycles of numbers of the four species were related, and the results are shown in Table 2.4. As with the lynx and sunspot relationship it is inappropriate to test these correlation coefficients with 71 degrees of freedom (they are each based on 73 pairs of observations) since the series

TABLE 2.4 Correlation coefficients between four series of game-birds. (Source: Moran, 1952)

	Ptarmigan	*Capercailzie*	*Blackgame*
Grouse	0·504	0·504	0·347
Ptarmigan	–	0·417	0·052
Capercailzie	–	–	0·189

are serially correlated. Using a lag of one year, the serial correlation coefficients calculated from Equation (2.3) are shown in Table 2.5. Moran argues that the number of degrees of freedom to use in testing the correlation coefficients in Table 2.4 is approximately

$$n^* = (n - 1)\frac{1 - \rho_s\bar{\rho}_s}{1 + \rho_s\bar{\rho}_s} \tag{2.5}$$

where the values of ρ_s and $\bar{\rho}_s$ are estimated from the values of r_1 for each of the two series being correlated. Moran shows that the values of r_1 tend

TABLE 2.5 Serial correlation coefficients (r_1) for the four species of game-bird. (Source: Moran, 1952)

Species	r_1	ρ_1 *(values used)*
Grouse	0·643	0·70
Ptarmigan	0·483	0·55
Capercailzie	0·592	0·65
Blackgame	0·427	0·50

to underestimate the values of ρ_s in Equation (2.5), and hence he increases the r_1 values by small arbitrary amounts, as indicated in the third column of Table 2.5. Thus, for the comparison between grouse and ptarmigan, the number of degrees of freedom is given by

$$n^* = 72\left(\frac{1 - 0{\cdot}7 \times 0{\cdot}55}{1 + 0{\cdot}7 \times 0{\cdot}55}\right) \doteqdot 32$$

Using 32 degrees of freedom it is found that the correlation between these two series is significant at the 1 per cent level. The correlation between the grouse and the capercailzie is also significant at the 1 per cent level, and that between the capercailzie and ptarmigan is very near the 1 per cent level, whilst that between grouse and blackgame is significant at the 5 per cent level. Thus, overall it can be seen that the series of grouse, capercailzie and ptarmigan are definitely related.

One word of caution in using Equations (2.4) and (2.5) is that they are only appropriate when trend has been eliminated from the data. If there is a trend in the same general direction in both the series this will show as a spurious positive correlation, whilst if the trends are in opposite directions this will provide a negative correlation. It is thus advisable to eliminate the trend by some regression technique as is indicated in the first section of this chapter.

Returning to the lynx cycle Moran (1953) further analyses the data (see Fig. 2.10 (*b*) for the correlogram), and proposes two models for the cycles. One is based on the smooth recurrence properties of the sine curve

$$x_t - m = a \sin (bt - c) + e_t \tag{2.6}$$

where x_t is the number at time t, m is the mean of the series, a, b and c are constants, and e_t is a sequence of independent random variables which can be considered as errors. The essential feature of this model is that the random element does not affect the future course of the cycle. The second model postulates that the population at time t is dependent upon the population at the k previous times, thus

$$x_t - m = a_1(x_{t-1} - m) + a_2(x_{t-2} - m) + a_k(x_{t-k} - m) + e_t \tag{2.7}$$

where $a_1, a_2, \ldots, a_k$ are constants. By a study of the correlogram there is evidence that the fluctuations are slightly damped, Fig. 2.10 (*b*), indicating that the first model would be inappropriate since it would generate peaks and troughs in the correlogram of approximately equal amplitude. A study of the partial serial correlation coefficients indicates that a value of k of 2 in Equation (2.7) would be appropriate for the second model, in which case the model becomes

$$x_t - 2{\cdot}9036 = 1{\cdot}4101(x_{t-1} - 2{\cdot}9036) - 0{\cdot}7734(x_{t-2} - 2{\cdot}9036)$$

and thus

$$x_t = 1{\cdot}0549 + 1{\cdot}4101x_{t-1} - 0{\cdot}7734x_{t-2} \tag{2.8}$$

The main interest in Equation 2.8 is its predictive value. Using the counts of 485 in 1929 and 662 in 1930 (taking logarithms $x_{t-2} = 2{\cdot}686$ and

$x_{t-1} = 2{\cdot}821$), Equation (2.8) estimates x_t as $2{\cdot}956$ or a total of 904 lynx (the actual catch was 1000 animals). If Equation (2.8) is repeatedly applied, using estimated values for x_{t-1} and x_{t-2}, it yields a series which shows damped oscillations, reflecting the increasing uncertainty of the phase cycle for an increasing number of years ahead.

Ecological succession

In the analysis of time series at the start of this chapter the term 'trend' was used to denote the directional changes that occur during a period of time, and for the purposes of ecological work trend can be taken as being related to succession. Succession can be defined in terms of the evolution of communities, and particular features of this process are the permanent change in the relative proportions of species, the arrival of species in a community and the extinction of other species.

In order to describe ecological succession several approaches to the problem are possible. We shall, however, reserve discussion of the recently developed analytic techniques till the end of this section and concentrate first on the descriptive studies. Areas that have had the community destroyed present a possible situation for observing succession, and in Chapter 1 we examined such a situation following the experimental defaunation of small islands. The method is, however, extremely slow if the whole succession is to be followed, and is particularly slow in the temperate regions of the world. In tropical regions growth is faster, and hence direct observation of destroyed communities is possible. Another line of approach is to find areas where some continuing process, such as the melting of a glacier or the deposition of sand, is continuously providing a new and uncolonized habitat. In these situations as one moves further from the present area of habitat addition one should pass through the stages of the succession until one reaches the climax or edaphic climax communities.

Descriptive methods

Richards (1957) describes the secondary succession to rain forest in Nigeria. In the Shasha Forest Reserve, the land economy is based on shifting cultivation. When land is cleared the debris is burnt, and tropical crops such as maize, yams, bananas and plantains are grown. After a few seasons the ground becomes unprofitable, is abandoned in favour of a new area, and secondary succession begins. Thus, within a fairly small Reserve it is possible both to watch the process of succession and in different areas to find different stages of the succession.

When cultivation is abandoned the ground is bare, except possibly for a few perennial crops. Within a few weeks it is covered by herbaceous species, Group (*a*) species, which may rapidly form a closed stand though their dominance is only temporary. Fairly soon, within a year or so, a few species of tree typical of such secondary habitats, Group (*b*) species, become dominant. Certainly within three years the species *Musanga cecropioides*, a species of Group (*b*), had become dominant in all the areas. This species is abundant in tropical Africa, but it is apparently unable to colonize bare ground, only establishing itself rapidly after there is a cover of vegetation. After five years the canopy is 4–5 m high. After 15–20 years the *Musanga* trees die when the canopy is 20–25 m high. Other Group (*b*) species persist for a while but at this stage many young trees of other high forest species become increasingly numerous, Group (*c*) species. The main characteristic of Group (*c*) species is that they are shade-intolerant high forest species whose seedlings are able to establish themselves under open conditions. After this in the succession there will be a decrease in the frequency of Group (*c*) species, and a gradual increase in the shade-tolerant high forest species. It is interesting to note that the complexity of the community, measured by the number of species present, increases during the succession.

Thus tropical secondary succession to high forest consists of three clearly defined stages. Firstly, there is the short-lived herbaceous stage which forms a closed canopy and provides the necessary environment for the second stage. The second or *Musanga* stage is characterized by the dominance of this one species, and the shade cast by the species forming this stage provides a suitable microclimate for the development of the high forest species. These two stages are also characterized by rapidly growing species that are intolerant of shade. The third stage lasts for a much longer period of time, and consists of the gradual replacement of secondary species by the slower growing species of the primary forest. Thus although the first stages of the succession are clear-cut, the final stage merges very slowly into the climax community. Succession is seldom a series of distinct stages but is usually determined as a series of stages that merge from one to the other, quicker at the start of the succession and slower towards the end of the succession.

Fungal successions on decaying plant material occupy a relatively short period of time. Frankland (1966) describes the succession of fungi on the decaying petioles of bracken, *Pteridium aquilinum*. She shows that although the normal succession on plant material is:

Weak parasites → Primary saprophytic sugar fungi → Cellulose decomposers and Secondary sugar fungi → Lignin decomposers and Associated fungi

the bracken succession is:

Cellulose and/or lignin decomposers → Sugar fungi

Harley (1971), however, asked if such successions were true successions. He suggested that the colonization of all fungi occurred at approximately the same time, but that the apparent succession represented the differential rates of development of the species.

These descriptions of succession have largely relied upon the observation of one habitat for a long period of time. The other means of establishing the successional stages is to observe many habitats at one period of time. A site which has been subjected to habitat accumulation has been the shingle beach at Dungeness in Kent (Scott, 1965). By examining the present state of the vegetation, Scott was able to postulate a relatively complicated successional sequence for the site. The younger stages in the succession tend to occur towards the shore on the more recent shingle, whilst the older stages were on the oldest established shingle to landward. Scott does, however, state that distance from the shore is not the only factor since the establishment of vegetation depends upon the presence of fine particle material amongst the shingle. Since such material was initially absent from Dungeness the gradual accumulation of mineral and humic material by wind transport is an important influence acting upon the initiation of the succession.

The first stage in the succession is bare shingle with encrusting lichens. Although the lichen contributes little to the fine particle content, it does indicate areas of stable shingle that are suitable for further colonization. The second stage comes with the establishment of prostrate broom, *Sarothamnus scoparius*. Broom can become established in shingle which has a very small content of fine particle material, but once it is established its roots penetrate to a depth of at least 1 m where they can find water.

Broom itself slowly improves the moisture holding properties of the shingle by providing humus and by trapping wind carried particles. The succession, illustrated in Fig. 2.11, enters a cycle of humus increase and of death and regeneration of the broom. At some time during this cycle, when the broom is declining and when sufficient humus has been accumulated, heath becomes established and broom no longer features in the succession. The floristic diversity increases during this process. At stage 3 (broom with pioneer species) the list contains seven species, at stage 4

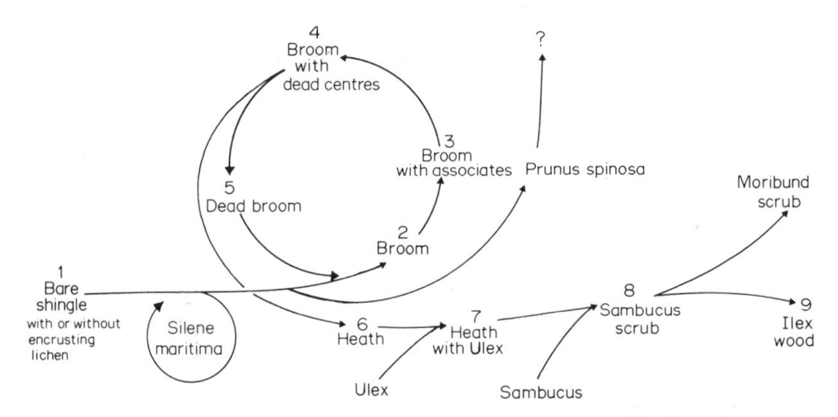

Fig. 2.11 A diagrammatic representation of the succession on shingle at Dungeness in Kent. (Redrawn from Scott, 1965)

(broom with dead centres) there are 14 species, whilst at the heath stage, stage 6, there are 25 species. It would appear that at this stage in the succession there is the maximum floristic diversity.

After the heath stage the succession enters a process whereby the character of the vegetation changes from being low (less than 1 m in height) to being high, with the development of a sort of scrub. First, gorse, *Ulex europaeus*, appears in the succession, and adds a large amount of humus to the shingle. In the next stage bushy species appear. The most abundant species is the elderberry, *Sambucus nigra*, but other species include the holly, *Ilex aquifolium*, blackthorn, *Prunus spinosa* and bramble, *Rubus fruticosus*, with willows, *Salix* spp., where the water table is high. The elderberry has its projecting shoots severely damaged by the wind, and the species would seem to need the protection of the gorse at least until it is well established. *Sambucus* bushes grow up to a height of about 3 m. At this stage in the succession there can be a halt to any further species change, with moribund elderberry bushes slowly degenerating, though perhaps occurring with *P.spinosa*, *R.fruticosus* and *Urtica dioica*. Alternatively, if *Ilex* is present, this becomes dominant and forms small thickets of one to three trees. Such thickets are associated with moribund or dead *Sambucus*, and they contain such climbing and trailing species as *R.fruticosus* and *Lonicera periclymenum*. Since no further development of these holly thickets are to be found, and since holly seedlings are abundant in the thickets, it is concluded that this forms the climax type of community. Fig. 2.11 shows diagrammatically the succession just described, but it also includes some other forms of succession that occur if the first stage is followed by *Silene maritima* or *P. spinosa* instead of broom.

More classical work on succession studies through the accumulation of new habitat has centred on sand dune systems. An area of exceptional interest, because of its rapid rate of growth of 7 to 14 m per year, is Tentsmuir on the south of the River Tay in eastern Scotland. Gordon (1963) shows that the succession starts with the foreshore and the embryo dunes, which in the southern part of the area contain isolated plants of *Agropyron junceiforme* and in the northern part a denser cover of *Elymus arenarius*. The second stage in the dune succession is the semi-stable dune which is dominated by the marram grass, *Ammophila arenaria*, but contains the floristic elements of the embryo dune. As the succession proceeds to stable dunes a mixture of species forming a completely closed ground cover replaces *A.arenaria*. Grasses such as *Festuca rubra* are dominant, but herbs such as *Trifolium repens*, *Plantago lanceolata* and *Lotus corniculatus* are abundant.

The parent material of the dunes consists of blown sand, and since this is derived from the sea it is likely to contain quantities of shell material. On the west of Scotland shell forms a large proportion of the sand, and on weathering it releases the calcium to give a base rich soil. However, on the east coast of Scotland the amount of shell in the sand is minimal, as, for example, at Aberlady Bay where Usher (1967) showed that the proportion of shell in the sand was less than two per cent. Thus, at the stable dune stage in the succession in eastern Scotland the soil contains little calcium and no formation of horizons is evident. Since there is no clay to hold the plant nutrients most of them are subsequently removed by leaching, leading the succession in the direction of acid heathland. Three stages can be noted at Tentsmuir. First there is the fixed moss dune where the vegetation is dominated by hypnoid mosses, particularly *Hylocomium splendens* and *Rhytidiadelphus triquetrus*. The next stage is the grey dune, so named because of the abundance of lichens of genus *Cladonia* and the relative infrequency of the hypnoid mosses. The final stage of the succession is the dune heath, dominated by ericaceous species such as *Calluna vulgaris*, *Erica cinerea* and *Empetrum nigrum*.

At the same time as the dunes are developing the slacks are also undergoing a succession. This succession is less obvious to a casual observer, but the process has been analysed by Crawford and Wishart (1966). Their analysis depends upon the numerical methods of classification described in Chapter 4. It demonstrates that the most forward dune slacks, the youngest in the succession, are dominated by *Honkenya peploides* and *Juncus gerardii*, which evolve into slacks dominated by *Salix repens* and *Juncus balticus*. This change requires a minimum of about 25 years. Then the succession depends upon the depth of the water table and the flooding frequency. In the wetter areas it evolves into a slack dominated by alder, *Alnus glutinosa*; in the drier areas into a *Hieracium pilosella* slack; and in intermediate areas into various types of *Erica tetralix* slacks. The first *E.tetralix* slack appears on land that is about 50 years old. It would seem that the final stage in the slack succession is a marsh slack, with *Filipendula ulmaria* and *Juncus effusus*, which is in the region of 100 years old.

Thus in the sand dune system we have twin developments in the dune and the slack successions. The former evolves towards a community that is dependent upon the concentration of shell material in the sand and the latter evolves towards a marsh, whose exact characteristics depend upon the water availability.

Analytic methods

Two recent developments in the analysis of successional data introduce new ideas. Working on a forest succession in America, Stephens and Waggoner (1970) and Waggoner and Stephens (1970) use a probability approach. The analysis depends upon estimating the probability that a community dominated by species A will become dominated by species B, C, . . . after a period of time. Using a three species situation as an example, let us assume that at the start of a period we have 300 sample plots, of which 200 are dominated by species A and 100 by species B. At the end of the period let us assume that 100 are still dominated by A, 160 by B and 40 by C. The changes in the series of sample plots are shown in Table 2.6, and the transition probabilities are shown in Table 2.7. Thus, of the 100 plots dominated by species B at the start, 20 evolved into

TABLE 2.6 The changes in dominance of species A, B and C over a period of time. The table gives the numbers of trees

		Start of period			*Totals*
		A	*B*	*C*	
End	A	100	0	0	100
of	B	80	80	0	160
period	C	20	20	0	40
Totals		200	100	0	

TABLE 2.7 The changes in dominance of species A, B and C over a period of time. The table gives transition probabilities

		Start of period		
		A	*B*	*C*
End	A	0·50	0	–
of	B	0·40	0·80	–
period	C	0·10	0·20	–

dominance by C, none into dominance by A, leaving 80 that remained dominated by B. The transition probabilities are thus:

B changing to A, $p = 0$
B remaining as B, $p = 0{\cdot}80$
B changing to C, $p = 0{\cdot}20$

Using a total of 327 plots, Stephens and Waggoner recorded dominant species in five categories, namely (i) maple, (ii) oak, (iii) birch, (iv) other major hardwood species and (v) minor species. All of these categories include more than one species. Several criteria for dominance can be used, but in a forest succession the number of stems per unit area or the basal area (sectional area of trees measured at breast height expressed on a unit area basis) can be used to determine dominance. With a 10-year period, the transition probabilities are given in Table 2.8. Maintaining the

TABLE 2.8 The transition probabilities for five groups of dominant species measured over the period 1927–1937. (Source: Stephens and Waggoner, 1970)

			1927		
	Maple	*Oak*	*Birch*	*Other*	*Minor*
1937					
Maple	0·82	0·16	0·13	0·07	0·07
Oak	0·07	0·72	0·02	0·03	0·07
Birch	0·02	0·08	0·83	0·07	0·07
Other	0	0	0·02	0·69	0·07
Minor	0·09	0·04	0	0·14	0·72

order of maple, oak, birch, other and minor, the data in Table 2.8 can be expressed in matrix form,

$$\mathbf{T} = \begin{bmatrix} 0{\cdot}82 & 0{\cdot}16 & 0{\cdot}13 & 0{\cdot}07 & 0{\cdot}07 \\ 0{\cdot}07 & 0{\cdot}72 & 0{\cdot}02 & 0{\cdot}03 & 0{\cdot}07 \\ 0{\cdot}02 & 0{\cdot}08 & 0{\cdot}83 & 0{\cdot}07 & 0{\cdot}07 \\ 0 & 0 & 0{\cdot}02 & 0{\cdot}69 & 0{\cdot}07 \\ 0{\cdot}09 & 0{\cdot}04 & 0 & 0{\cdot}14 & 0{\cdot}72 \end{bmatrix}$$

The matrix **T** can be used for three analytic techniques in relation to the forest succession that it describes. **[The reader not familiar with matrices is referred to Appendix A for a definition of a matrix and a summary of matrix operations.]**

First, it can be used to predict what is likely to happen during the next period of time. If we consider that the structure of the community at time t is represented by the vector $\mathbf{v}_t$, such that the elements of the vector give the number of plots dominated by maple, oak, birch, other and minor species in that order, then the structure after a further period of time, $\mathbf{v}_{t+1}$, is estimated as

$$\mathbf{v}_{t+1} = \mathbf{T}\mathbf{v}_t \tag{2.9}$$

As an example, if we had 113 maple, 128 oak, 42 birch, 29 other species and 15 minor species plots, then after a further period of 10 years we should expect there to be

$$\mathbf{T}\begin{bmatrix} 113 \\ 128 \\ 42 \\ 20 \\ 15 \end{bmatrix} = \begin{bmatrix} 122 \\ 103 \\ 50 \\ 22 \\ 30 \end{bmatrix}$$

plots of each of these groups. It will be seen that in both these vectors the number of plots adds up to the correct total of 327.

Secondly, we can make predictions further into the future. If we repeatedly use Equation (2.9) it can be seen that

$$\mathbf{v}_{t+2} = \mathbf{T}\mathbf{v}_{t+1} = \mathbf{T}^2\mathbf{v}_t$$

and hence

$$\mathbf{v}_{t+k} = \mathbf{T}^k\mathbf{v}_t \tag{2.10}$$

where k is any whole number of periods of time in the future.

Thirdly, we can ask if there is any stable end point and if all of the groups of species will be represented. For this we need a latent root and vector of **T**. Since all of the elements of **T** are probabilities and are thus non-negative (i.e. zero or positive) and since all the column totals are one, the theorem of Brauer (1962) states that the maximum latent root is one. Other theoretical considerations lead us to realize that this is the only latent root that we are interested in. Thus, solving the equation

$$\mathbf{T}\mathbf{v} = \lambda\mathbf{v} \tag{2.11}$$

where $\mathbf{v}$ is the latent vector and λ the latent root, gives

$$\mathbf{v} = (128, 54, 72, 17, 56)' \tag{2.12}$$

Hence, we should expect that after a considerable length of time there would be 128 plots of maple, 54 of oak, 72 of birch, 17 of other species and 56 of minor species, and that there would be no more change in the overall structure of the series of 327 plots. The latent vector given in Equation (2.12) is not exactly correct since the elements have been rounded to the nearest whole number such that their sum is 327.

Stephens and Waggoner discuss the relationship between prediction based on the 1927–1937 transition probabilities and the actual observed transitions up to 1967. In general the matrix model agrees very closely with the observed changes in the forest.

The second recent advance in the analysis of successional data is described by Williams, Lance, Webb, Tracey and Dale (1969). Their mathematical treatment is complex, relying first upon the numerical recognition of distinct groups in the succession. Their analysis is of particular interest since during the first year of the tropical succession the groups refer to temporal changes in the succession, whilst after this period of time the groups refer to the spatial distribution of the sites. After the initial numerical determination of the groups the analysis continues with the calculation of transition matrices in a similar manner to that of Stephens and Waggoner (1970).

Random fluctuations

The final or residual feature of the analysis of time series, described at the start of this chapter, was the random fluctuation. After removing all other sources of variation, trend, seasonal and cyclical, this one component was left. Similarly, in this chapter it has been left until last since its effect is universal in all biological studies. Random variability affects all biological observations, making replication of experiments essential and the statistical treatment of data imperative.

Perhaps the most interesting random fluctuations acting upon populations and communities are the catastrophes. These are essential components of some forest ecosystems where seral stages in the succession occur where fire has destroyed the climax forest type. An example of this is the stands of lodgepole pine, *Pinus contorta*, in the climax forests of North West America, where the fire which destroyed the original forest was caused either by human negligence or by the result of lightning. Flooding is a similar catastrophe which affects ecosystems. Reference to the list of floods in the Vale of York (Radley and Sims, 1970) shows that there have been 23 serious floods in the last 200 years, eight of these in the last 100 years and 15 in the previous 100 years. There is no periodicity in these floods, and their effect on a very wide area of Yorkshire must be one of the random influences acting upon communities living in these areas. In general, though, most random variability cannot be assigned to such definite causes.

3. The Concept of the Ecosystem

Definition

In the previous two chapters we have investigated the spatial and temporal heterogeneity of organisms, using from time to time the terms 'population' and 'community'. A 'population' was assumed to be all of the individuals of one of the species in a multi-species 'community', or to be a community of only one species. However, this does not define either term rigorously. We must either make a precise definition or investigate why a more casual definition is the more useful.

In talking about pattern (Chapter 1) and cyclical changes (Chapter 2) we used as one example the work carried out by Watt (1960) on *Galium saxatile*. He collected data from an area measuring 10 cm by 160 cm on Lakenheath Warren in the Breckland, and the results, based on this one small portion of the Warren, were taken to characterize the whole area. Which then is the population of *G.saxatile*? In some cases we could consider it to be those plants within the 10 cm by 160 cm study area, whilst in other cases we might wish it to be all the *G.saxatile* on the Warren.

It appears then that the casual definition of population or community is the most appropriate for ecological and conservation usage. Sometimes the terms will be used to refer to the organism(s) within the actual study area and sometimes to refer to those within a wider area of which the study area is just a sample. The term population will be used for one species, and community for two or more species. But these two terms relate only to the species that inhabit a given area, and we require a more general term for the totality of this area. Such a term is 'ecosystem'. Once again a casual definition will be used, implying either a restricted study or sampling area, or, and more frequently, a wider area that has been sampled.

We have, however, to fix some limits to the spatial size of the ecosystem. A boundary is fixed in a completely arbitrary manner, but there are certain conventions on the nature of such a boundary. First, the ecosystem must be large enough for all the functional processes to be represented. These processes include the flow of energy and the cycling of nutrients, which are both discussed in later sections of this chapter. Secondly, the boundaries should be sited so that the inputs and outputs across the boundaries are relatively easily measured.

An ecosystem is an arbitrary class in a hierarchy of classifications of the sum of the biotic and non-biotic environment. Each stage in this hierarchy is called an ecosystem. Thus, we could consider a single oak tree with the space around it as an ecosystem, though with the neighbouring trees overlapping with it in the canopy and in the soil it would be very difficult to measure inputs and outputs. We could consider an oak forest to be an ecosystem, and this is the scale that would more usually be referred to as an ecosystem. We could, however, take larger and larger units until we consider the whole biosphere to be one ecosystem. The size of the ecosystem is thus purely a function of its use.

This arbitrary concept is a means whereby comparisons can be made between areas. The concept is useful initially, being a framework on which to build studies and compare them with other studies, but there it ceases to be useful. The classification of ecosystems and the allocation of areas to particular ecosystems is not a research tool and cannot advance our knowledge of the functions of the biosphere. However, to the conservationist the concept of the ecosystem is of tremendous importance. It is seldom possible to expend the effort required to elucidate all the processes occurring within an area when plans for management are being drawn up. By allocating the area to an ecosystem it can be compared with other similar ecosystems, and extrapolation of the results of conservation management in other areas may be possible. In this chapter we will be concerned with the functional aspects of the ecosystem, analysing the processes that occur and the methods whereby these processes can be quantified and modelled. In the next chapter we will be concerned with subjective and objective methods of their classification.

The flow of energy

One of the unifying concepts in ecology, bringing together studies on both plants and animals, has been the energy flow in ecosystems. In order to carry out such studies we require climatological knowledge to determine whence the energy comes and how much arrives, and we have to partition the ecosystem so as to trace where the energy goes. This later factor has been the stimulus for a deeper understanding of biological production, since food for an increasing population of man has to be manufactured by a member or members of various ecosystems.

Input of energy

Light is one of the most important physical factors since it is the ultimate source of energy for all ecosystems and since it governs the circadian and seasonal rhythms of many species (see Chapter 2). Odum (1963) separates three components of light that are of importance. First, he considers intensity, and shows that plants can become genetically adapted to differences in light intensity. Thus, the alpine sorrel, *Oxyria digyna*, from the Yukon where the light intensity is low reaches its maximum rate of photosynthesis at a relatively low light intensity (about 2000 foot candles), whilst plants from Colorado, where the light intensity is higher, do not reach their maximum rate of photosynthesis till over 6000 foot candles. The maximum rate of the Colorado plants is, however, well in excess of that of the Yukon plants, and even at 2000 foot candles the rate of the Colorado plants is the greater. Secondly, Odum considers wavelength. The sun's radiation emission ranges from radio waves (10^4 microns (μ) in wavelength) to gamma waves with a wavelength of $10^{-6}\,\mu$, but about 99 per cent of the radiation is in the range 0·2 to 4 μ (Kormondy, 1969). About half of this energy is in the range 0·38 to 0·77 μ, the visible spectrum. Thirdly, Odum considers duration. In the temperate regions of the world this has a considerable influence upon the diurnal and seasonal activity of plants and animals.

Besides these three features of radiation mentioned by Odum we also need to be able to quantify the amount of energy available, and to investigate its geographical and seasonal variability. The solar flux is the amount of radiant energy per unit area per unit time to reach the earth's upper atmosphere. This has been estimated as averaging $1{\cdot}53 \times 10^9$ cal m^{-2} an^{-1} (calories per square metre per annum). This is an average figure, since the flux for any particular location on the upper atmosphere obviously varies seasonally and diurnally.

Collingbourne (1966) analyses the destination of the solar flux. First, some of the energy will be scattered by the air molecules and particles smaller than the wavelength of the energy. Secondly, there will be absorption by the atmospheric gases. Thirdly, there will be scattering and absorption by particles in the atmosphere that are larger than the wavelength of the radiation. Finally, some of the radiation will reach the earth's surface. On a clear day Collingbourne estimates that about 22 per cent of the radiation will be absorbed, 17 per cent reflected back into space and 61 per cent will reach the earth's surface. However, on an overcast day the amount reaching the earth's surface will be reduced to 28 per cent, whilst 19 per cent and 51 per cent are absorbed and reflected respectively.

Thus the prevailing weather conditions can cause seasonal disturbances in the amount of solar energy reaching the earth's surface. Also important can be the pollution in the atmosphere, as it is known that the amount of solar radiation reaching the ground surface is reduced in heavily industrialized areas. The latitude is also of considerable importance as this determines the sun's position in the sky relative to the site. Usher (1970a) has developed a computer programme for use in ecological studies of the sun's position. Besides these factors the aspect of the site and the degree of slope also affect the amount of solar radiation received. Measurements of solar energy are standardized by always being carried out on a horizontal surface. Using the theoretical considerations of solar flux and latitude, curves can be drawn showing the seasonal distribution of the amount of radiation received at any point on the earth's surface (e.g. Kormondy

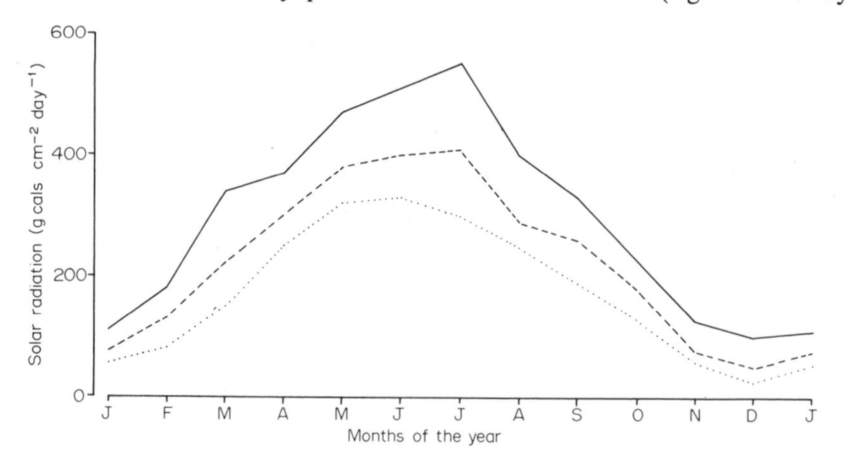

Fig. 3.1 The seasonal variation in solar energy. The localities are
———————— Central France
— — — — — — Southern England
· · · · · · · · · · · · · Northern Scotland

1969). These curves demonstrate the general pattern of radiation availability, but they do not show how local factors of climate and pollution affect the conditions. Fig. 3.1 shows the energy received at three sites, one near the Massif Centrale in France (46°N), one on the south coast of England (51°N) and one on the Moray Firth in Scotland ($57\frac{1}{2}$°N). These show that the symmetrical annual cycle of the theoretical curves is not in fact observed on the ground. The most nearly symmetrical curve in Fig. 3.1 is that from France, an area of a more continental type of climate with generally clearer skies.

General features of solar radiation availability are that the energy is reduced by an oceanic as opposed to a continental type of climate, and that energy is reduced by heavily polluted atmospheres. Near to the Tropics the amount of available solar energy is relatively constant throughout the year, being greatest at the time of the spring and autumn equinox and least at the summer and winter solstice. As one moves further into the temperate regions the seasonal pattern becomes more pronounced, with the maximum at the summer solstice and the minimum at the winter solstice. Beyond 70°N or 70°S no radiation is received for a part of the year, although very large quantities of radiation can be received during the short summer period in arctic and antarctic regions.

Movement of energy in the ecosystem

In order to discuss the flow of energy through the ecosystem we must separate out the two main pathways of energy flow. Most importantly there is the grazing food chain of plants and animals. An example of this would be:

grass → cattle → man

though it can be more complicated with more links, as:

phytoplankton → zooplankton → herring → man.

These pathways of energy flow can be generalized as:

plant → herbivore → carnivore_1 → carnivore_2 → . . .

and will be referred to as the grazing food chain or grazing food web. At each stage of such a system death and waste organic products means that organic material is available to a different pathway. An example to this pathway would be:

dead grass → fungi → Collembola → predatory mite

which can be generalized as:

dead organic material → decomposer_1 → decomposer_2 → . . .

and will be referred to as the decomposer food chain or web. It will be seen that some of these decomposer organisms can also be classified as herbivores and carnivores.

Phillipson (1966) discusses the background of such concepts as the food chain and the food web. Each step in such a chain is referred to as a trophic level. Thus, within an ecosystem all organisms of similar feeding habits are grouped together, and this group forms a trophic level. Let us assume that there are n trophic levels (in practice n is often four, the species being assigned to categories of producers, herbivores, carnivores and top carnivores). Now we can define the energy content of the ith trophic level as Λ_i. Λ_1 is the standing crop, the energy content at a given time, of the producers, Λ_2 of the herbivores, and so on. We know that the energy content of a trophic level is not static, but is continuously changing due to assimilation from lower trophic levels and due to loss both to higher trophic levels and in the form of heat. Let us define the energy flow from the $i-1$th level to the ith level per unit of time as λ_i. Let us also define the energy lost as heat (due to respiration and metabolism) from the ith level per unit of time as R_i. If we consider the ith level, the energy passing from it to the $i+1$th level is λ_{i+1}.

We can now write an equation for the rate of change of energy in the ith trophic level. This equation is:

$$\frac{d\Lambda_i}{dt} = \lambda_i - R_i - \lambda_{i+1} \qquad (3.1)$$

which simply states that the rate of change of energy in the standing crop is equal to the input (λ_i) minus the output ($R_i + \lambda_{i+1}$). Equation (3.1) is one of the most fundamental relations in discussing the flow of energy through ecosystems. Phillipson poses a number of questions about this trophic–dynamic structure, principally:

(1) Is Λ_i/Λ_{i-1} constant? This is both of interest within an ecosystem, to see if $\Lambda_2/\Lambda_1 = \Lambda_3/\Lambda_2$, as well as for comparisons between ecosystems, to see if Λ_2/Λ_1 is the same for all ecosystems.

(2) λ_i is a measure of the production of the $i-1$th tropic level, since it represents the food available to the ith level. How can this parameter be maximized?

(3) The efficiency of the ith level can be measured as λ_i/λ_{i-1}. Is this ratio constant, or how can it be maximized?

A direct result of the trophic–dynamic analysis is the plot of the data for a visual comparison of the trophic structure of ecosystems. The data for Silver Springs, Florida, is plotted as a pyramid in Fig. 3.2. On the left side of the pyramid are the amounts of energy accumulated by each trophic level as net production. On the right side of the pyramid are the

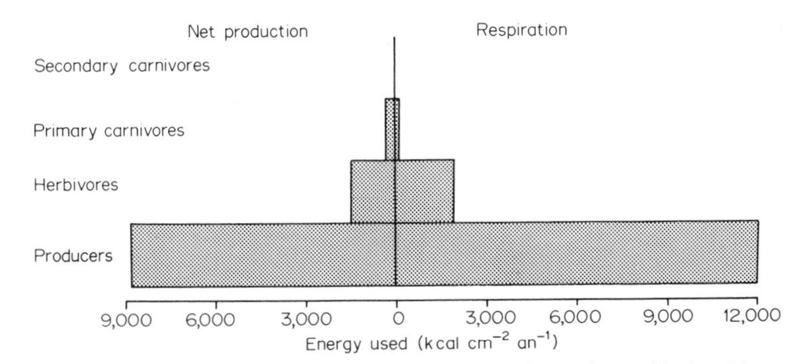

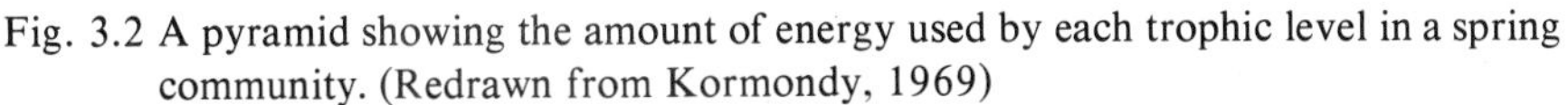

Fig. 3.2 A pyramid showing the amount of energy used by each trophic level in a spring community. (Redrawn from Kormondy, 1969)

amounts of energy used in respiration. Thus, each block of the pyramid represents the total amount of energy utilized by each trophic level per year.

Following on from Phillipson's questions two efficiencies can be defined, namely:

food chain efficiency

$$= \frac{\text{calories of } i\text{th level consumed by } i+1\text{th level}}{\text{calories of food supplied to } i\text{th level}} \times 100$$

Maximizing this efficiency implies that the food of the ith level is being exploited to the best advantage of the $i+1$th level.

gross ecological efficiency

$$= \frac{\text{calories of } i\text{th level consumed by } i+1\text{th level}}{\text{calories of food consumed by } i\text{th level}} \times 100$$

This latter efficiency is equal to λ_{i+1}/λ_i, expressed as a percentage. In order to maximize the food chain efficiency it must be that no food from the ith level is wasted, and thus that food supplied to the level is equal to the food consumed by the level. Thus the maximum food chain efficiency is equal to the gross ecological efficiency. A fuller analysis of these two efficiencies, and of a number of other definitions of efficiencies, is given by Kozlovsky (1968), who extends the concepts of the transitions between pairs of trophic levels to the overall production of the community.

Values of the gross ecological efficiency are obviously important in management practice as they set an upper limit on the food availability, either for wildlife or for man. Phillipson suggests that the most probable value of this efficiency in natural ecosystems would be in the order of 10 per cent. Edwards, Reichle and Crossley (1970) show that values for invertebrate herbivores and oribatid mites vary from 6·8 to 30 per cent.

One of the strongest criticisms of such trophic–dynamic studies is that they ignore the species composition of the trophic levels. This has certainly been realized by all writers, since analyses have been based on food webs rather than on the simplified food chains. One complication is that many species cannot be allocated to only one trophic level, since they are omnivorous to a greater or lesser extent. The main argument in favour of the trophic structure is that it would be an impossibly complex task to sort out a food web for each individual species and for each species to work out its feeding relationships and efficiency.

However, it is interesting to consider the number of species making up each trophic level. In energy flow studies this has probably not been investigated, though Kormondy (1969) states '. . . at the "top" of the energy hill, i.e. at the carnivore level, there is not enough energy to support more than very few carnivores'. A theorem of population dynamics will, if it is accepted, clarify the speculation. Macarthur and Wilson (1967) state and prove the theorem that 'there can be no more predator species than prey species in a given habitat, provided the predator species are resource limited'. In this theorem the term predator is taken to be any species belonging to any trophic level except the first. The theorem thus shows that if there are m prey species the number of predators will be fewer than m, particularly so when the predator is larger than the prey (a fine-grained situation, as explained on pp. 96–97 of Macarthur and Wilson's book). Since many top carnivores are large animals one would expect the number of species to decrease in the higher trophic levels. A further analysis of these theorems is given by Levins (1968).

An example

In the previous sub-section we have taken a theoretical look at the flow of energy through ecosystems. In order to illustrate the concepts we will examine the Cone Spring in Iowa, studied by Tilly (1968), which demonstrates the complexity of the analyses in what is essentially a rather simple ecosystem. The Cone Spring covers an area of approximately 140 m^2 with a slope of 3·2 cm m^{-1}. There is an average water depth of 3 cm, but this varies between 0·5 and 16·5 cm. Seasonal averages for pH, temperature and rate of flow are 7·2, 12·6°C and 1 l sec^{-1} respectively. There are only three major producer species, the perennials *Bacopa rotundifolia* (Scrophulariaceae) and *Lemna minor* (Lemnaceae), and the annual *Impatiens capensis* (Balsaminaceae). During the winter *Impatiens* completely dies, and *Bacopa* dies back to form submerged or barely emergent rosettes. *Lemna* is found throughout the year between the leaves of *Bacopa* and in mats on the water surface in zones of weak flow. During the summer *Impatiens* grows in height and its shade causes *Bacopa* to die back.

Of the 30 or so species of consumer organisms only a few have much impact on the energy flow. The most important species is *Gammarus pseudolimneus* since its population was supplying the bulk of the energy to the higher trophic levels. *Gammarus*, an amphipod, is a scavenger, feeding on all types of detritus that are available. A trichopteran, *Frenesia missa*, is also a scavenger feeding on dead vegetable matter and twigs. It has a strong seasonal periodicity since there is an adult emergence during September and October. *Physa integra*, a pulmonate snail, is in general a detritus feeder, though it will occasionally feed on live plant material. Two genera of Diptera were also present in relatively large numbers. *Cardiocladius* is a detritus feeder, but with a preference for large particles (5 to 30 mm in diameter). *Pentaneura* is generally predatory, though it can be omnivorous. In the laboratory it would survive to maturity on detritus, and gut analyses of field specimens showed that some detritus was taken. In calculations 80 per cent of their food is assumed to be other animals and 20 per cent detritus. *Phagocata vetata* is a top carnivore. This planarian does not, however, completely occupy such a position in the trophic structure since it has been observed to feed on both herbivores

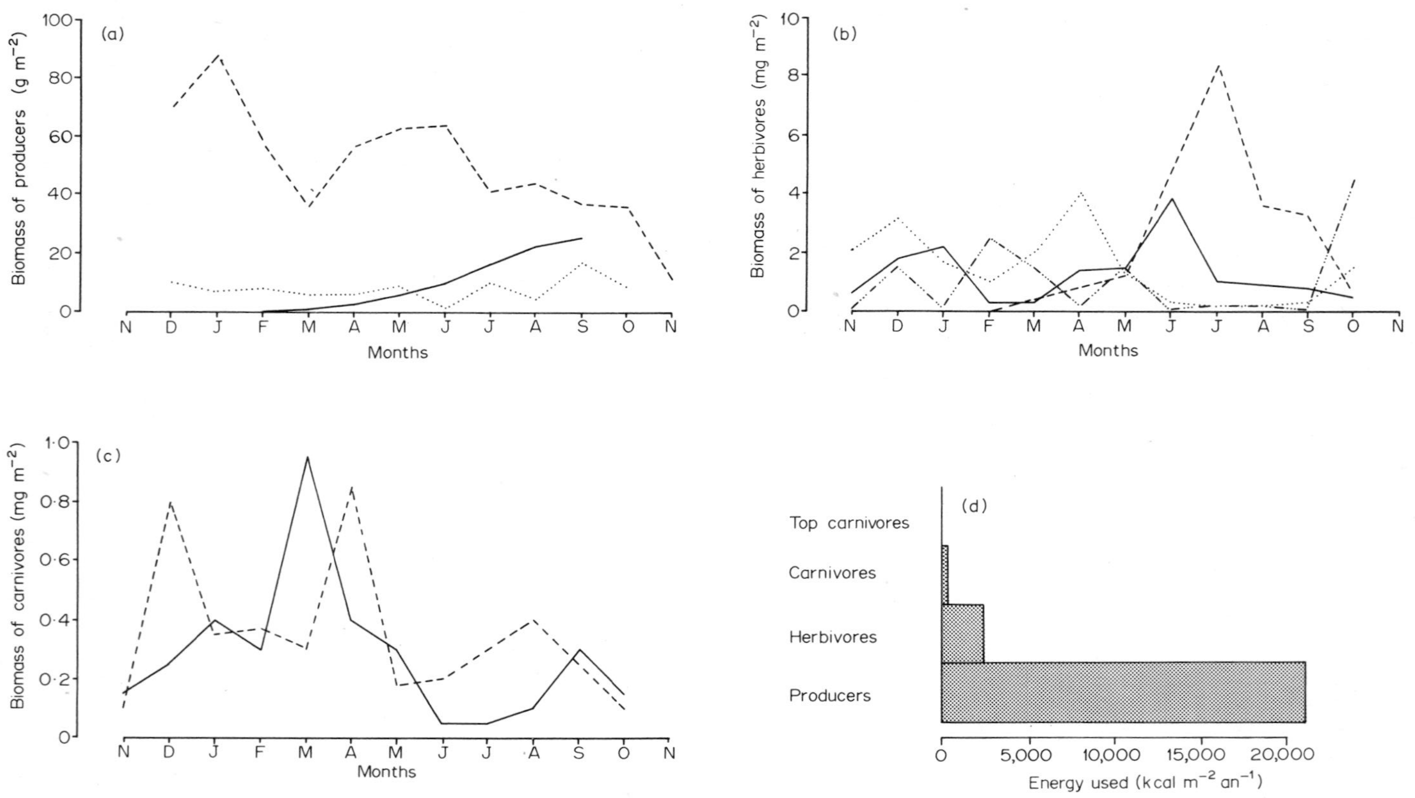
(a)
Biomass of producers (g m^{-2})
100
80
60
40
20
0
N D J F M A M J J A S O N
Months
(b)
Biomass of herbivores (mg m^{-2})
10
8
6
4
2
0
N D J F M A M J J A S O N
Months
(c)
Biomass of carnivores (mg m^{-2})
1·0
0·8
0·6
0·4
0·2
0
N D J F M A M J J A S O N
Months
(d)
Top carnivores
Carnivores
Herbivores
Producers
0 5,000 10,000 15,000 20,000
Energy used (kcal m^{-2} an^{-1})

and carnivores. It is estimated that it gains approximately 60 per cent of its energy from herbivores and 40 per cent from other carnivores. The seasonal distribution of the standing crop biomass of all of these species is shown in Figs 3.3 (*a*)–(*c*).

We can thus consider this ecosystem to have four trophic levels, the producers, the herbivores, the carnivores and the top carnivore. Tilly does not calculate an energy pyramid though this can be done from his data. If we use the amounts of energy used at each level, Tilly's data can be used directly for the second, third and fourth trophic levels. For the first level we have an estimated net annual production of 9508 kcal m^{-2} an^{-1}, but there is no assessment of the respiration at this level. Using the data for another spring ecosystem quoted by Kormondy (1969), where respiration = $1 \cdot 27 \times$ net production, we can estimate that the respiration of plants in Cone Spring was in the order of 12,000 kcal m^{-2} an^{-1}. Thus, the amount of energy used per annum by the first trophic level is of the order of 21,000 kcal m^{-2}. The energy pyramid is shown in Fig. 3.3 (*d*).

Having looked at the community broadly to establish the energy pyramid it is useful to break it down into its various component parts. It is here that we see the difficulty of quantifying the whole of the ecosystem. Sampling procedures were required to estimate the input of energy (net primary production and import). The results of this analysis are given in Table 3.1. The figure for the utilization of the standing crop is derived from changes in the ecosystem during the year. Thus, the standing crop of *Bacopa* was reduced whilst the standing crops of *Lemna* and the consumers increased slightly.

This is the amount of energy entering the ecosystem at the first trophic level, as estimated by sampling procedures. Tilly also estimated the utilization of energy by measuring the respiration of the consumers and

---

Fig. 3.3 Data for the Cone Spring ecosystem. (Drawn from data given by Tilly, 1968).
(a): The seasonal variability in the standing crop of the producers.
———— *Impatiens* — — — — *Bacopa* · · · · · · · · · · · · *Lemna*
(b): The seasonal variability in the standing crop of the herbivores.
———— *Gammarus* — — — — *Frenesia*
· · · · · · · · · · · · *Physa* — · · · — · · · *Cardiocladius*
(c): The seasonal variability in the standing crop of the carnivores.
———— *Pentaneura* — — — — *Phagocata*
(d): The ecological pyramid showing the amount of energy used by each trophic level of the Cone Spring ecosystem

TABLE 3.1 The input of energy into Cone Spring. (Source: Tilly, 1968)

Source	*Energy* (kcal m^{-2} an^{-1})
Net primary production	
Bacopa	204
Impatiens	730
Lemna	134
Import	626
Utilization of standing crop	247
Total	1941

TABLE 3.2 The utilization of energy in Cone Spring. (Source: Tilly, 1968)

Source	*Energy* (kcal m^{-2} an^{-1})
Respiration	
Consumers	2008
Micro-organisms	3400
Export	4100
Total	9508

micro-organisms, as well as measuring the export of organic material in the water leaving the spring. Details of the utilization of energy are given in Table 3.2. There is a very wide discrepancy between the two sides of the energy budget given in Tables 3.1 and 3.2. Tilly analyses this discrepancy, and is able to show how his sampling procedures would underestimate the net primary productivity and import terms on the credit side of the balance. For further study of this ecosystem we will have to assume that the input of energy is of the order of 9508 kcal m^{-2} an^{-1}.

The next step in the analysis is to build up a food web for the ecosystem and to quantify the relationships. This has been attempted by Tilly, and a shortened form of the web to include only those species previously mentioned is shown in Fig. 3.4. The other species of herbivore and carnivore are lumped together for the sake of convenience. The web is not strictly quantified since the net secondary productivity is consider-

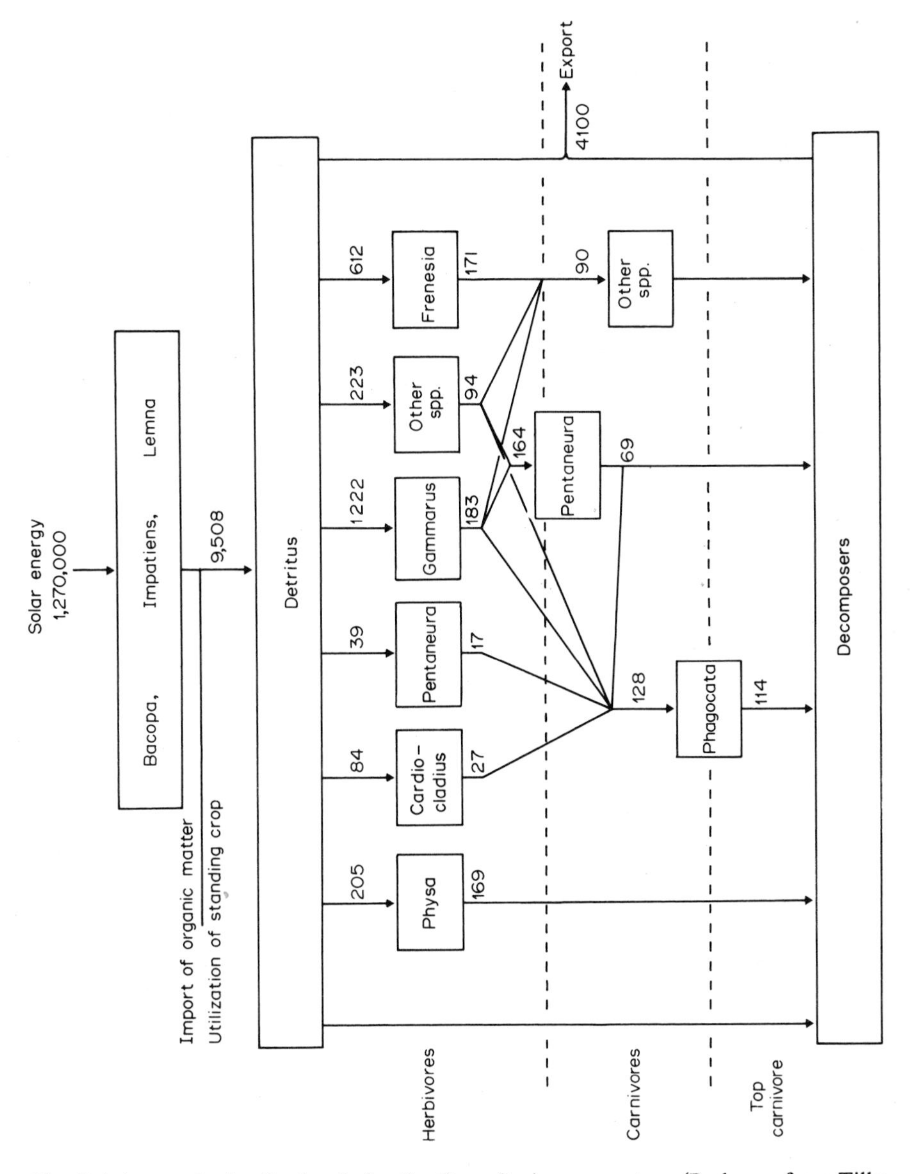

Fig. 3.4 A quantitative food web for the Cone Spring ecosystem. (Redrawn from Tilly, 1968)

ably greater than the assimilation of the carnivores. To illustrate this point the group of other carnivores assimilate 90 kcal m^{-2} an^{-1}. Observations showed that these species were feeding on *Frenesia*, *Gammarus* and the group of other herbivores. The net secondary production of these groups was 171, 183 and 94 kcal m^{-2} an^{-1} respectively. The latter two groups were also used by other links in the food web, but *Frenesia* was preyed upon only by *Chauliodes*, a member of the group of other carnivores. The group only assimilated 90 kcal m^{-2} an^{-1} although over 171 kcal m^{-2} an^{-1} were available for its exclusive use. The web does not quantify what happened to the missing 81 kcal m^{-2} an^{-1}, and we can only assume that it has contributed towards the decomposer food chain.

We are now able to synthesize the whole community and construct an energy flow diagram. Such diagrams have been prepared for generalized ecosystems (e.g. Odum, 1959, and Kormondy, 1969). Summation of Tilly's figures shows that 2385 kcal m^{-2} an^{-1} enters the herbivore (detritus feeder) part of the food web, and from the energy budget we know that 2008 kcal m^{-2} an^{-1} is lost in the respiration of the consumer organisms. Earlier we have postulated that the respiration of the producers would account for about 12,000 kcal m^{-2} an^{-1}. Using these figures given by Tilly, and by making a number of other calculations and estimates, the diagram shown in Fig. 3.5 can be built up. In this diagram all estimates and calculations not given by Tilly are enclosed in brackets, and the widths of the lines are proportional to the logarithm of the energy flow. In this ecosystem the detritus food chain is directly equivalent to the grazing food chain of terrestrial ecosystems. There has been no attempt to split up the separate components of the decomposer chain, and this is shown in the diagram with a loop to indicate that carnivorous activity occurs here as well as in the main feeding chain. There has been no attempt to quantify the flow from any of the consumer groups to either the decomposer or export systems. The main effects would be the migratory activity of insects that have bred in the ecosystem, particularly *Frenesia*, a member of the herbivore category. The majority of the flow from the top carnivore category would be towards the decomposer system. The only species concerned in this category is *Phagocata*, and Tilly shows that of an assimilation of 128 kcal m^{-2} an^{-1} (apportioned 79 to the carnivore level and 49 to the top carnivore level) a total of 82 kcal m^{-2} an^{-1} was mucus, and respiration accounts for only 14 kcal m^{-2} an^{-1}.

Thus, Tilly's work demonstrates the use of ecosystem analyses, but it also demonstrates the difficulties of obtaining data that is accurate. For conservation management this sort of approach will become essential as one wishes to establish the role of every species in the community. However, to conclude the consideration of the Cone Spring ecosystem, it

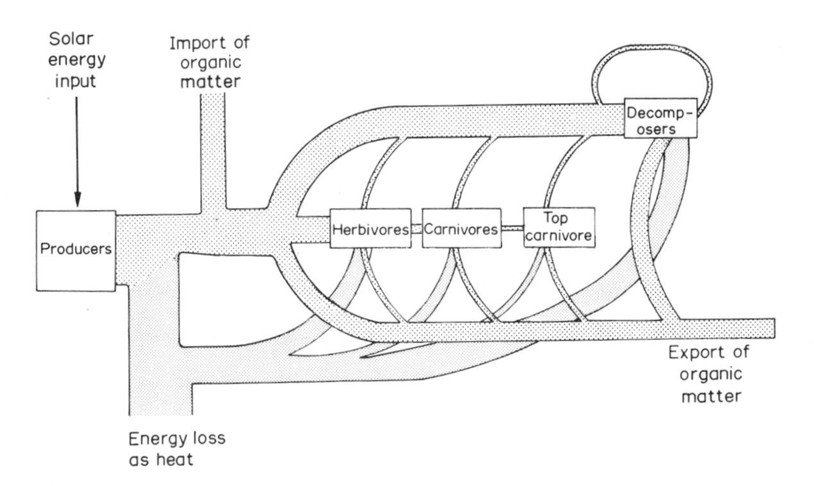

Fig. 3.5 A diagrammatic representation of the energy flow in the Cone Spring ecosystem. For a further discussion see the text. The widths of the pathways in the illustration are proportional to the logarithm of the amount of energy flowing. (Drawn from data given by Tilly, 1968)

would be useful to investigate its efficiency. The trophic–dynamic parameters, estimates based on Tilly's data, are shown in Table 3.3. One can see that the standing crop is relatively small compared with the annual net production. Thus, the animals are producing in the order of 10 times their own standing crop of energy per annum, whilst the net production of the plants is over 20 times their standing crop. Calculating the ratios of the standing crop we obtain

$$\Lambda_2/\Lambda_1 = 0{\cdot}215$$
$$\Lambda_3/\Lambda_2 = 0{\cdot}230$$
$$\Lambda_4/\Lambda_3 = 0{\cdot}210$$

The amount of variability in these three ratios is very small, indicating that there is a relatively constant relationship between the standing crops of energy in the trophic levels.

TABLE 3.3 The trophic–dynamic parameters for Cone Spring. The parameters have been estimated from the data of Tilly (1968). All figures are expressed as kcal m^{-2} an^{-1}

Trophic level i	*Standing crop (average annual)* Λ_i	*Energy passed to next trophic level* λ_{i+1}	*Energy lost as heat* R_i	*Net production* P_i
1	280	2385	12000	9508*
2	60·1	333	1817	568
3	13·8	49	236†	147
4	2·9	–	37†	12

* Includes 3000 kcal m^{-2} an^{-1} organic matter import
† Includes mucus production of *Phagocata*

The efficiency of the system is shown in Table 3.4. The gross ecological efficiency has been calculated as

$$\text{G.E.E.}_i = \frac{\text{Food passed to } i+1\text{th level}}{\text{Food consumed}} \times 100 = \frac{100\lambda_i}{\lambda_{i-1}} = \frac{100\lambda_i}{(P_i + R_i)}$$

and the food chain efficiency as

$$\text{F.C.E.}_i = \frac{\text{Food passed to } i+1\text{th level}}{\text{Food supplied}} \times 100 = \frac{100\lambda_i}{P_{i-1}}$$

where P_i is the production of the ith level and R_i is the respiration of that level. In order to calculate the efficiencies for the first trophic level I have used the estimated utilization of 21,000 kcal m^{-2} an^{-1} of solar energy, and the stated total of 1,270,000 kcal m^{-2} an^{-1} of solar radiation received by the site.

TABLE 3.4 The efficiency of the trophic levels of Cone Spring. Estimated from data given by Tilly (1968). All figures are percentages

Trophic level	*Gross ecological efficiency*	*Food chain efficiency*	*Production efficiency*
1	11·36	0·19	31·0
2	13·96	3·50	23·8
3	14·71	8·63	29·1
4	–	–	24·5

For natural ecosystems Phillipson suggested that a probable value of the gross ecological efficiency would be 10 per cent. Judged by this criterion the efficiency of Cone Spring was high. However, such semi-aquatic ecosystems are known to be more productive than many terrestrial ecosystems since water availability never becomes a limiting factor for plant production. The food chain efficiency of 0·19 per cent for the producers also compares very favourably with average values of 0·10 per cent for potatoes and 0·05 per cent for grain crops (Watt, 1968).

The final column of Table 3.4 shows the production efficiency of the various trophic levels, and is calculated as

$$\text{P.E.}_{\cdot i} = \frac{\text{Production of level}}{\text{Food consumed by level}} \times 100 = \frac{100P_i}{\lambda_{i-1}} = \frac{100P_i}{P_i + R_i}$$

This calculation is of interest as it shows what would be the absolute maximum value of the gross ecological efficiency. Such a maximum value could only be achieved if there was no loss of energy to either the decomposer or the export system, and if all the net production of the ith level was assimilated by the $i+1$th level.

The cycling of nutrients

We have seen how energy flows through the ecosystem – there is an input in the form of gross primary production, and a loss throughout the biotic part of the ecosystem due to respiration and other heat losses. Import and export of energy in the form of organic material is another feature of ecosystems, though this tends to form a smaller proportion of the total input and output with larger ecosystems. Energy is thus definitely a flow, with a continuous input and losses throughout the system. Nutrients, however, have a definite cycle, resulting from interactions between the grazing food chain, the decomposer food chain and the abiotic environment. However, in all of these cycles some driving force is needed, and this is provided by the flow of energy, ensuring the carriage of nutrients through part of their cycle.

Nutrient cycles can be either open or closed. Nitrogen and carbon cycles can be considered as open since there is a direct interchange between these elements and the atmosphere. Calcium, sulphur and potassium cycles can be considered as closed, since input of these nutrients is largely through the physical and chemical weathering of the parent material, and output is either through a harvest or through drainage in ground water. There is no direct exchange between the ecosystem and the atmosphere, though it is known that some input might occur with rainwater.

Most books that give an introduction to ecology (e.g. Odum, 1963; Kormondy, 1969; Whittaker, 1970) discuss the importance of the nitrogen cycle. A diagrammatic representation of the cycle, rather less elaborate than that given by Odum (1959), is shown in Fig. 3.6. The illustration omits any reference to the source of the nitrogen or to the destiny of the nitrogen since we shall be concerned with a single ecosystem. Looking at the input side of the ecosystem, we can see that apart from artificial fertilization the only source of nitrogen is the atmosphere. The atmospheric nitrogen concentration is 79 per cent, and with this amount it would seem that the atmosphere would act as a nitrogen

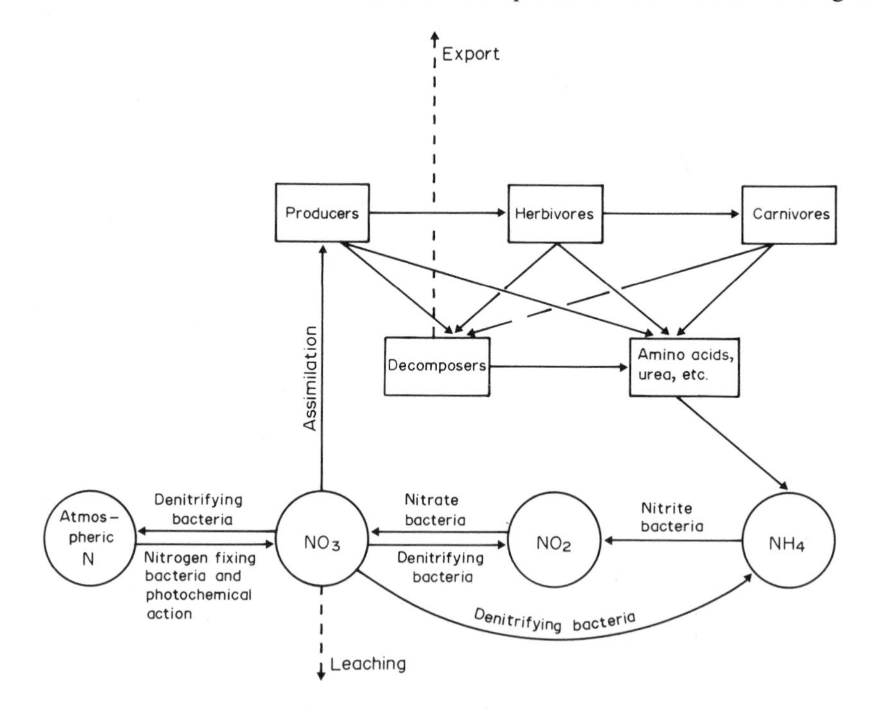

Fig. 3.6 The nitrogen cycle

reservoir for the ecosystem. However, very few organisms are capable of directly using this nitrogen source, and hence the reservoir is actually in the organic part of the ecosystem. A long list of bacteria and blue-green algae is now known to be capable of fixing atmospheric nitrogen, but the best known are in the bacterial genus *Rhizobium*. These bacteria, which show strong host-specificity, are associated with leguminous plants (e.g. clover, peas, etc.) and with a few plants in other families (e.g. alder). These bacteria, once they have penetrated the root hairs, establish an increasing population resulting in the plant's response of enlarging the root and forming a characteristic nodule. Nitrogen fixation occurs within this nodule. This symbiotic relation between bacteria and plants accounts for a large proportion of the terrestrial fixation of atmospheric nitrogen, but there are also some free-living bacteria that are capable of nitrogen fixation. *Azotobacter* (an aerobic species) and *Clostridium* (an anaerobic species) are known to be widely distributed in most soils as well as in fresh-water and marine ecosystems. The amounts of nitrogen fixed by all bacteria, symbiotic and free-living, range between 140 and 700 mg m^{-2} an^{-1}, though in clover fields the amount can be very greatly in excess of these values. The amount of nitrogen entering the ecosystem as a result of electrochemical or photo-chemical activity is relatively small, being of the order of 35 mg m^{-2} an^{-1}.

Output of nitrogen from the ecosystem generally occurs in three ways, by leaching in ground water, by export (either in organic matter or by a harvest) or by denitrification. Leaching usually affects the soluble nitrates, but the export of nitrogen depends upon the nature of the ecosystem and its management. In many natural ecosystems this term would be negligible, being balanced by an input of migratory animals and detritus. However, in managed ecosystems such as forests and wildlife parks there will be an export of nitrogen in the crop that is removed. Denitrification occurs as a result of the activity of certain bacteria, such as *Pseudomonas*, and by fungi using nitrate as an oxygen source. The full reduction of nitrate to nitrogen usually only occurs under anaerobic or partially anaerobic conditions, and so it can be expected to have the greatest impact on poorly aerated soils, and where there is a large build-up of humus and hence considerable demands on oxygen.

The inputs and losses usually operate at the nitrate point of the nitrogen cycle. Within the ecosystem we can see that the main nitrogen uptake into the grazing food chain is in the form of nitrate. Within this food chain the general pathway of nitrogen is from nitrate towards ammonia, which is derived from the decomposition of protein, amino acids and the products of excretion (urea and uric acid). Bacterial action completes the cycle of nitrogen back from the ammonia form to the nitrate form, a process referred to as nitrification. *Nitrosomonas* converts ammonia to the toxic nitrite form, which is further oxidized either by the atmosphere or by *Nitrobacter* to nitrate. These bacteria gain their energy from this nitrification process, using some of this energy to assimilate carbon from carbon dioxide or bicarbonate. The amount of ammonia or nitrite converted is very much greater than the amount of carbon assimilated. Thus *Nitrosomonas* has been found to convert 35 units of ammonia to nitrite for every unit of carbon dioxide assimilated, and *Nitrobacter* 76 to 135 units of nitrite to nitrate for every unit of carbon dioxide. The movement of nitrogen in this decomposer part of the ecosystem is not always in the direction from ammonia to nitrate. The denitrifying bacteria, such as *Pseudomonas*, in more aerobic conditions do not reduce nitrate all the way to gaseous nitrogen, but the process stops either at the nitrite or the ammonia stage.

Thus, in Fig. 3.6 there are three components of the nitrogen cycle. Firstly, there is the exchange component between the atmosphere and the nitrate pool, with the flows of nitrogen operating in both directions. Secondly, there is the main biotic component, that of the grazing food chain, where the general movement of nitrogen is from the nitrate to the ammonia form. Thirdly, there is the micro-organism or decomposer component where the general direction of movement of nitrogen is from the ammonia to nitrate form.

Studies on the energy flow in ecosystems were quantified such that the relative proportions of all of the links in the system were known. Unfortunately, there seems to have been relatively little work on the quantification of nutrient cycles in ecosystems. Nutrient budgets have been drawn up to show the amount of nutrient taken by the producers, the amount returned to the soil in litter, the input from the atmosphere and the output in harvest (all per unit of time). Studies do not seem to have been made at the depth of the energy studies. Whittaker (1970) illustrates the nitrogen cycle and attempts to show the average amounts of nitrogen for the earth's surface, and the average rates of some of the links in the cycle. Unfortunately this study is very general and does not relate to specific ecosystems.

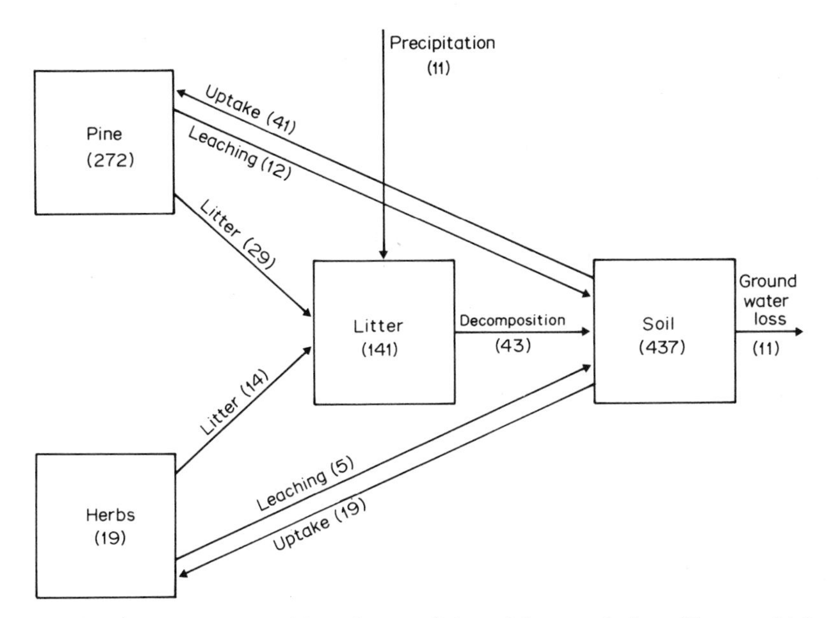

Fig. 3.7 Quantitative relationships of part of the calcium cycle in a 55 year old Scots pine forest. The standing crop data, within the boxes, is expressed as kg ha^{-1}, and all transfers between parts of the ecosystem are expressed as kg ha^{-1} an^{-1}. (Drawn from data given by Ovington, 1962)

An example of the partially quantitative studies on nutrient cycles is given by Ovington (1962, 1965). Based on this work, a schematic representation of the calcium cycle is shown in Fig. 3.7. The rates of flow and the standing crops assume an equilibrium state, i.e. that there is no net gain or loss from any part of the ecosystem. It shows the relative importance of the links in the calcium transport, and how the main reservoir of calcium is in the soil. Ovington also shows the distribution within the tree (not shown in the illustration) where 122 kg ha^{-1} are in the stems, 93 kg ha^{-1} are in the roots, and 57 kg ha^{-1} are in the leaves and branches. The only input is 11 kg ha^{-1} in the precipitation, and the only output is 11 kg ha^{-1} in the ground water.

Ovington (1965) states that the quantitative nature of the nutrient flow is determined, to a large extent, by the nutrient status of the soil. The soil acts as a nutrient reservoir, and the amount released into the biotic part of the ecosystem is a relatively constant proportion of this reservoir. Thus, the larger the reservoir the larger the amount released. However, Ovington also shows that the nature of the biotic part of the ecosystem affects the nutrient flow. He presents data for two different species, the pedunculate oak *Quercus robur* and the Scots pine *Pinus sylvestris*, both 47 years old and growing on the same soil type. Fig. 3.8 shows the two flow diagrams for a part of the ecosystem, and one can see that the potassium uptake by the pine crop is less than that of the oak crop. However, in the pine forest the bracken, the predominant species in the ground flora, is taking up a large amount of the potassium.

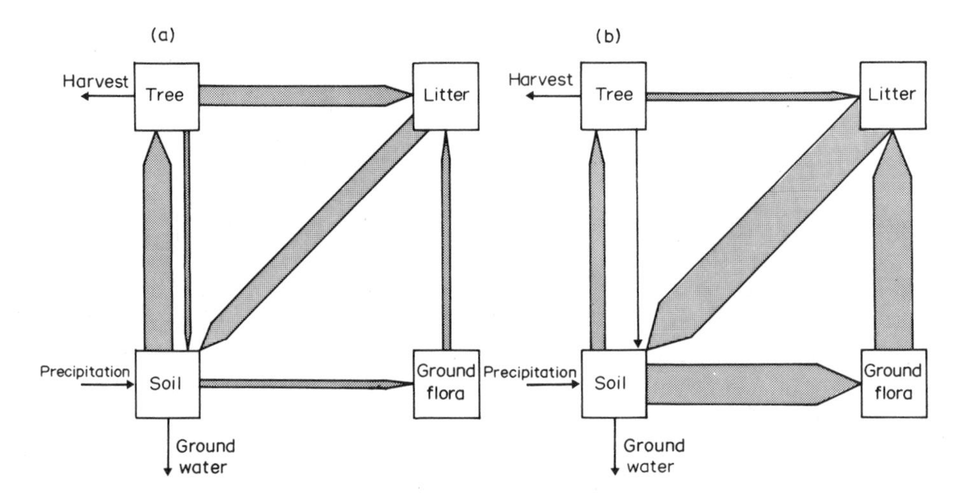

Fig. 3.8 The potassium flow in two adjacent woodlands, both aged 47 years and on the same soil type. The width of the arrows refers to the amount of potassium forming the transfer. (a): Oak. (b): Scots pine. (Redrawn from Ovington, 1965)

Nutrient cycles are a feature of all ecosystems, and for conservation and management one will want to know the location of the reservoir, the characteristics of release from the reservoir to the grazing food chain, and the effect on the ecosystem of the export of nutrients in a harvest. Most of the non-gaseous nutrients have a reservoir in the litter and soil portion of the ecosystem, and these will be characterized by being the parts of the ecosystem with the greatest nutrient standing crop. With carbon there is a free exchange of carbon dioxide between the biotic part of the ecosystem and the atmosphere, and hence the atmosphere acts as the reservoir. With nitrogen, although the atmosphere has the greatest amount of the element, exchange is difficult and slow, and hence the reservoir lies in the speed of production and in the quantity of nitrate.

When the reservoir has been located one wishes to know what will be the effect of management. Porter (1969) has considered the effect of nitrogen in grassland ecosystems. He shows that the nitrogen concentration in cultivated fields slowly decreases during cultivation, until after about 60 years the concentration has dropped to about 65 per cent of the concentration in uncultivated fields. However, various crops have different impacts on this decrease in concentration. Thus corn will decrease the concentration by more than other rotational crops, whilst the concentration will be slightly increased under legumes.

The literature is full of descriptions. The majority of basic ecological text-books illustrate at least the nitrogen, carbon, phosphorus and sulphur cycles, and discuss the various links in these cycles. More and more research effort has been put into understanding the processes involved, particularly the nitrification process of the nitrogen cycle and the photosynthetic process of the carbon cycle. Although these studies have given us a knowledge of the physiological and biochemical mechanisms, the application of this research has hardly been attempted. The recent interest in production ecology has resulted in more research into primary production of the ecosystem. Particularly for forest ecosystems there have been quantitative studies determining how much of each nutrient is taken from the soil each year, how much is returned in litter, how much is removed in the crop and how much is stored in different parts of the tree. But we know very little about the nutrients in the consumer organisms. The ecosystem concept has hardly yet found its way into nutrient studies. There seem to be no answers to questions about the rate of nutrient flow between trophic levels of the ecosystem, or to the efficiency of the food chain in respect of each of these nutrients. Also, there seems to be little knowledge of the effects of management practice on the nutrient resources of the site. The only indication is that the level of nutrients in the reservoir determines the rate of output from the reservoir to the grazing food chain. In forests the continued export of nutrients in the crop can only deplete the reservoir – will fertilizers have to be used to restore the reservoir, or can the forest function adequately on the depleted reservoir? It seems therefore that nutrient studies on ecosystems should follow two lines of development. First, the standing crop and the rates of flow of nutrients in the whole ecosystem will have to be measured and quantified. Secondly, and this is 'applied ecology', one will want to know the effects of intervention in the ecosystem on the nutrient status and nutrient dynamics.

Ecosystem modelling

In the previous sections of this chapter we have investigated some of the fundamental flows and cycles that occur within an ecosystem. A considerable amount of research was required to evaluate any particular flow or cycle, and even research aimed at quantifying each link does not relate

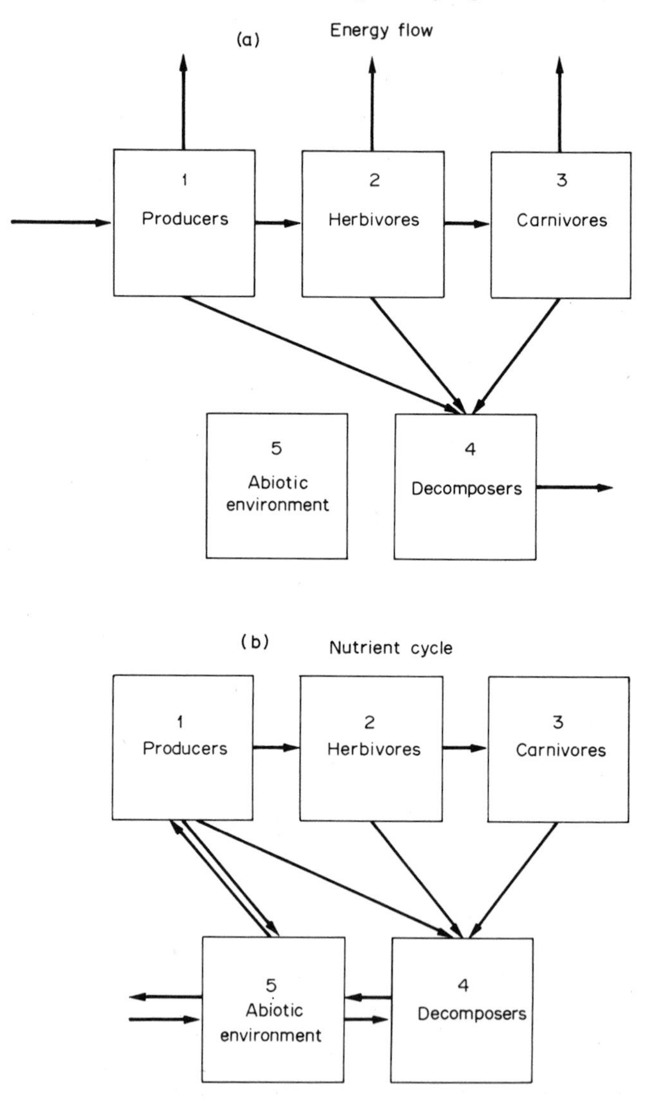

Fig. 3.9 A diagrammatic representation of a five compartment ecosystem. For an explanation see the text. (a): The energy flow. (b): The cycling of a nutrient.

these links one to another and to the various standing crops. Research described in the examples is of a 'here-and-now' nature, showing what happens with a given complement of species, each with their own ecological efficiencies, and with given inputs and outputs to and from the ecosystem. It could not be used to answer such questions as 'What would happen if management of the ecosystem required a specified harvest?', 'What would be the effect on the ecosystem if the consumers were more efficient?' or 'What would happen if the input were increased?'. Obviously these questions do not apply to all the flows and cycles, since, for example, the input of solar energy cannot be altered, though the input of nitrogen could be doubled by fertilization. Research of this 'here-and-now' kind is essentially static, whereas the ecosystem is a dynamic collection of various parts, and hence it will require a more dynamic approach for its full analysis. Such an approach not only quantifies the links in the flow or cycle, but also shows in a functional way how one link depends on the other links and components of the ecosystem.

The basic idea in the systems analysis approach to modelling is to divide the ecosystem into a number of compartments, and to investigate the relations between each pair of compartments. The division is arbitrary, and depends upon the aims of the research or the management application of the model. If, for example, we were interested in a very rough model for energy flow we could divide the ecosystem into five compartments, namely producers, herbivores, carnivores, decomposers and abiotic environment, see Fig. 3.9 (*a*). Only the energy flows between compartments that are likely to contribute much to the system are shown in the illustration (predatory green plants, for example sundews, are not included by an arrow from herbivores to producers). A similar simplified system for the cycle of nutrients is shown in Fig. 3.9 (*b*).

If we define x_i as the standing crop of energy or nutrient in the ith compartment, then for each of the models shown in Fig. 3.9 we have five values of x_i. There is an input into some compartments, as well as an output, resulting in an exchange of energy or nutrients between the ecosystem and its surrounding ecosystems. Let these, per unit of time, be a_i and z_i respectively. Table 3.5 shows the non-zero values of a_i, z_i and x_i for the two models in Fig. 3.9. But a system requires more parameters in order to describe the transfers of energy or nutrients between compartments. Thus f_{ij} is defined as the transfer parameter for energy or nutrients moving

TABLE 3.5 The non-zero parameters for standing crop, input and output for energy flow and nutrient cycle models (Fig. 3.9). 1 = producers, 2 = herbivores, 3 = carnivores, 4 = decomposers and 5 = abiotic environment

Energy flow model *Compartments*	*1*	*2*	*3*	*4*	*5*
Standing crop	x_1	x_2	x_3	x_4	0
Input	a_1	0	0	0	0
Output	z_1	z_2	z_3	z_4	0
Nutrient cycle model *Compartments*	*1*	*2*	*3*	*4*	*5*
Standing crop	x_1	x_2	x_3	x_4	x_5
Input	0	0	0	0	a_5
Output	0	0	0	0	z_5

TABLE 3.6 The non-zero transfer parameters for the diagrammatic ecosystems in Fig. 3.9. The compartment numbers are listed in Table 3.5

Energy flow model		*Transfer to compartment* 1	2	3	4	5
	1	–	f_{12}	0	f_{14}	0
Transfer	2	0	–	f_{23}	f_{24}	0
from	3	0	0	–	f_{34}	0
compartment	4	0	0	0	–	0
	5	0	0	0	0	–
Nutrient cycle model		*Transfer to compartment* 1	2	3	4	5
	1	–	f_{12}	0	f_{14}	f_{15}
Transfer	2	0	–	f_{23}	f_{24}	0
from	3	0	0	–	f_{34}	0
compartment	4	0	0	0	–	f_{45}
	5	f_{51}	0	0	f_{54}	–

from the ith to the jth compartment. Again, recording only the non-zero values in Fig. 3.9, these parameters are shown in Table 3.6.

Using this notation we can write equations for the rate of change of the energy or nutrient content of any compartment. Thus, for compartment 2, the rate of change of energy or nutrient content is given by

$$\frac{dx_2}{dt} = a_2 - z_2 + \sum_{\substack{i=1 \\ i \neq 2}}^{5} f_{i2} - \sum_{\substack{i=1 \\ i \neq 2}}^{5} f_{2i}$$

or for the general ecosystem where there are n compartments, the rate of change in the rth compartment is given by

$$\frac{dx_r}{dt} = a_r - z_r + \sum_{\substack{i=1 \\ i \neq r}}^{n} f_{ir} - \sum_{\substack{i=1 \\ i \neq r}}^{n} f_{ri} \tag{3.2}$$

For the flow of energy through the system shown in Fig. 3.9 (a) we can prepare the set of equations

$$\left.\begin{aligned} dx_1/dt &= a_1 - z_1 - f_{12} - f_{14} \\ dx_2/dt &= f_{12} - f_{23} - f_{24} - z_2 \\ dx_3/dt &= f_{23} - f_{34} - z_3 \\ dx_4/dt &= f_{14} + f_{24} + f_{34} - z_4 \end{aligned}\right\} \tag{3.3}$$

and we can prepare a similar set of equations for the cycling of nutrients. From these equations it is simple to compute the energy content of any

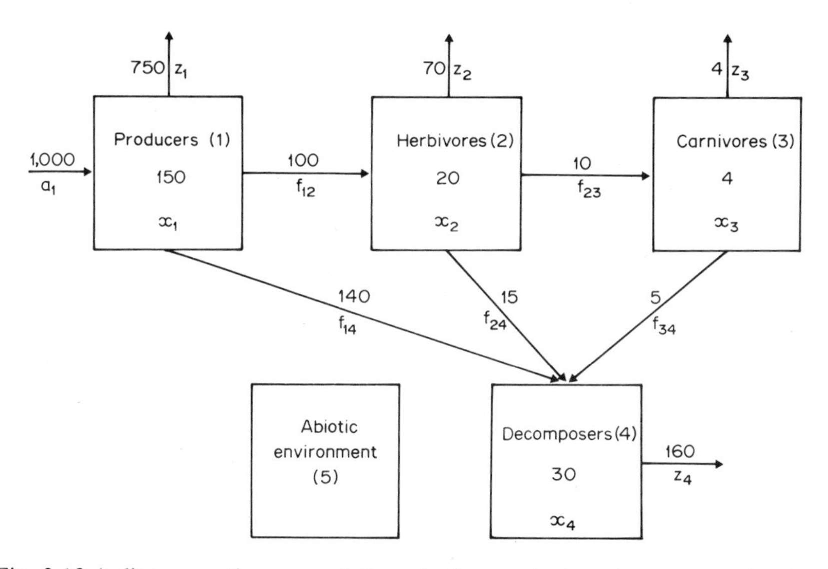

Fig. 3.10 A diagrammatic representation of a five compartment ecosystem, showing the various parameters (symbols and hypothetical values) for the energy flow. For a discussion of this system see the text

compartment after one, two, three, ... periods of time. If we use the parameters given in Fig. 3.10, then application of Equations (3.3) gives

$$dx_1/dt = 10 \qquad dx_2/dt = 5$$
$$dx_3/dt = 1 \qquad dx_4/dt = 0$$

implying that there are increases in the standing crop of producers, herbivores and carnivores, but no change in the decomposers.

Two features are faulty with such a simple model. First, on the mathematical side we have used calculus notation (dx/dt) to denote the rate of change. Calculus assumes infinitesimal time, which is certainly not the case in ecological work where rates of change are given on a per day or per annum basis. It is more mathematically realistic to talk about differences, using difference equations, thus

$$\Delta x_{i,t} = x_{i,t+1} - x_{i,t}$$

where $x_{i,t}$ is the energy or nutrient content of the ith compartment at time t. It will be seen that dx_i/dt in Equations (3.3) has been replaced by Δx_i. However, ecological literature has used the calculus notation for these changes, and in order to avoid confusion with other published accounts the calculus notation will be used here. Secondly, the model as described by Equations (3.3) is biologically unsound since the energy or nutrient content of a compartment cannot increase for an infinitely long period of time as indicated by the model. Hence we need to further develop the model.

In order to demonstrate this next stage of the modelling a simpler situation with only three compartments will be used (Smith, 1970). Smith's example consists of an aquatic plant being eaten by a herbivore. The parameters x_i, a_i, z_i and f_{ij}, as previously defined, are shown in Fig. 3.11. It will be seen that the ecosystem has reached equilibrium since the input of phosphorus (100 mg day^{-1}) equals the output (19 mg day^{-1} in the water flowing out of the system and 81 mg day^{-1} due to emigration of the herbivores). This equilibrium state of the ecosystem can be represented by the equations

$$dx_1/dt = dx_2/dt = dx_3/dt = 0$$

It will also be seen in Fig. 3.11 that there are 10 non-zero parameters, three for standing crop, three for input and output, and four for transfers of phosphorus within the ecosystem. In a real ecosystem with far more compartments it will be seen that a lot of research will be required just to determine these parameters for one nutrient. Smith poses one set of questions concerning these parameters, namely 'How important are the parameters in understanding the ecosystem and how accurately should they be measured?'. He also poses a second set of questions about the

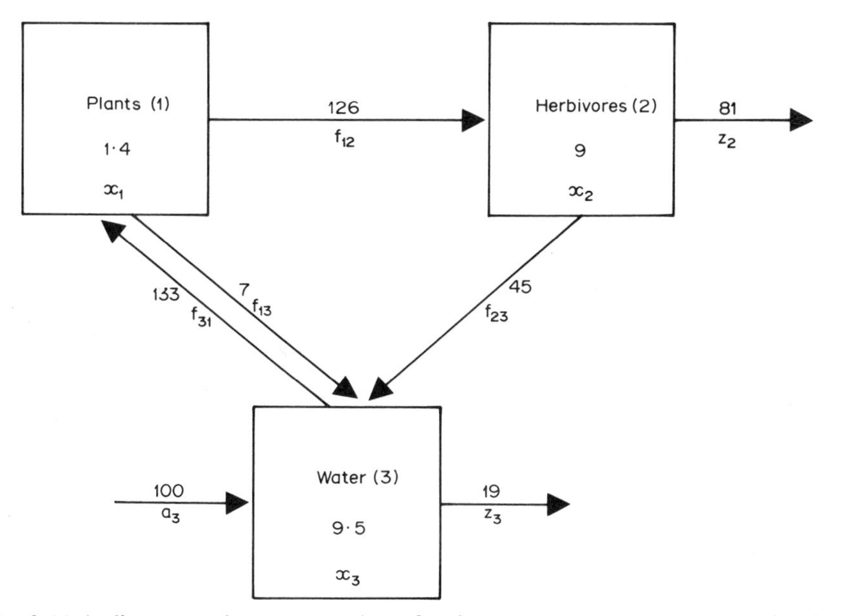

Fig. 3.11 A diagrammatic representation of a three compartment ecosystem, showing the amount of phosphorus in each compartment (mg), and the rates of input, output and transfer (mg day^{-1}). (Based on a model developed by Smith, 1970)

dynamic nature of the ecosystem, asking what would be the responses if the parameters were to be changed.

In order to investigate the dynamic nature of the ecosystem, let us assume that not only is the system run with an input of 100 mg day^{-1}, but also with 25 and 400 mg day^{-1}. The parameters for these three trials on the hypothetical ecosystem are listed in Table 3.7. It can be seen that as the phosphorus input increases so all the parameters increase. The data in Table 3.7 give us a measure of the response of the ecosystem processes to differences in the other parameters. In real ecosystems it might be impossible to alter one set of parameters by so large an extent, and one will have to rely upon climatic changes from year to year, or upon management practices altering the level of one or more of the parameters. Also, in real ecosystems the values of parameters in a table such as Table 3.7 are all estimates, derived from sampling, and hence statistical techniques will be required to move from this stage of the investigation to the next stage in model building.

It is perhaps useful to break the development of Smith's model at this point in order to ask what features a biological model should possess

TABLE 3.7 The parameters for three trials on a three compartment ecosystem. (Source: Smith, 1970)

a_3	x_1	x_2	x_3	z_2	z_3	f_{31}	f_{13}	f_{12}	f_{23}
100	1·4	9	9·5	81	19	133	7	126	45
25	0·9	4	4·5	16	9	40·5	4·5	36	20
400	2·4	19	19·5	361	39	468	12	456	95

Models take the form of one or more equations that each or collectively describe a phenomenon. Watt (1960), introducing a mathematical model for the effect of population density on fecundity, considers that a model should possess four attributes. As these are generally required by all biological models they can be stated as:

(1) The equation should provide a satisfactory statistical fit to the data that it describes.

(2) The equation should be no more complex than is necessary for such a statistical fit.

(3) The equation should incorporate insight into the biological mechanism of the phenomenon that it describes.

(4) The equation should be general, so that by altering parameter values it can be made to fit any set of measurements of the phenomenon.

Thus, Watt concludes that 'the equation should be steeped in, and a product of, the lore and data of biology'.

In modelling an ecosystem such as Smith's simplified example it is perhaps most appropriate to use the simplest possible functional relationships. This can often be achieved by linear regression analysis, relating the ecosystem parameters to one or a combination of values of the standing crop. Such model building is slightly dangerous since regression equations can only be used for interpolation and not extrapolation. However, regression equations describe the data within its limits, and hence can provide a useful first approximation at a functional relationship.

Smith's model building can be conceived as being partly based on a regression fit and partly on known and simple biological models. The most simple of these functional relationships, those that can be determined by linear regression analysis, are for the output of phosphorus in water,

$$z_3 = c_1 x_3 \text{ where } c_1 = 2 \tag{3.4}$$

for the transfer of phosphorus to the water from the plants,

$$f_{13} = c_2 x_1 \text{ where } c_2 = 5 \tag{3.5}$$

and for the transfer of phosphorus to the water from the herbivores,

$$f_{23} = c_3 x_2 \text{ where } c_3 = 5 \tag{3.6}$$

In a real ecosystem, where sampling errors were affecting the estimates, such parameters as c_1, c_2 and c_3 would have been estimated by regression analysis. Using a physical model, the emigration of the herbivores from the ecosystem has been assumed to be proportional to the rate of contact between individuals, i.e. the social interaction. Estimating the constant of proportionality by regression gives

$$z_2 = c_4 x_2{}^2 \text{ where } c_4 = 1 \tag{3.7}$$

The ideas of predatory prey relations have been used to model the transfer of phosphorus from water to plants and from plants to herbivores. Putting in the constants of proportionality we have

$$f_{31} = c_5 x_3 x_1 \text{ where } c_5 = 10 \tag{3.8}$$

and

$$f_{12} = c_6 x_1 x_2 \text{ where } c_6 = 10 \tag{3.9}$$

The set of six Equations (3.4)–(3.9) inclusive now represent the functional relationships within the ecosystem. In the differential calculus notation we can see that

$$\left.\begin{aligned} dx_1/dt &= f_{31} - f_{13} - f_{12} \\ dx_2/dt &= f_{12} - f_{23} - z_2 \\ dx_3/dt &= a_3 + f_{23} + f_{13} - f_{31} - z_3 \end{aligned}\right\} \tag{3.10}$$

We have now reached the position where we can experiment with our ecosystem model, using a computer to carry out the calculations. Smith analyses the sensitivity of the six parameters c_1 to c_6 inclusive by seeing the result of slight changes, in the order of one per cent, in their values on the ecosystem processes. He also analyses what will happen if the input of phosphorus is fixed at some new amount, say 200 mg day^{-1}, or if there are other major changes in the parameters of the ecosystem. However, in conservation management these might not be the questions that are asked.

The analysis of sensitivity will tell us where to put the greatest research effort in modelling the component parts of the ecosystem, but it is unlikely that energy inputs or efficiency of herbivores can be substantially changed. The sort of question that might be asked would be, for example, 'What would happen to the phosphorus cycle if we harvested a proportion of the plants?'.

Substituting Equations (3.4)–(3.9) in Equation (3.10) we obtain

$$\left.\begin{aligned} x'_1 &= x_1(10x_3 - 10x_2 - 4) \\ x'_2 &= x_2(10x_1 - x_2 - 4) \\ x'_3 &= 5(x_1 + x_2) + a_3 - x_3(10x_1 + 1) \end{aligned}\right\} \qquad (3.11)$$

where x'_1, x'_2 and x'_3 give the amounts of phosphorus in each compartment after one period of time, the initial amounts being x_1, x_2 and x_3. Equations (3.11) can now be used to compute a solution to the management question posed above. Within the computer Equations (3.11) are repeatedly applied and after each application a harvest of a specified intensity carried out, until such a time as the equilibrium situation has been reached. This then is the normal method of model operation. A series of simulation equations similar to those in Equation (3.11) are set up, and the computer run until such time as an equilibrium has been reached or the system has broken down. When an equilibrium situation has been reached we can term it the stable state of that ecosystem under a particular management regime, an analogous concept to the stable state of populations (see Chapter 5). In the model that we are considering here, since there are so few compartments, there is an algebraic solution to Equations (3.11). If we define the proportion of the standing crop of plants harvested as p $(0 < p < 1)$, it can be shown that the equilibrium solution is given by

$$\left.\begin{aligned} x_2^2 + \frac{20 + p}{10}x_2 + \frac{7p - 990}{10} &= 0 \\ x_1 &= 0{\cdot}5 + \frac{x_2}{10} \\ x_3 &= x_2 + \frac{5 + p}{10} \end{aligned}\right\} \qquad (3.12)$$

where the positive root of the first equation is chosen for x_2. The effects of exploitation on the processes of the ecosystem are thus predicted by the model to be:

(1) to increase the standing crop of phosphorus in the water;
(2) to decrease the standing crop of phosphorus in the plant and herbivore populations;
(3) to increase the output of phosphorus in the water;
(4) to decrease the output of phosphorus from the herbivores; and
(5) to decrease the transfer of phosphorus between all pairs of compartments of the ecosystem.

The results of repeated computer simulation of such an ecosystem, each simulation having a different harvest rate p, would allow for a manager to evaluate the optimal harvesting rate for that ecosystem. Also, the computer results quantify all of the increases and decreases listed above so that it would be possible to have a management constraint such that the harvest was maximized whilst not reducing the herbivore population below some clearly defined limit.

The extent to which the model accords with reality is a measure of the skill of the model builder, and a measure of the generality of the functional relationships, as well as being a reflection of the quality of the data on which the model was based. It is important to stress that a model can be no better than the data on which it is founded.

We have so far been concerned with the simple models where there is only one species of plant and one species of herbivore. It may be that in an actual model several species are involved, in which case each species forms a compartment. The table of transfer coefficients, as for example Table 3.6, then specifies the links in the food web, and it has been referred to as the 'who-eats-whom' matrix. As the systems become more complicated matrix terminology is useful. These models are beyond the scope of this brief introduction to the modelling of ecosystems, but the methods have been described by Usher (1972). Texts on the subject of ecosystem modelling include those of Watt (1966) and Van Dyne (1970). Such analytic techniques in the planning of the management of renewable resources are considered by Watt (1968).

Van Dyne (1969) points out that any model building is liable to two sorts of error. First, ecological processes are extremely complicated, and are partially determined by many random, or apparently random, variables, for example weather. If one builds a model that is completely realistic the model will probably be mathematically intractable. The result tends to be that gross simplifications are made in order to achieve a model that can be dealt with by usual mathematical techniques. This is unacceptable biologically. Secondly, one can rely for simplicity upon empirical models, which include no analysis of the mechanism in the ecosystem, and hence they lack depth and fail to have much predictive value. This approach may be mathematically simple, but it is equally biologically unsound. These two sorts of errors stress the importance of the basis of models, and of the criteria that have been used by the model builder (Watt, 1960).

4. Classification of Ecosystems

Introduction

A feature of human thinking during the processes of scientific development has been the categorization of the objects with which one is dealing. This is reflected in the development of such disciplines as taxonomy and systematics, but the classification process has also been applied to communities of plants, and, to a very much lesser extent, to animal communities. Initially classification may have been an aim in itself, although there has usually been an essentially practical side of the work: classifications provided the framework within which different communities could be compared and contrasted, and from which predictions could be made about similar communities. The work of the European foresters amply illustrates this point. They classified forests by the tree species and the ground flora, and then having erected the forest classes they were able to apply the production data from sample plots to other forests in the same class.

During more recent years the classificatory process has dichotomized. On the one hand there is a search for more objective methods of deciding the criteria on which the classification is based. The development of this thinking has led to theoretical and computer based studies, a parallel process to the development of 'numerical taxonomy' in the classification of organisms. On the other side of the dichotomy there is the search for ecological significance. This has manifested itself in the theory of ecological gradients, and the analysis of these gradients into categories which form the basis of land-use planning or resource production estimates.

There are, however, various features common to all classification schemes. Perhaps the most notable is that discrete categories are formed from what is, biologically, a continuum or something fairly close to a continuum. This will always result in a few communities not fitting into a category, or perhaps being wrongly assigned to a category. A second feature is that there is no universal classification. Any classification is erected for communities in a defined area, even if this is the whole Earth, and for a particular purpose, such as forestry, an academic study of vegetation, or land-use planning. When considering examples of classifications it is therefore essential to bear these two features in mind – whence was the classification derived, and for what was its purpose?

European forestry classifications

Although much of the impetus for the classification of forest types can be traced back to the plant geographers of the latter half of the nineteenth century, the work of Cajander in Finland, published in its initial form in 1909, has become particularly well known. Cajander (1943) defines a forest type in terms of the ground flora at a time when the forest has reached *an age of early maturity* and when the trees have *an approximately normal density.* [The italics have been added by the author since these two terms cannot be defined precisely.] All forest areas fulfilling these conditions which have more or less the same species composition of the ground flora belong to the same forest type. Cajander realized that management practice, in relation to age, density and species composition, would influence the nature of the ground flora, and hence he defines his classification as an ecological expression of the site and not as an expression of the site's history and management.

TABLE 4.1 The five classes of forest types erected for Finland by Cajander

Class	*Forest type*
I	Heath forests
II	Fresh moss-rich forests
III	Grass-herb forests
IV	Fen forests
V	Dwarf-shrub bog forests

Cajander's classification for Finland erected five classes of forest types, shown in Table 4.1. These classes reflect the general structure and composition of the ground flora. For example, in class (I) lichens are nearly always present and dwarf shrubs (*Ericaceous* plants) are abundant. The abundance of the mosses is inversely proportional to that of the lichens, grasses and herbs are scarce and the most frequent forest species is the Scots pine (*Pinus sylvestris*). The class is divided into six forest types

TABLE 4.2 The division of the class of heath forests into six forest types

Forest type	*Symbol*	*Geographical range in Finland*
Cladonia-type	ClT	mainly in north
Myrtillus–Cladina-type	MClT	north
Calluna-type	CT	most frequent in southern half
Empetrum–Vaccinium-type	EVT	north
Empetrum–Myrtillus-type	EMT	north
Vaccinium-type	VT	most frequent in southern half

which are based on the individual component species. These types are shown in Table 4.2 together with an indication of their geographical range in Finland. Köstler (1956) discusses the rationale behind Cajander's classification and the historical development of this branch of forestry. He also discusses schemes that have been erected for Sweden and Switzerland, both of which show similarities to Cajander's approach, though the Swiss classification relies more upon the forest tree species than on the ground flora.

Although in Finland Cajander could base his classification on mature forests, Anderson (1950) in Britain faced the difficulty of attempting to classify sites for their forestry potential by using the 'waste land' plant communities. Anderson defines 'waste land' in this context as any land available for tree planting which has been devoid of a stand of trees for many years, or which has never carried such a stand. The classification is based on two locality factors, namely soil fertility and soil moisture, factors which Anderson considered to be important in affecting the growth of young trees. He hypothesizes that the existing vegetation, even if shallow rooted, will reflect the nature of the mineral and water supply in the upper soil horizons.

In the classification he proposes six arbitrary categories of soil fertility, represented by the letters *A* to *F* in order of decreasing fertility. In each category there is a further subdivision into three sub-categories, one, two and three, based on an increasing availability of water in the soil under normal summer conditions. This combination of fertility and soil water categories yields 18 'site-classes' or 'locality units', two of which are further subdivided according to the presence or absence of peat. Anderson identifies these 20 site-classes according to the dominant plant species, giving the classification set out in Table 4.3. In descriptions of each of these site-classes Anderson records the geographical distri-

TABLE 4.3 Anderson's classification of waste land communities

Fertility class	*Soil Moisture*			
	Dry *1*	*Moist* *2*	*Wet without peat* *3a*	*Wet with peat* *3b*
A	Dry grass-herb	Moist grass-herb	Grass-rush (hard rush)	Willow-reed
B	Dry grass	Fern	Sedge	Soft rush
C	Grass-heath	Rush-grass (Jointed rush)	–	*Molinia*
D	*Erica cinera*	*Nardus-Molinia*	–	Cottongrass
E	*Calluna*-heath	*Vaccinium*	–	*Myrica*
F	Lichen	*Erica tetralix*	–	*Sphagnum*-moor, *Calluna*-moor

bution in Great Britain, the normal geological strata with which it is associated, and the associated plant species that are indicative of the site-class when found together with the dominant species.

In putting forward this classification of waste land communities there are four points which emerge as important in Anderson's work:

(1) There is the problem of which species to use as indicators. *Calluna vulgaris*, for example, occurs in all the infertile communities irrespective of the water conditions, whilst *Molinia caerulea* occurs in most of the wet communities irrespective of the fertility. Hence, one needs to record not only the dominant species but also the species of plants associated with the dominants. Some species which can be dominant, for example bracken, *Pterydium aquilinum*, indicate very little about either the fertility or the soil water conditions of the site.

(2) As with most classifications there are anomalous communities which fail to fit into a site-class since either the vegetation is intermediate between two or more classes, or the species composition is unlike any of the classes. For example, the dominance of the bog-rush, *Schoenus nigricans*, in places on the Island of Rhum in the Inner Hebrides probably denotes a community that would be ecologically closest to the *Calluna*-moor (*F*3) community.

(3) The total number of site-classes is arbitrarily chosen. In producing his classification Anderson's site-classes had to be small enough in number to make recognition practicable, but yet large enough in number to render the classes ecologically meaningful. In planning the afforestation of land he considers that 20 classes is about the maximum number,

though he realizes that the ecologist might wish to have a finer subdivision of the sites.

(4) It must be emphasized that Anderson's classification is forestry orientated, since it was derived for land that was to be afforested. Anderson's discussion of the site-classes relates to the ease of afforestation, to the need for land management such as drainage and ploughing, and to the species of trees that will grow commercially upon these sites.

The European forestry classifications have thus aimed to provide a forestry orientated classification that is essentially practical and not academic. The fact that such schemes have worked is a tribute both to the experience of the men who originated them and to their ability to subjectively choose the species on which the classifications are based. Ecologists have realized that subjectivity can lead to errors, and hence there have been many searches for objective criteria on which to erect new classifications.

The search for objective criteria

Dichotomies are commonplace in scientific research where the effort is divided between different lines of approach. It is evident that in the field of the classification of vegetation a dichotomy results from the quantitative approach, the numerate school, and from the descriptive approach, or the innumerate school. The latter is by far the older, resulting in an extensive literature of the plant sociology of the European Continent. Although the study of plant sociology is strictly beyond the scope of a book on 'ecosystems' or 'wildlife conservation', yet the principles and practice of the Braun-Blanquet school have spread so widely that their classifications must be considered briefly. The methods have been described by Braun-Blanquet (1932, 1951), and they have been summarized, criticized and adapted by Poore (1955a, 1955b, 1955c, 1956, 1962). The influence of the phytosociological thinking has been so profound that it is reflected in zoological studies such as Gisin's (1943) study of the Collembola (Springtails) of a Swiss National Park, and da Gama's (1964) study of the Collembola communities of Portugal. Gisin (1955) carried the process further in that he erects a classification of vineyard soils based on their populations of micro-arthropods.

A subjective approach

The Braun-Blanquet system essentially develops a hierarchical classification of vegetation based on the grouping of 'associations' into 'alliances', 'alliances' into 'orders', and so on. The system thus attempts to parallel the hierarchical structure of classical taxonomy, whilst the development of this classification has depended upon the collection of a very large amount of descriptive data by the Zurich-Montpellier School. Poore (1955a) analyses five components of the Braun-Blanquet system, stating that:

(1) The fundamental unit of vegetation is the 'association', which is an abstraction, obtained by the comparison of a number of lists made in selected sites in the field. The association is defined entirely by the floristic composition and not by the habitat.

(2) The description and foundation of the association is the association table, which consists of species lists, together with an estimate of the quantitative presence of each species, of all communities thought to belong to the association.

(3) Each association is characterized by a number of species of 'high fidelity' to that particular association – fidelity implies species of narrow ecological amplitude, and thus they characterize an association, an alliance (species of wider amplitude occurring in several associations), etc. Fidelity and dominance are unrelated.

(4) The hierarchy develops from the association until it includes all the communities and species in a natural vegetation region.

(5) The associations are arranged according to the 'principle of sociological progression', the simplest communities first and those with the highest internal integration (woodlands) last.

To use the system sample plots are chosen in uniform vegetation stands, though it is impossible to define 'uniform' precisely, and statistical tests of homogeneity are virtually unused as they are thought to be too time consuming. If no uniformity can be found the Braun-Blanquet system breaks down. Having decided upon a uniform stand the environmental features are recorded and a species list made, all species being recorded for their cover or abundance as well as for their sociability (i.e. whether individual plants occur singly, in tufts, in patches, etc.). It is important that the sample plot should not be smaller than the 'minimal area' of the association, a situation that can be empirically tested by repeatedly doubling the quadrat size and recording the addition of new species. When the doubling yields few or no new species for the list the sample can be said to approach the minimal area of the association – working with a 1 m^2 quadrat Poore (1955c) found this to be about 4 m^2 on a Scottish mountainside.

The field lists are arranged into association tables, but this is a subjective process open to circular argument. For example, of this process Poore (1955a) states: 'The bases for recognizing similarity may be many and various, and may include, for instance, dominance by the same species, constancy of certain species, similarity of physiognomy and habitat and the regular occurrence of species which are already suspected to have a narrow amplitude.' Thus, although 'dominance' may be important in erecting the association, it is then disregarded in favour of 'fidelity' in characterizing the association. The process of determining fidelity is one that is subjective, relying upon a considerable degree of personal experience and intuition.

Thus, the Braun-Blanquet system starts with a thorough collection of vegetational data, though the location of the sample plot is not randomly determined and so the application of any statistical analysis is open to question. Thus, in the search for objective criteria on which to base classification, the system is more developed than those of the foresters – no *a priori* assumptions have been made about which species to look for and record. Despite the collection of quantitative data about the species and the step towards objectivity, the system gets no further as there is the subjective process of shuffling the lists of communities into associations, and then the assessment of fidelity in order to characterize associations. A search for the rationale of this latter process is long overdue. Poore (1956, 1962) has hinted that such a rationale may not exist, but that as classificatory experience accumulates a process of successive approximations may yield a final solution.

A numerical approach

It was the formulation of classificatory rationale that has led to the concepts of numerical classifications. It should be emphasized that the majority of this work is not statistical since there is no attempt to formulate an *a priori* hypothesis and then to collect data to test it. Rather, the processes are mathematical, using the results of algebra and numerical analysis, coupled with the speed of electronic digital computers, to probe the complex and multi-dimensional situations.

The methods of numerical classification can be divided into two groups, namely agglomerative and divisive. The agglomerative techniques start with a large number of units which are successively clustered together until all of the units are included in the classification. The divisive techniques start with a whole population of units, which is then divided successively until there is some cut-off point in the process or until every unit has been separated from all other units. Each of these techniques can,

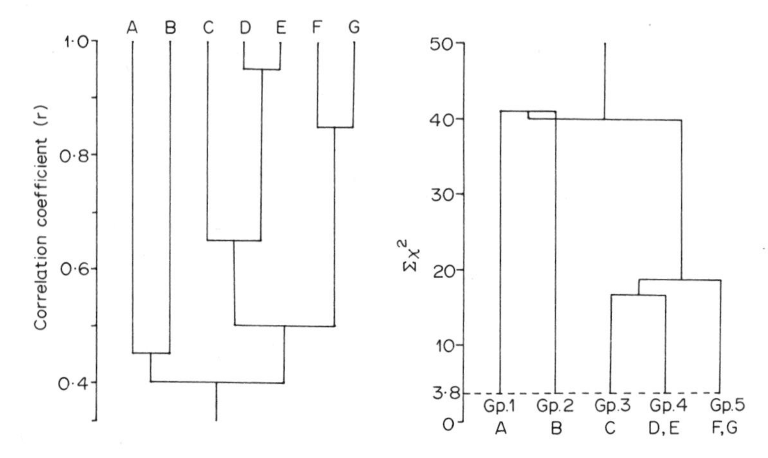

Fig. 4.1 The dendrogram on the left shows a polythetic agglomerative classification of seven units. The analysis proceeds until all seven units have been clustered ($r = 0{\cdot}4$). The dendrogram on the right shows a monothetic divisive classification. The analysis proceeds until χ^2 falls below some arbitrary level, in this example the level is 3·8. The seven units in each dendrogram can be considered as identical

theoretically at least, have two variants, monothetic when fusion points are determined by one character, and polythetic when the fusion points are determined by several characters. In practice polythetic agglomerative techniques have been used most frequently in numerical taxonomy where the correlation coefficient or some other measure of association is used to assess the similarity between the units being classified. Fig. 4.1 (*a*) shows a typical dendrogram resulting from such an analysis. The methods of analysis and their application in systematics are discussed by Sokal and Sneath (1963), and Moss (1968) discussed the results of the experimental application of several of these techniques.

In the study of plant communities the monothetic divisive techniques have received the greatest attention. Williams and Lambert (1959, 1960) describe a technique known as 'association-analysis'. As with the Braun-Blanquet School the analysis uses as its data complete lists of plant species within sample quadrats. However, the conditions of 'uniform stand' and 'minimal area' are relaxed, and sampling of the vegetation can either be at random or on a systematic grid. Since the technique is pseudo-statistical, the estimation of a χ^2 criterion of interspecific association requires that the sample size should be large. A dendrogram using a monothetic divisive technique is shown in Fig. 4.1 (*b*). With a cut-off value of 3·8 the seven units have been divided into five groups. By changing the cut-off value of the χ^2 criterion the analysis can produce more or less groups. Anomalies do sometimes occur where a χ^2 criterion for a division is greater than at the previous division, as is shown for the separation of groups one and two from the remainder in Fig. 4.1 (*b*).

In monothetic divisive techniques the simplest rationale to employ is that of Williams and Lambert's association-analysis. The interspecific association between two species is measured by a χ^2 value for a contingency table of the presence and absence data. Thus, if we have a sample size of 200, and we consider three species A, B and C, we can set out three contingency tables thus ('+' denotes presence, and '–' absence):

		Species A +	Species A –	
Sp. B	+	65	15	80
	–	69	51	120
		134	66	200

		Species B +	Species B –	
Sp. C	+	22	52	74
	–	58	68	126
		80	120	200

		Species C +	Species C –	
Sp. A	+	52	82	134
	–	22	44	66
		74	126	200

The probability that species A will occur in a quadrat is estimated from our data as 134/200 or 0·67 (frequency of 67 per cent). Similarly the probabilities for species B and C are 0·40 and 0·37 respectively. If we make the *a priori* hypothesis that there is no association between species A and B, then we can expect the probability that both A and B occur in the same quadrat to be $0{\cdot}67 \times 0{\cdot}40 = 0{\cdot}268$. Since our sample contains 200 quadrats, on the hypothesis that there is no association between species A and B we would expect 53·6 quadrats to contain both species. If this number is exceeded we can suspect that the two species are positively associated, and if there are fewer joint occurrences we can suspect negative association. χ^2 can be used to test the departure of the observed data from the hypothesis expectation. In the general case

		Species X		
		+	−	
Species Y	+	a	b	$a+b$
	−	c	d	$c+d$
		$a+c$	$b+d$	$a+b+c+d=n$

the value of χ^2 is

$$\chi^2 = \frac{n(|ad - bc| - \frac{1}{2}n)^2}{(a+b)(c+d)(a+c)(b+d)} \tag{4.1}$$

after carrying out Yate's correction (see, for example, Bailey (1959)). In our example the values of χ^2 are 11·195 (53·6), 4·505 (29·6) and 0·358 (49·6) for the associations between species A and B, B and C, C and A respectively (the expected frequency of the joint occurrence of the two species is shown in brackets). Since χ^2 has 1 d.f. it can be seen that there is a strong association between species A and B ($p < 0{\cdot}001$), that the data are indicative of a negative association between species B and C ($p < 0{\cdot}05$), and that there is no association between species A and C. In using these contingency tables it should be remembered that the validity of χ^2 is dubious when any one of the expected frequencies are less than two.

Williams and Lambert (1959) applied the technique of association analysis to a Callunetum in the New Forest, Hampshire. The site lies over Barton Sand, except for a ridge capped by Plateau Gravel towards the south of the sampling area. The area is subjected to rotational heather burning on an approximately seven year cycle. Only six vascular species were present, though one of these, *Trichophorum caespitosum*, occurred in less than two per cent of the 615 quadrats and was not included in the analysis. The other five species are *Calluna vulgaris*, *Molinia caerulea*, *Erica tetralix*, *Erica cinerea* and *Pteridium aquilinum*. Using the χ^2 method of estimating interspecific associations, Table 4.4 sets out the observed associations. Williams and Lambert discuss the possible methods for dividing the community into groups, and they argue from a consideration of the vegetation that *M.caerulea* gives the best separation, although *C.vulgaris* is the most abundant species. They postulate that the

TABLE 4.4 Significant associations are shown by their χ^2 values, (calculated with Yate's correction) negative associations being in italics. x denotes a non-significant value of χ^2. (From Table 1 of Williams and Lambert (1959))

	C.vulgaris	M.caerulea	E.tetralix	E.cinerea	P.aquilinum
C.vulgaris	–	51·31	45·66	*x*	*x*
M.caerulea	51·31	–	93·76	*12·62*	*68·64*
E.tetralix	45·66	93·76	–	*4·84*	*14·08*
E.cinerea	*x*	*12·62*	*4·84*	–	6·92
P.aquilinum	*x*	*68·64*	*14·08*	6·92	–
$\Sigma\chi^2$	96·97	226·33	158·34	24·38	89·64

TABLE 4.5 The χ^2 values for all quadrats in which *M.caerulea* is present. x denotes a non-significant association, and o denotes associations that are indeterminate in the sample. (Derived from Table 2 of Williams and Lambert (1959))

	C.vulgaris	E.tetralix	E.cinerea	P.aquilinum
C.vulgaris	–	24·8	*x*	*o*
E.tetralix	24·8	–	*x*	*o*
E.cinerea	*x*	*x*	–	*o*
P.aquilinum	*o*	*o*	*o*	–
$\Sigma\chi^2$	24·8	24·8	0	0

TABLE 4.6 The χ^2 values for all quadrats in which *M.caerulea* is absent

	C.vulgaris	E.tetralix	E.cinerea	P.aquilinum
C.vulgaris	–	*x*	4·34	*x*
E.tetralix	*x*	–	*x*	*x*
E.cinerea	4·34	*x*	–	*x*
P.aquilinum	*x*	*x*	*x*	–
$\Sigma\chi^2$	4·34	0	4·34	0

value of $\Sigma\chi^2$ for each species is a useful decision-parameter on which to make a division. In the New Forest example *M.caerulea* has the greatest $\Sigma\chi^2$. The community is thus divided into those quadrats in which *M.caerulea* is present (474 quadrats), and those in which it is absent (141 quadrats). Considering only the quadrats in which *M.caerulea* is present, the significant associations are shown in Table 4.5. Here it can be seen that an ambiguity arises – does one divide on the presence–absence data of *C.vulgaris* or *E.tetralix*? Reference to Table 4.4 shows that *E.tetralix* has

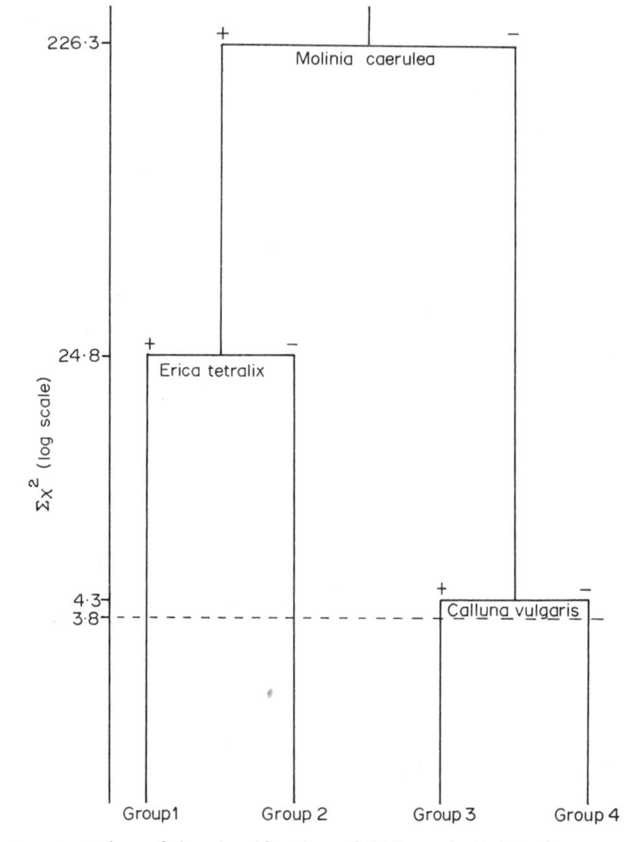

Fig. 4.2 A representation of the classification of 615 quadrats into four groups, based on the presence (+) or absence (−) of the species named. (Drawn from data given by Williams and Lambert, 1959)

a greater $\Sigma\chi^2$ than *C.vulgaris*, and hence division is based on *E.tetralix*. Similarly, in the absence of *M.caerulea* (Table 4.6) the ambiguity arises as to division on *C.vulgaris* or *E.cinerea*. Since the former has the greater value of $\Sigma\chi^2$ (Table 4.4), division is based on that species. The division of the whole community is shown schematically in Fig. 4.2.

One essential feature of any analysis is that the groups should be ecologically meaningful. Fig. 4.3 shows the distribution of the four groupings, demonstrating that the first division, based on the presence and absence of *M.caerulea*, reflects the geological/pedological change from Barton Sands to Plateau Gravel. The more obvious burning pattern is reflected in the second-order divisions. The mathematical relationship between the summation of the columns of the χ^2-matrix and the methods

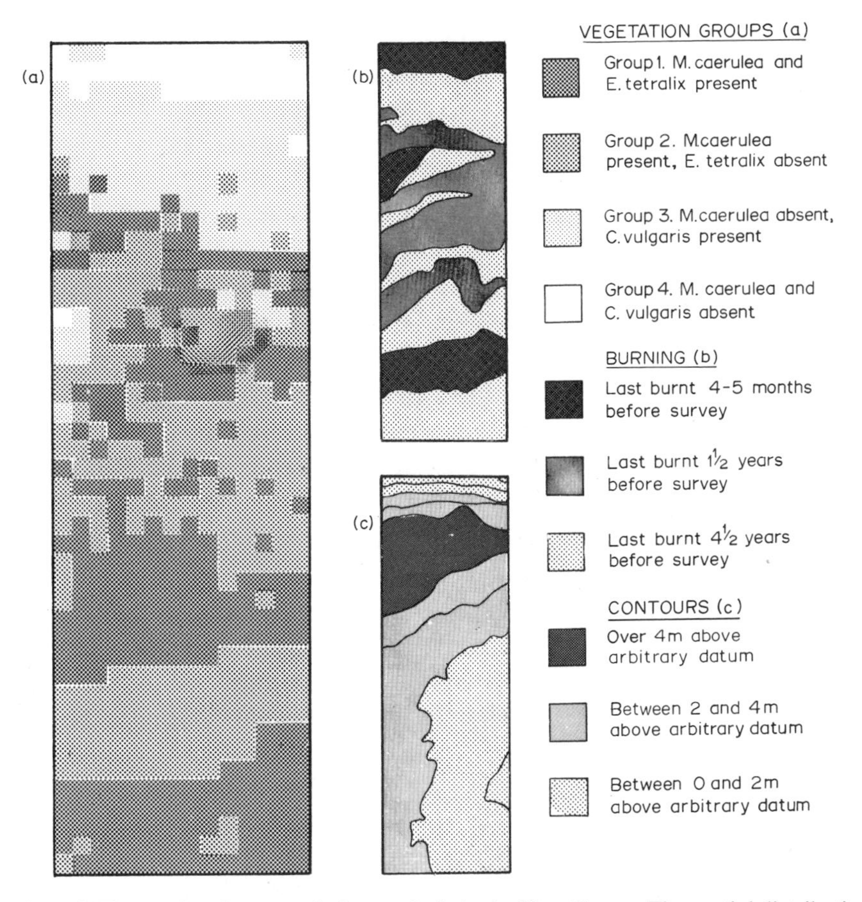

Fig. 4.3 The results of an association analysis in the New Forest. The spatial distribution of the four groups distinguished in the analysis is shown on the left, whilst on the right the two main ecological influences of burning and topography (related to geology) are shown. (Redrawn from Williams and Lambert, 1959)

of principal component analysis suggests that the first-order division will be related to the principal axis of variation. It is therefore most likely that the first divisions in the classification will reflect ecological influences acting upon the horizontal distribution of the species in the community.

The treatment of the χ^2 values in the association matrix is a difficult problem. Williams and Lambert (1959) point out the disadvantage, from theoretical considerations, of adding corrected χ^2s, and they experiment (1960) with other criteria based on the χ^2. The practice of ignoring non-significant values of χ^2 can be criticized, and in cases where ambiguities occur it might be better to include in $\Sigma\chi^2$ all of the non-significant values. The significance level of $\chi^2 = 3{\cdot}841$ is, after all, a subjective level at which to work. The cut-off level for terminating the analysis is also subjectively chosen. Williams and Lambert (1960) discuss two such points, one in which the population of quadrats is successively divided until there are no significant associations remaining, and the second where $\Sigma\chi^2$ for each of the species falls below $N.2^{-5}$ (where there were N quadrats originally). A third point of conflict is in determining which species to include in the analysis. Williams and Lambert (1959) excluded *T.caespitosum* as it occurred in less than two per cent of the quadrats. However, when they include it they show (1960) that the division in the absence of *M.caerulea* is on the presence or absence of *E.cinerea* rather than *C.vulgaris*. Despite this division on a different species the distribution of the four groupings of quadrats is similar in both analyses, and the ecological interpretations of the analyses are identical.

Bunce (personal communication) has used the techniques of numerical classification in a framework of conservation management characterizing woodlands so that the conservation efforts could be directed to preserving as wide a spectrum of woodland types as possible. During a Nature Conservancy review in Britain, 2463 woodland sites were visited, and 136 tree, shrub and herbaceous species were recorded for their presence or absence from the whole woodland. An association analysis was performed by calculation of the χ^2 values for the 136 square association matrix. Using all significant χ^2 values the analysis yielded 557 groups of woodlands, a number of categories too large to be included in an overall conservation plan. Thus a cut-off point for the analysis of $\Sigma\chi^2 = 20$ was used, giving 103 groups.

Bunce considers the ecological characteristics of each of the groups. He discusses, as an example, the 68 woodlands that comprise group 25, and the 42 woodlands comprising group 47. The mean values of the environmental variables are shown in Table 4.7, and the representation of the sites in the Nature Conservancy regions in Table 4.8.

The key characters for group 25 woodlands are set out in Table 4.9. It can be seen that the group consists of woodlands particularly characteristic of the East Anglian boulder clays. The woods are usually coppices of varying age, with oak, *Quercus* spp., and ash, *Fraxinus excelsior*, forming the canopy with an understory of hazel, *Corylus avellana*.

TABLE 4.7 The mean values of environmental variables for woodlands in Bunce's classification

Variable	*Group 25*	*Group 47*
Area	90·3 acres	83·0 acres
Altitude (top)	277 ft	527 ft
Altitude (bottom)	201 ft	319 ft
Horizontal/vertical distance	47·2	13·4
Aspect: Degrees from south (in either direction)	96·0°	98·3°
Aspect: Degrees from east (in either direction)	80·0°	81·8°
Distance from west coast	140 miles	35 miles
Distance from south coast	91 miles	157 miles
Annual rainfall	37·4 in	56·0 in
Days with no sunshine	73 days	78 days
Days with snow lying	16 days	21 days
Saturation deficit (July)	8·0 mb	5·8 mb

The characters of group 47 are set out in Table 4.10. These woodlands occur on steep slopes, extending over a wide range of contours. Many of the sites in this group are in narrow valleys often with rock outcrops or scree present and with ungrazed ledges. Oaks, *Quercus* spp., are the dominant species, though ash, *F.excelsior*, and alder *Alnus glutinosa*, may be locally important.

TABLE 4.8 Distribution of woodlands in Nature Conservancy regions

Region	*Group 25*	*Group 47*
England		
South	10	1
South East	10	0
South West	11	7
East Anglia	27	0
Midlands	3	0
North East	2	7
North West	0	2
Wales		
South	4	15
North	1	2
Scotland		
South	0	1
East	0	2
West	0	5

TABLE 4.9 Diagnosis of group 25 woodlands

Species	*Condition*	*Notes*
Acer campestre	present	Distributed in south, central and east England. Indicates a moist base-rich soil
Potentilla erecta	absent	Indicates the absence of wet, acid soils
Oxalis acetosella	absent	Indicates the absence of mildy acid areas and patches of raw humus
Primula vulgaris	present	Generally distributed in Britain, and has no indicative value about the soil base-status or moisture
Euonymus europaeus	absent	Indicates the absence of calcareous, usually well-drained soils
Teucrium scorodonia	absent	Indicates the absence of skeletal and disturbed soils
Viburnum opulus	absent	The occurrence is related to management practices in the past

TABLE 4.10 Diagnosis of group 47 woodlands

Species	*Condition*	*Notes*
Acer campestre	absent	No marked affinities for southern base-rich sites
Filipendula ulmaria	present	Generally distributed in Britain. Soil usually eutrophic, wet to very wet
Galium hercynicum	absent	Absence of dry acid areas
Lonicera periclymenum	present	Generally distributed in Britain. Soil mildly acid, little grazing of sites
Sanicula europea	present	Infrequent in Scotland and west Wales. Loamy soils, generally well-drained to moist

The main use of such a wide scale analysis is in planning the conservation effort locally, and deciding which sites require the greatest conservation effort. Thus any site can be fitted into the National classification, and comparison with existing conservation areas will show if this site would fit into a series of similar reserves, or whether this is a woodland type previously given no protection. Bunce (personal communication) has prepared a dichotomous key to all of the 103 groups of woodlands that the association-analysis has determined, showing the distribution of woodlands into these groups.

The search for ecological significance

Throughout the history of community classification interest has always focussed on the ecological significance of the group or hierarchy that is produced. Thus, a test of the usefulness of the numerical techniques has been the question 'Are the results ecologically meaningful?'. In general the answer has been 'Yes'. Much then of the classification research has been concerned with the *a posteriori* assessment of the results.

It was shown that Anderson's classification was partially *a priori* in concept. He considered that young trees were influenced by fertility and soil water conditions, and based on these two criteria the classification of 20 groups was erected. The concepts of this *a priori* approach to classification are most fully developed in land-use planning projects in Canada. Farrar (1962) states his six basic premises as:

(1) Land is a multiple continuum. All land classifications and all subdivisions of land are arbitrary. Every parcel of land is unique and differs in some way from all other land.

(2) Regardless of size, any area of land can be treated as an ecosystem if included with it are all organisms, the soil below and the atmosphere above.

(3) Regardless of size, any area of land can be subdivided arbitrarily into any number of ecosystems.

(4) A site classification is a classification of land for a particular purpose. To be useful it must be based on ecological principles.

(5) The site classification designed for a particular area is unique.

(6) Advances in scientific aspects of forestry do not depend on site classification. A site classification is solely an aid to management in its early stages.

Farrar's concept of an ecosystem thus differs from many that consider the ecosystem to be bounded by dissimilar ecosystems. Also, Farrar's sixth premise is in direct contradiction to Poore's (1962) statement that classification is essential to the advancement of the biological sciences. The methods of gradient analysis are not, however, entirely new. Champion and Brasnett (1958) review many of the indices of site and climate that have been postulated.

Hills (1961) applies this concept of ecological gradients. His scheme is summarized in Table 4.11. The geographical area under consideration is Canada, which is divided into climatic regions, based on broadly similar patterns of climate. Within each climatic region several *landtypes* are recognized, these being distinguished as areas of similar geological

TABLE 4.11 A schematic representation of Hills' methods of land classification

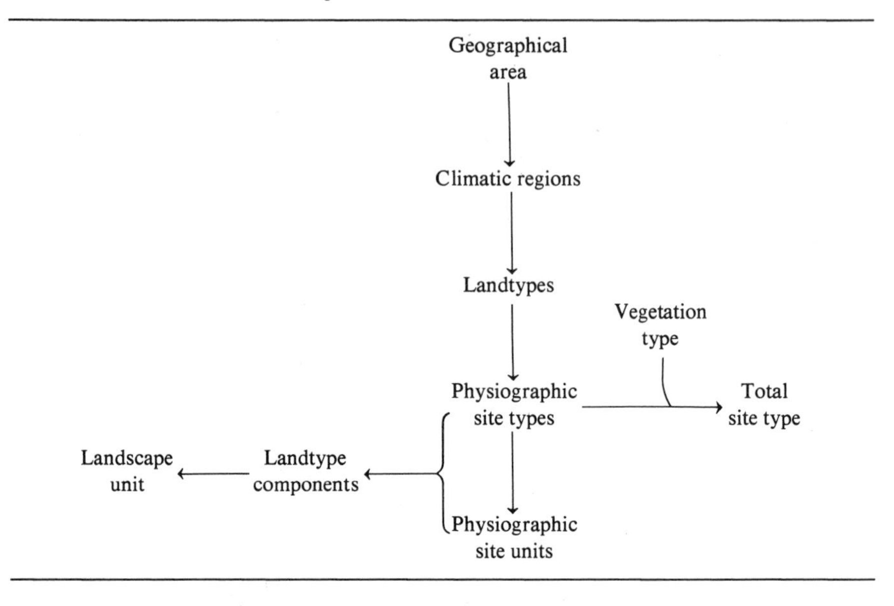

material in respect of texture and petrography. Thus each landtype is defined in terms of its climate and geology.

Consideration of three gradients then divides the landtype into *physiographic site types*. The gradients of prime ecological significance are considered to be soil moisture, depth of parent material and local climate. An example of a physiographic site type would be 'dry site, moderately shallow soil, with a hot and dry local climate'.

Other gradients of lesser general ecological significance can be used to subdivide the site types into *physiographic site units*. These gradients can be chosen such as factors of the soil profile development, stoniness, degree of slope, aspect, slight variation in soil texture, small temperature variations, etc. The amount of division depends on the fineness of the separation of the sites that is required. Integration of the site units or site types with the *vegetation type* gives the *total site type*, which Hills equates with the 'ecosystem'.

The process of division is matched in the classification by a process of synthesis. The divisive process can yield a very large number of categories, and the synthetic process puts together physiographic site types or units that have similar ecological characteristics, depending upon the object of the classification. Such pooling can be of a more general kind,

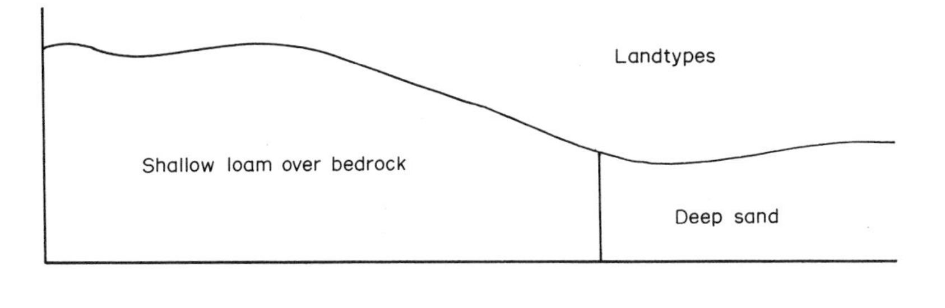

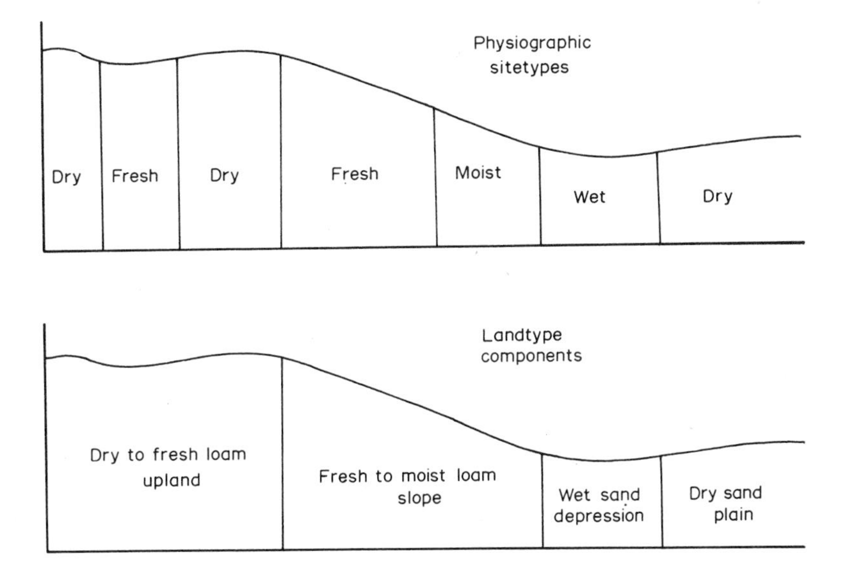

Fig. 4.4 A diagrammatic representation of Hills classification. (Redrawn from Hills, 1961)

for example a warm dry site may have similar growth potential to a cool moist site. The processes are illustrated by reference to Fig. 4.4, which shows the division of an area into two landtypes, and the subsequent division into seven physiographic site types. The synthesis produces four landtype components for inclusion in land-use planning.

It must be emphasized that both the divisive and synthetic processes are subjective, and rely upon the experience of the planner. The separation of gradients is often based on judgements such as 'cold, cool, average, warm, hot' without a definition of the precise meaning of these terms. It might also be that too many categories would be produced, since if three gradients are each divided into five states, 5^3 or 125 groups are produced. However, many ecological factors are correlated with each other, and it often happens that only a small percentage (20–30 per cent) of the possible classes are actually found in the field. The whole process of the analysis is one that awaits the application of quantitative techniques so that the *a priori* reasoning can be treated in an objective manner.

The application of the classification depends on the scale at which the land planning is to be carried out. For example, Hills (1961) maps the landtypes for an extensive scheme of approximately 250 square miles. For each of the 27 landtypes he maps gradings of the area's potential for use in agriculture, forestry, wildlife conservation, fish production (in the water bodies) and recreational use. If a more intensive land-use plan were required the landtype components or the physiographic site types would be mapped. Again the potential of each mapped category would be assessed for the five aspects of land management described above, and maps drawn for each form of land-use. These six maps form the basis for the integrated land-use plan, balancing the economic needs of the community, the predicted development of the community and the biological potential of the sites.

The method has been used extensively in Canada. Hills and Pierpoint (1960) discuss forest site evaluation in Ontario, and Pierpoint (1962) shows the application of the technique to a more restricted area, The Kirkwood Management Unit. The methods have been criticized by Rowe (1962) as 'difficult to comprehend and apply by any excepting those who have worked in close association with him [Hills]'.

A similar system to that used in Canada has been developed in Australia (Christian and Stewart, 1953, 1964). Their method consists of dividing a large area of land into units, known as land systems, which are areas or groups of areas which have a recurring pattern of topography, soils and vegetation. The system is orientated towards the planning of land-use development, working in part from aerial photographs. It is now being extensively used in tropical countries, as, for example, in Nigeria (Bawden and Tuley, 1966) and in The Gambia (Hill, 1969). Adaptation of these methods to forest use (Duffy, 1969) requires a subdivision of the land system into land units, each of which can be assessed and mapped for its forest capability.

The Canadian and Australian methods of land classification provide an important development in our thinking on the classification of ecosystems by introducing *a priori* reasoning into the range of available methods.

5. The Response of Ecosystems to Exploitation

Introduction

In the previous chapters we have seen that an ecosystem is built of a number of components, each component being linked to at least one other by a transfer of energy or nutrients. In modelling the ecosystem we were concerned with the equilibrium or steady state, a state that exists when the output from every compartment exactly balances the input. This is an idealized situation, since in nature the random influences, such as weather, will be continually modifying this equilibrium. However, more drastic disturbances may result from the management practices to which the ecosystem is subjected. Some link in the system may be broken, and one wishes to know what effect this will have on the whole ecosystem and in particular on one or two of the species in the ecosystem. Such a broken link may result from the removal of the top carnivore since this animal may be dangerous not only to the other animal species but also to man himself. How do the populations of herbivores respond in this new situation? A link may not be completely broken. It is possible to exploit a population heavily without exterminating it completely. What are the characteristics of such a population so that we can recognize the symptoms and carry out prophylactic treatment before the species is added to *The Red Book* of animals threatened with extinction?

In this chapter the concepts of plant and animal ecology will mostly have to be separated one from the other. First, some models used in population mathematics will be developed, and then laboratory and field populations will be examined in order to elucidate the general characteristics of populations subjected to exploitation. Finally some indices of plant production will be investigated, and examples will be given to demonstrate the effects of over-grazing and under-grazing on the community structure.

Introduction to population mathematics

Within the space of an introduction it is impossible to explore the whole subject of population dynamics, and hence only the development of models that will be useful in understanding the management of populations will be considered. For a much fuller treatment of the subject reference should be made to text-books such as Andrewartha (1961), Slobodkin (1961), Clark, Geier, Hughes and Morris (1967), Solomon (1969) or Williamson (1972).

Looking first at the simplest of models, we can assume that a population doubles itself in a fixed period of time. This is the same as saying that the rate of increase in population size is proportional to the size of the population, or mathematically,

$$dN/dt \propto N \tag{5.1}$$

where N is the size of the population and t is the time. Putting in the constant of proportionality, Equation (5.1) becomes

$$dN/dt = rN \tag{5.2}$$

In order to solve Equation (5.2) we want to be able to express N as a function of t. Integration of Equation (5.2) gives

$$N_t = N_o e^{rt} \tag{5.3}$$

where e is the base of natural logarithms (2·7182818284) and N_o is a constant, being the size of the population at the start. The constant r is a measure of the speed of increase of the population, and it is referred to as the 'intrinsic rate of natural increase'. Returning to the initial statement that such a model represented a population that doubled itself in a fixed length of time, we can see from Equation (5.3) that this period of time, T, is given by

$$T = 1n2/r \tag{5.4}$$

where $1n$ denotes a natural logarithm.

Equation (5.3) is often referred to as exponential or logarithmic growth, since if the logarithm of population size is plotted against time the graph is a straight line with slope r. This is a common feature during the initial growth of a population. For example, bacterial populations have a logarithmic growth phase before the supply of nutrients begins to run out and the growth slows down. The model is thus based on an over-simplification, since in nature populations do not continue to grow exponentially, but are regulated. The nature of this regulation, and the mechanisms involved, need not concern us during the building of these models. The only feature of concern is that some sort of regulation exists.

Let us now assume that the regulatory factor, be it the supply of nutrients to a bacterial population or some social interaction, would maintain an equilibrium population size of N_E. Let us also assume that the rate of increase in the size of the population is proportional to the difference between this equilibrium size and the actual population size, i.e. to $(N_E - N)$. By analogy with Equations (5.1) and (5.2) we have

$$dN/dt = K(N_E - N)N \tag{5.5}$$

where K is a constant. Integration of this differential equation gives

$$N_t = \frac{N_E}{1 + e^{a-bt}} \tag{5.6}$$

where a and b are constants related to the steepness and curvature of the graph. [The steps in the integration of Equation (5.5) to a form similar to Equation (5.6) are shown by Maynard Smith (1968).] Equation (5.6) is referred to as the logistic equation, and the graph is sigmoid or S-shaped (see Fig. 5.1). The graph is symmetrical about the point on the curve for which $N_t = 0{\cdot}5\ N_E$. In many populations there has not been a good fit, in a statistical sense, between the population and the model. This has been due to the symmetric property of the model, since populations have shown a very short exponential phase before the rate of increase starts to decline. For these cases other assumptions will have to be made in formulating the differential equation, and integration will result in other forms of sigmoid curve. One of these non-symmetrical sigmoid curves is described by the equation

$$N_t = \frac{N_E}{1 + e^{a-bt-ct^2}} \tag{5.7}$$

where a, b and c are all constants. It will be seen that the maximum population size predicted by Equations (5.6) and (5.7) is N_E, and that the population size at the start ($t = 0$) is given by $N_E/(1 + e^a)$.

Another method of building similar models is by the application of matrix algebra. This method overcomes some of the theoretical objections to differential calculus since the matrix operates over discrete periods of

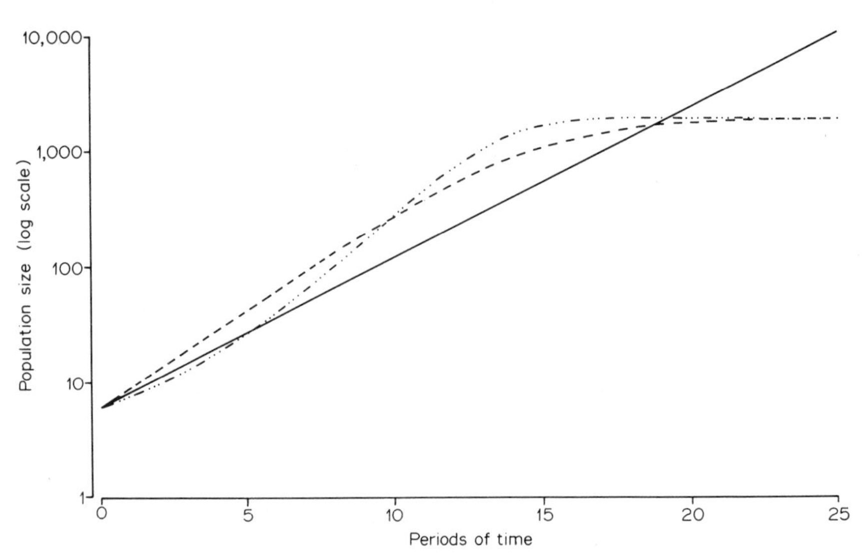

Fig. 5.1 Three curves describing the increase of population size with time. The curves are:
———————, Equation (5.3) with $N_O = 6$ and $r = 0{\cdot}3$;
— — — — —, Equation (5.6) with $N_E = 2000$, $a = 5{\cdot}8061$ and $b = 0{\cdot}4$;
— · · · — · · ·, Equation (5.7) with $N_E = 2000$, $a = 5{\cdot}8061$, $b = 0{\cdot}2$ and $c = 0{\cdot}02$

time, since the age or size structure of the population is taken into consideration, and since matrices are more suited to numerical methods and computation. The model was originally described by Lewis (1942), independently by Leslie (1945) and extended by Leslie (1948). In order to use the model we divide the population into $n + 1$ $(0, 1, 2, \ldots, n)$ equal age groups, and assume that the oldest group possible, the age at which all surviving animals die, is represented by an age of n. The data concerning the population are used to calculate the elements of a matrix that will link the numbers in the various age groups at successive times. This link is provided by the matrix equation

$$\begin{bmatrix} f_0 & f_1 & f_2 & \cdots & f_{n-1} & f_n \\ p_0 & 0 & 0 & & 0 & 0 \\ 0 & p_1 & 0 & & 0 & 0 \\ 0 & 0 & p_2 & & 0 & 0 \\ \vdots & & & & & \\ 0 & 0 & 0 & & p_{n-1} & 0 \end{bmatrix} \begin{bmatrix} n_{t,0} \\ n_{t,1} \\ n_{t,2} \\ n_{t,3} \\ \\ n_{t,n} \end{bmatrix} = \begin{bmatrix} n_{t+1,0} \\ n_{t+1,1} \\ n_{t+1,2} \\ n_{t+1,3} \\ \\ n_{t+1,n} \end{bmatrix} \tag{5.8}$$

In the matrix f_i ($i = 0, 1, 2, \ldots, n$) refers to the fecundity of a female in the ith age group, and p_i ($i = 0, 1, 2, \ldots, n-1$) is the probability that a female in the ith age group will be alive in the $(i+1)$th age group. It is evident that $p_i \leqslant 1$, since a female must either die or become one age group older during the period of time over which the matrix operates. The vector with elements $n_{t,i}$ represents the structure of the population in its various age groups at time t, and the elements $n_{t+1,i}$ are the predicted numbers in each of the age groups at time $t+1$. The matrix can thus be broken down into the sum of two matrices, one stochastic matrix which gives the probability that an individual will be in another class at the end of the period of time, and the second matrix that gives the reproductive data for all the classes. For readers who are not familiar with matrix algebra a synopsis of matrix operations is given in the first appendix.

As an example of the application of this model, Williamson (1967) uses the matrix

$$\begin{bmatrix} 0 & 9 & 12 \\ \frac{1}{3} & 0 & 0 \\ 0 & \frac{1}{2} & 0 \end{bmatrix}$$

on an initial population of just one old animal, represented by the column vector

$$\{0, \quad 0, \quad 1\}'$$

Repeated application of the matrix predicts a population of 12 young after one period of time, of four middle-aged animals after two periods of time, of 36 young and two old after a third period of time, and so on. This results from the fact that each old female produces on average 12 young before dying, and each middle-aged female produces on average nine young before either dying (probability one half) or becoming one age group older (probability one half). The young animals are not fecund, and have a probability of surviving to become middle-aged of one third. A graph of the numbers in each of the age classes is shown in Fig. 5.2 together with the total population size which is calculated by adding up the sum of the vector elements.

If the matrices in Equation (5.8) are represented by symbols, then

$$\mathbf{Mn}_t = \mathbf{n}_{t+1} \tag{5.9}$$

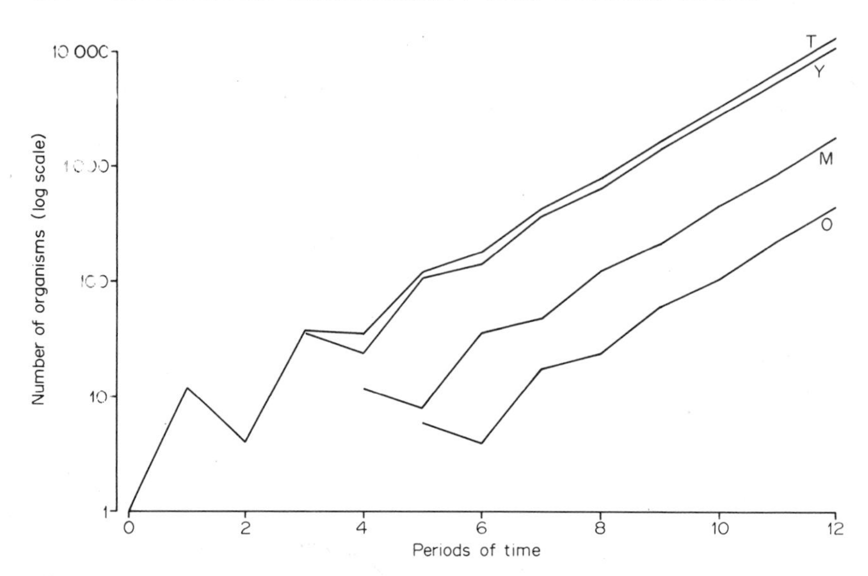

Fig. 5.2 Graphs showing the increase of a population governed by the matrix model of Williamson (1967). The symbols indicate the total population size (T), and the numbers of young (Y), middle-aged (M) and old-age (O) animals

where $\mathbf{n}_t$ represents the population structure at time t. In calculating the population structure for Fig. 5.2 it can be seen that

$$\mathbf{n}_1 = \mathbf{M}\mathbf{n}_0$$
$$\mathbf{n}_2 = \mathbf{M}\mathbf{n}_1$$

and so on. In general we can see that after k periods of time, the population is estimated as

$$\mathbf{n}_k = \mathbf{M}^k\mathbf{n}_0 \tag{5.10}$$

It is evident from Fig. 5.2 that the total population size and the numbers of each of the three classes show damped oscillations. After a few periods of time it is also evident that all three classes are increasing at the same steady rate, and thus that the proportion of each class to the others remains constant. This is due to the property of the dominant latent root of $\mathbf{M}$, which in this case is two. Thus, during each period of time the population size is doubled. More generally, if the dominant latent root is λ, then

$$\mathbf{M}\mathbf{n} = \lambda\mathbf{n} \tag{5.11}$$

where **n** is the stable population structure, measured by proportions rather than by actual numbers. In a graph where the logarithm of the population size is plotted against time the slope of the line after the stable population structure has been reached is $\ln\lambda$. It will be recalled that the slope of the graph for the logarithmic growth of populations was r, the intrinsic rate of natural increase. Thus, the relation

$$r = \ln\lambda \tag{5.12}$$

serves to link the results of the two models. Williamson (1967) has shown that although the rate of increase is a useful ecological parameter, its application is restricted since it is based on experimentally determined death rates, which are likely to be lower than the death rates in natural populations, and since it does not show the speed of approach to the stable state. Using the matrix model the approach to the stable state can be estimated by the ratio of the two latent roots of largest modulus (absolute value). For the matrix that is illustrated in Fig. 5.2 the three latent roots are 2, -1 and -1. The ratio of the first to the second roots $|\lambda_1/\lambda_2|$ is equal to two, and is large enough for the oscillations to be heavily damped. As explained in the appendix, these matrices have a single root of largest modulus, and hence the second root will always have a smaller modulus. Thus, the ratio can never equal one, and the oscillations will always be damped. However, the closer the ratio becomes to one the more unstable the population, and the oscillations will be virtually undamped.

The matrix model suffers from the same disadvantage as the logarithmic growth model – it assumes that the population size will increase for ever. However, field studies have shown that the elements of the matrix, the age-specific fecundity and the age-specific survival, are both influenced by the population density. Thus a more realistic model would be derived by making all the matrix elements functions of the total population size, or functions of the size of one or more age groups. Such a model has been described by Pennycuick, Campton and Beckingham (1968), where the age-specific parameters were each multiplied by a factor for the total population size. Using a population of ten age groups their matrix was

$$\begin{bmatrix} 0 & 0{\cdot}1F & 1{\cdot}2F & 1{\cdot}0F & 0{\cdot}8F & 0{\cdot}6F & 0{\cdot}3F & 0{\cdot}1F & 0 & 0 \\ 0{\cdot}3S & 0 & 0 & 0 & 0 & 0 & 0 & 0 & 0 & 0 \\ 0 & 0{\cdot}95S & 0 & 0 & 0 & 0 & 0 & 0 & 0 & 0 \\ 0 & 0 & 0{\cdot}9S & 0 & 0 & 0 & 0 & 0 & 0 & 0 \\ 0 & 0 & 0 & 0{\cdot}8S & 0 & 0 & 0 & 0 & 0 & 0 \\ 0 & 0 & 0 & 0 & 0{\cdot}8S & 0 & 0 & 0 & 0 & 0 \\ 0 & 0 & 0 & 0 & 0 & 0{\cdot}7S & 0 & 0 & 0 & 0 \\ 0 & 0 & 0 & 0 & 0 & 0 & 0{\cdot}65S & 0 & 0 & 0 \\ 0 & 0 & 0 & 0 & 0 & 0 & 0 & 0{\cdot}3S & 0 & 0 \\ 0 & 0 & 0 & 0 & 0 & 0 & 0 & 0 & 0{\cdot}1S & 0 \end{bmatrix}$$

where
$$F = \frac{15000}{2500 + N_t}$$

$$S = \frac{1}{(1 + \exp(N_t/1389 - 5))}$$

$$N_t = \sum_{i=0}^{n} n_{t,i}$$

Thus, the fecundity falls off in a more or less exponential fashion with increasing population size, and the survival decreases in a sigmoid manner.

Starting with an initial small population of 48 animals, an equilibrium

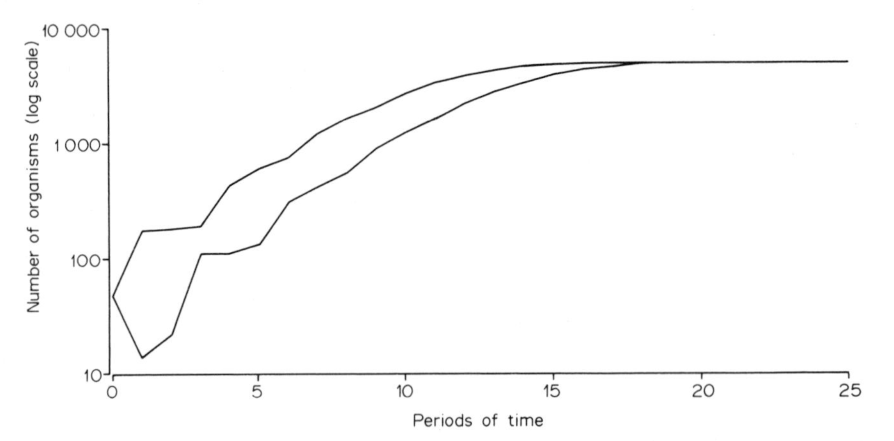

Fig. 5.3 The increase in size of a population according to a matrix model that has density dependent survival and fecundity elements. The upper line represents a population in which all of the age classes were initially equally abundant, and the lower line a population in which, intially, all the organisms were in the youngest age class

population size of 4953 animals with a stable age structure was achieved after about 20 generations of the computer simulation. The results of the simulation are shown in Fig. 5.3, where it can be seen that the matrix model has produced an asymmetric sigmoid curve. The main difficulty, however, in applying such a model is in determining the functional relationships between survival and fecundity and the population size.

So far we have been concerned with developing a matrix model that divides the population according to its age structure. It is, however, useful to be able to divide a population in other ways. Lefkovitch (1965) considers the case of insect populations where there are four life stages, namely egg, larva, pupa and adult, and where these form the most suitable basis for division of the population. It is obvious that the length of time spent in each phase is different, and hence the matrix will tend to have fewer zero terms. If we consider an insect where the egg stage lasts for three days, the larval stage for three weeks, the pupal stage for five days and the adult stage for a month, then with a time period of one week the matrix of fecundity and survival will be

$$\begin{bmatrix} 0 & 0 & f_3 & f_4 \\ p_1 & r_2 & 0 & g_4 \\ 0 & p_2 & 0 & 0 \\ 0 & q_2 & p_3 & r_4 \end{bmatrix}$$

In this matrix the p terms refer to animals that have moved on one stage, the q term animals that have moved on two stages, and the r terms the animals that have remained in the same stage. Thus larvae that were ready to pupate at the start of the period would be adults at the end of the period (q_2), whilst the young larvae would still be larvae at the end of the period (r_2). The f terms refer to the production of eggs, and the g term to eggs that were laid and hatched during the period of time. It will be observed that, as there are no elements f_2 or g_3, it has been assumed that at least four days elapse between emergence from the pupa and the first production of eggs.

The model has further been developed by Usher (1966, 1969c) for studying populations of trees measured by size classes. The matrix is essentially very similar to that used by Lefkovitch, with the p elements denoting trees that were becoming larger, r elements representing the trees that remained in the same size class and the f elements being functions of the latent root, as regeneration of the forest could only occur in the gaps caused by exploitation of the tree canopy. The aim of Usher's matrix models has been to determine the forest structure that would give the optimum yield of trees.

Adaptations of the basic model to investigate the effect of exploitation on animal populations have been attempted by Lefkovitch (1967) and Williamson (1967). Lefkovitch's method is to multiply the vector of the population structure by another square harvesting matrix. Thus the operation

$$\begin{bmatrix} \theta_0 & 0 & 0 \\ 0 & \theta_1 & 0 \\ 0 & 0 & \theta_2 \end{bmatrix} \begin{bmatrix} n_{t,0} \\ n_{t,1} \\ n_{t,2} \end{bmatrix}$$

will effectively carry out a harvest of $100\,(1-\theta_0)$ per cent of age class 0, $100\,(1-\theta_1)$ per cent of age class 1, and so on. If we wish to harvest part of the population at each period of time, then Equation (5.9) becomes

$$\mathbf{MHn}_t = \mathbf{n}_{t+1} \tag{5.13}$$

where **H** is the square harvest matrix. One can follow this approach and say that one wishes to harvest in such a manner that the absolute size of the population does not change from one period of time to another. This means that the latent root of (**MH**) is unity, and Equation (5.11) becomes

$$\mathbf{MHn} = \mathbf{n} \tag{5.14}$$

It may be possible to solve this equation to estimate values of θ_0, θ_1, and so forth.

Williamson (1967) uses a similar approach with a harvest matrix that is used to premultiply the vector. Since his basic model doubles the size of the population during each period of time, he asks if the population size would be held steady if half of it were to be harvested each time. He also asks if the harvest can be taken in any arbitrary way over the size classes. It is clear from the definition of the latent root that a harvest of half of each class will lead to a constant population size, but what happens if the harvest is made up completely of young? The harvesting matrix for such a situation is

$$\begin{bmatrix} \frac{1}{2} & -\frac{1}{2} & -\frac{1}{2} \\ 0 & 1 & 0 \\ 0 & 0 & 1 \end{bmatrix} \begin{bmatrix} n_0 \\ n_1 \\ n_2 \end{bmatrix} = \begin{bmatrix} n_0 - \frac{1}{2}(n_0 + n_1 + n_2) \\ n_1 \\ n_2 \end{bmatrix}$$

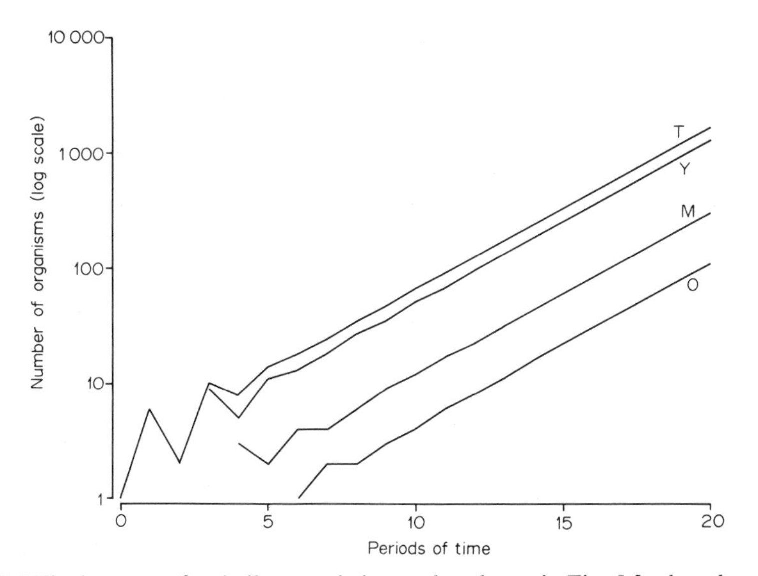

Fig. 5.4 The increase of a similar population to that shown in Fig. 5.2 when there is an exploitation of half of the total population size removed entirely from the youngest age class. The symbols *T*, *Y*, *M* and *O* are as in Fig. 5.2

The results of this harvest are shown in Fig. 5.4, where it can be seen that there is a steady increase in the population size after about the eighth operation of the matrix. The latent root of the matrix

$$\mathbf{MH} = \begin{bmatrix} 0 & 9 & 12 \\ \frac{1}{6} & -\frac{1}{6} & -\frac{1}{6} \\ 0 & \frac{1}{2} & 0 \end{bmatrix}$$

is 1·53, implying that although half of the population has been harvested there is still a substantial increase in the population size. Williamson states that an exploitation of 73 per cent of the young would be required to stabilize the size of the population such that the harvest accounts for all of the production.

These are thus the two basic approaches to modelling of animal populations. The one whereby rates of change are recorded in calculus notation, from which a differential equation is formed and integrated to give an algebraic solution. The other, which probably requires a greater biological understanding since more parameters are required, being in matrix form lends itself well to experimental computation and simulation. Both these techniques will be required later in this chapter when laboratory and field populations of animals are considered.

The responses of populations to harvesting: laboratory studies

Harvesting of laboratory populations has usually had two aims, the determination of the harvesting rate that will give maximum yield, and the investigation of the effect of harvesting on the age or size structure of the population. Intuitively there should be a maximum yield, since if there is no exploitation there is no yield, and if there is 100 per cent exploitation the population will be exterminated and there will still be no yield. Between these two extremes there will be at least one maximum, and the determination of this is obviously important for the management of the species concerned. This maximum is, however, subject to certain constraints, principally that the yield must be sustained at this maximum level for a very long period of time. The yield may or may not be made up of all parts of the population of the exploited species. Thus, harvesting may result in only the young being culled (e.g. three year old males of the Pribilof fur seal), or in only the older or larger individuals being taken (e.g. forest trees below a certain size are not thinned, and fishing nets permit the younger and smaller members of the population to escape). Laboratory studies must thus direct the harvest at different parts of the population so as to give a generalized result of the effects of exploitation. Studies must also take into account seasonal periodicity of the organism, as well as its ecological efficiency of food utilization.

It is because of the avoidance of some of the latter complications that most laboratory studies have relied upon invertebrate herbivores as experimental animals. These animals are characterized by continuous breeding throughout the year, and by having a large capacity to increase their population size because of the large numbers of eggs that are laid. The species used all have a simple population structure since they are not long-lived. In field studies the groups of the population are usually year-classes, and hence in modelling field populations one may need between six and 50 age or size classes to characterize the population structure. In the laboratory fewer age or size classes will be needed, and hence the model building and the interpretation of the results is very much simplified. Watt (1955) experimented with the flour beetle *Tribolium confusum*, Nicholson (1954) used the blowfly *Lucilia cuprina*, Slobodkin and Richman (1956) used the water flea *Daphnia pulicaria* and Usher, Longstaff and Southall (1971) have experimented with the Collembola *Folsomia candida*.

In considering the effect of harvesting on the population age structure we will have to look at studies that have aimed the exploitation at different age or size classes. Thus Slobodkin and Richman (1956) aimed their exploitation at the youngest size class of *Daphnia*, since the only individuals that could be aged in the population were those less than four days

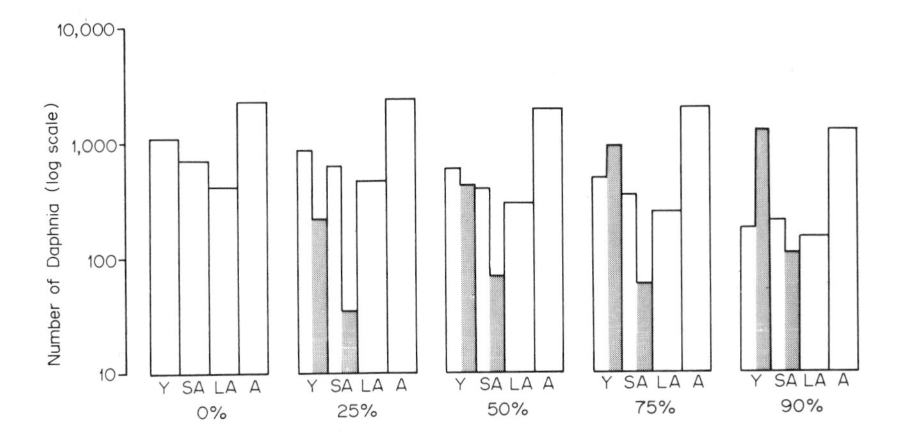

Fig. 5.5 The cumulative totals of the numbers of *Daphnia* in experiment H–15 of Slobodkin and Richman (1956). Each set of histograms represents an exploitation, by the percentage indicated, of the smallest size class. The symbol *Y* denotes young, *SA* small adolescents, *LA* large adolescents and *A* adults. The shaded histograms represent the *Daphnia* that were removed during the experiment. (Redrawn from Slobodkin and Richman, 1956)

old. The young class is thus based on age, but the other three classes that they recognized in their populations were characterized by size. The environmental factors were kept constant since there was a regular supply of food and a constant temperature. Harvesting occurred every fourth day. Fig. 5.5 show the effects of removing various proportions of this youngest size class. It will be noticed that the number of animals in each category decreases with increasing harvesting intensity (note that the illustration shows the numbers *after* each harvest), and thus the population is not only reduced in numbers but also in biomass. As well as this reduction in the overall population size there is also a change in relative abundance of the size classes. When there is no harvest there is a preponderance of young animals and a minimum in the large adolescent category in the three smaller size classes. As the intensity of the harvest increases the discrepancy between the numbers in these classes decreases until at the 90 per cent harvest rate the numbers in each category are more or less the same. It can be predicted that a size distribution similar to that for the 90 per cent harvest would be found in equilibrium populations. This is based on two assumptions. The first assumption is that the mortality of the smaller size classes is negligible, as indeed other laboratory work quoted by Slobodkin and Richman has shown. The second assumption is that the size of each individual *Daphnia* increases in time in an asymmetric sigmoid manner. Individual *Daphnia* are known to follow such a growth curve, with a short logarithmic growth phase when they are young followed by a long decelerating increase in size approaching the asymptotic adult size.

Nicholson (1954) aimed his exploitation at the adults of his blowfly populations. We will examine the results of one of his experiments which was maintained at constant temperature and the food supply to the larvae was provided in excess. For the adults there was an excess of water and sugar, but the protein food was limited to a small but constant daily amount. Nicholson's results are shown in Table 5.1. The restricted food supply to the adults did not influence their longevity, as confirmed by other experiments in which flies were maintained on only sugar and water. It did, however, influence fecundity since adults are incapable of developing eggs when they have had insufficient protein. Since it is known that larval mortality is very small when they have an excess of food, the

TABLE 5.1 Data for the removal of blowflies from a population where the only limiting factor was in the adult protein food supply. (Source: Nicholson, 1954)

Daily exploitation rate a	*Pupae produced per day* b	*Adults emerged per day* c	*Adults exploited per day* d	*Adults entering population per day* e	*Mean adult population size* g	*Mean adult life span (days)* $k = g/e$	*Mean birth-rate (per individual per day)* $l = b/g$
0%	624	573	0	573	2520	4·4	0·25
50%	782	712	356	356	2335	6·6	0·33
75%	948	878	658	220	1588	7·2	0·60
90%	1361	1260	1134	126	878	7·0	1·55

number of pupae being produced is directly proportional to the fecundity. Thus it can be seen in Table 5.1 that the birth-rate or fecundity is related to the exploitation rate. It can also be seen that the average life span of a fly increased with the exploitation rate. Nicholson shows that although the percentage of flies reaching maturity varied approximately directly with the percentage of flies not exploited, the mean adult population size varied to a lesser degree, since surviving adults lived longer and produced more offspring. The fly population is thus capable of replacing the flies that are removed in the exploitation.

Usher, Longstaff and Southall (1971) divided their populations of *Folsomia* into five size classes, and spread their exploitation over all classes. The effect of exploitation on the size structure of the population is shown in Fig. 5.6, where it can be seen that the response of the population to harvesting was dependent upon the food supply. The parameter that was used to characterize the size structure was a weighted mean size class, defined by

$$m = \sum_{i=1}^{k}(in_i)/\sum_{i=1}^{k}n_i \qquad (5.15)$$

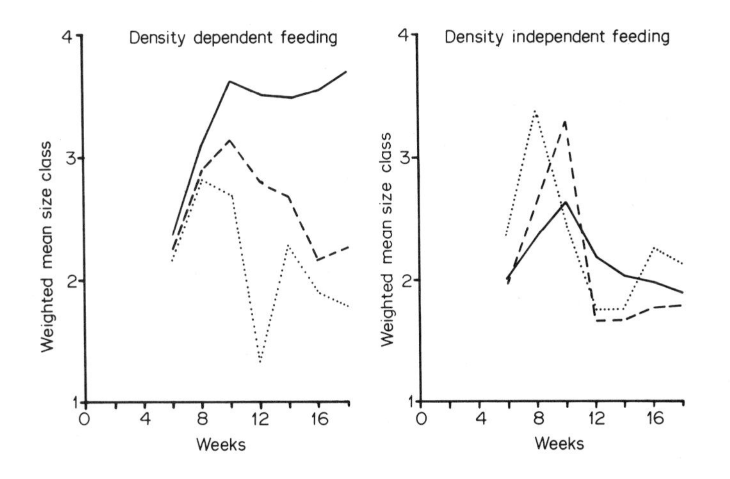

Fig. 5.6 The effect of exploitation on the size structure of a population of *Folsomia candida*. The weighted mean size class is defined in Equation (5.15). The lines represent:

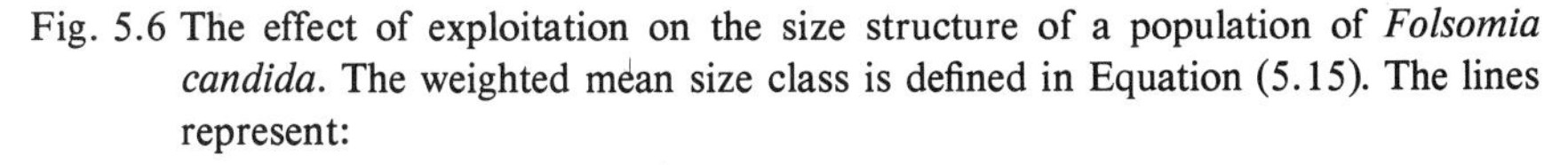

————————, no exploitation;
— — — — —, 30 per cent exploitation rate;
. . . ., 60 per cent exploitation rate.
(Redrawn from Usher *et al.*, 1971)

where there are k size classes (the smallest being 1 and the largest k), and where there are n_i individuals in the ith class. When the food supply was related to the population size, i.e. the larger the population the greater the amount of food supplied, the mean size class was about 3·5 after the tenth week, and it remained at this value when there was no harvesting. As the rate of harvesting increased so the mean size class of the population decreased, implying that there were more young (small) animals present. Similarly, when the food was supplied in a small fixed amount, after the tenth week the populations had a relatively stable structure with a mean size of approximately 2·0. When the populations were exploited the competition for food was still so intense that there was no significant change in the size structure of the population. It is, however, clear that the stability of the size structure in harvested populations is less than in the unexploited populations. Fig. 5.6 shows that the variance of the mean size of the exploited populations is considerably greater than the variance of the unexploited population.

Thus, these three studies on *Daphnia*, *Lucilia* and *Folsomia* have demonstrated the effects both on the population structure and on reproduction of aiming the impact of exploitation at the juveniles, at the adults and at all the size or age classes. One very obvious effect has always been to reduce the population size to a new level, but in no case so to reduce the size as to exterminate the population. Also, we can see that the effect of exploitation has been to reduce competition amongst the survivors. The relation between the rate of exploitation and the reduction of competition is dependent upon the totality of the influences contributing to the competition. Thus in the experiments with *Folsomia* the competitive influences acting upon the population came from two sources, the competition for oviposition sites and the competition for food. When competition for oviposition sites is severe the influence of exploitation upon the population structure is very obvious since the number of small animals in the population increases with increasing exploitation (implying that there has been more breeding). When, however, food has been the dominant source of the competitive influence, *Folsomia* population structure shows little change under exploitation.

It is also important to determine the maximum yield from these populations. Data from the H–15 experiment of Slobodkin and Richman (1956) are shown in Fig. 5.7, and represent the same experiment as shown in Fig. 5.5. The experiments were started with 20 animals and were run for just

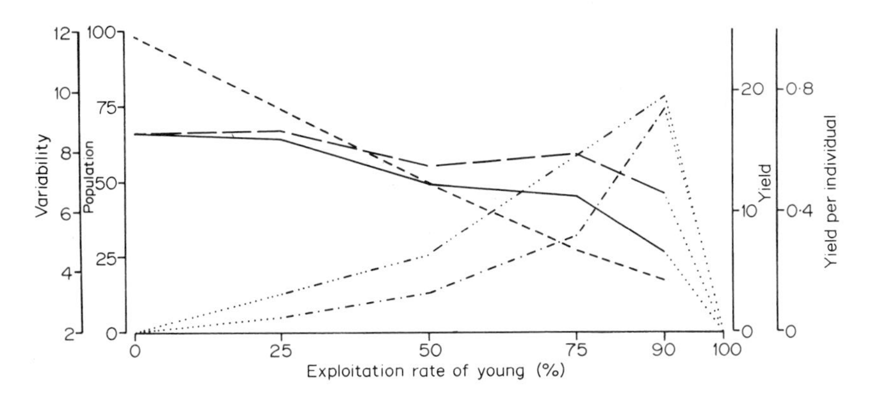

Fig. 5.7 The effect of exploitation on populations of *Daphnia*. The data are derived from the same experiment as those shown in Fig. 5.5. The lines represent:

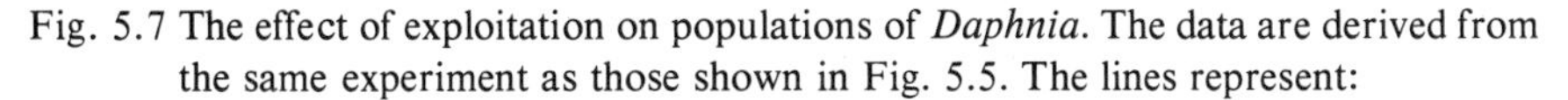

— — — — —, mean population size before harvest;
————————, mean population size after harvest;
– – – – – – – – – –, variability in the population size;
— · · · — · · ·, total numerical yield;
—·—·—·—·, yield per individual.

(Drawn from the data of Slobodkin and Richman, 1956)

over nine months. The average yield from the populations increased with the rate of exploitation, and as the experiments were carried out over a relatively long period of time these results can be accepted as sustained yields. The maximum yield occurred with an exploitation rate of 90 per cent of the juveniles every four days. This represents a rate of approximately 23 per cent of the juveniles every day, or about 11 per cent of the total population every day. It is obviously dangerous to express rates measured over a long period of time on a daily basis, as illustrated by the following example. If 100 per cent of the juveniles were removed every four days, after the adults had died of old age there would be no recruitment to the larger size classes and the population would become extinct. However, if 25 per cent or even 30 per cent of the juveniles were to be removed every day it might well be that the yields that Slobodkin and Richman found could have been increased slightly. It will be seen from Fig. 5.5 that a few of the young adolescent class were included in the harvest. However, since the exploitation was aimed almost entirely at one age class the biomass harvest was directly proportional to the numerical harvest.

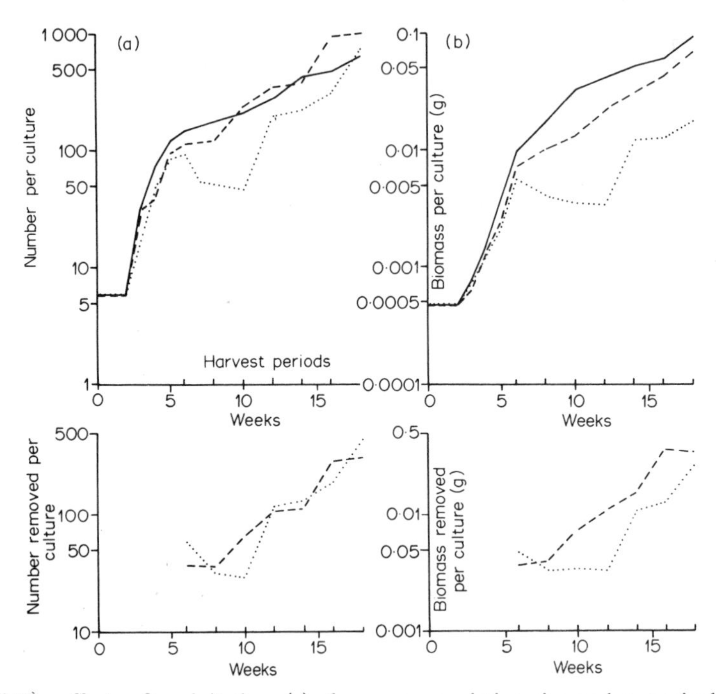

Fig. 5.8 The effects of exploitation. (a): the mean population size and numerical harvest and (b): the mean biomass of the standing crop and the biomass harvest, for a population of *Folsomia candida*. The graphs are discussed in the text. The lines have the same representation as those in Fig. 5.6. (Redrawn from Usher *et al.*, 1971)

This was not, however, the case in the exploited populations of *Folsomia*, described by Usher, Longstaff and Southall (1971), as becomes clear when the experiments with excess food are examined (Fig. 5.8). In Fig. 5.8 (*a*) it can be seen that the number of animals in the population is more or less independent of the level of exploitation, except during part of the time when statistical analysis showed there to be no significant difference between the zero and 30 per cent harvested populations, and the 60 per cent harvested population to have significantly fewer animals. Similarly, if the number of animals harvested are statistically examined there is no significant difference between the 30, 40, 50 and 60 per cent exploitation rates. However, a study of the population biomass shows that the biomass of the 60 per cent exploited population is significantly smaller than the 30 per cent exploited population which is significantly smaller than the unexploited population. Analysis of the biomass harvest shows that the harvest of the 60 per cent exploited population is significantly smaller than that of the 30 per cent exploited population. The results of these experiments are shown in detail in Fig. 5.8.

Thus, when the harvest is aimed at just one age or size class we need not distinguish between numerical and biomass yield. However, when animals from more than one age or size group are harvested we must define the term 'maximum harvest' either in terms of numbers or in terms of biomass. *Folsomia* studies have shown that numerical harvest will be maximized by removing between one-third and two-thirds of the total population every two weeks, whilst the biomass harvest is maximized by removing between 30 and 40 per cent of the population every two weeks.

The populations discussed above have all been of invertebrates, whereas in conservation management vertebrate species are of considerable importance. The laboratory study of guppies (Silliman and Gutsell, 1958) shows that vertebrate populations have many similarities with arthropods populations, although often the capacity to increase is less and hence experimental work is required over a much longer time period. Silliman and Gutsell's experiments lasted for 172 weeks, and some of the results are shown in Fig. 5.9. There was a deliberate attempt to make the guppy experiment resemble a commercial fishery. Thus, their populations were fully inter-breeding and self-reproducing, and as the food supply was constant there was an asymptotic maximum population size. Thus competitive influences for food affected the population, and cannibalism acted as a density-dependent control on the survival of the young fish. Fishing was carried out at intervals of three weeks, which corresponded to the reproductive period of the fish. The catch, to resemble a commercial fishery, was selective. Thus the smaller fish escaped whilst for the larger fish the probability of capture was independent of the age or size. We can see from Fig. 5.9 that there has been an effect on the size structure of the population, since when 10, 25, 50 and 75 per cent of the population was harvested the mean length of the catch was 26·5 mm, 25·9 mm, 23·6 mm and 22·2 mm respectively. Thus, the exploitation has decreased not only the total size of the population but also the mean size of the individual fish in the population. One can also see in Fig. 5.9 that the catch, either as numbers or biomass, is least when 10 per cent are harvested, is greater when 25 per cent are harvested (although the graphs are decreasing they level out before week 76 when the exploitation rate was changed), is decreasing steeply when 50 per cent are removed, and the population becomes extinct when 75 per cent are harvested. Silliman and Gutsell

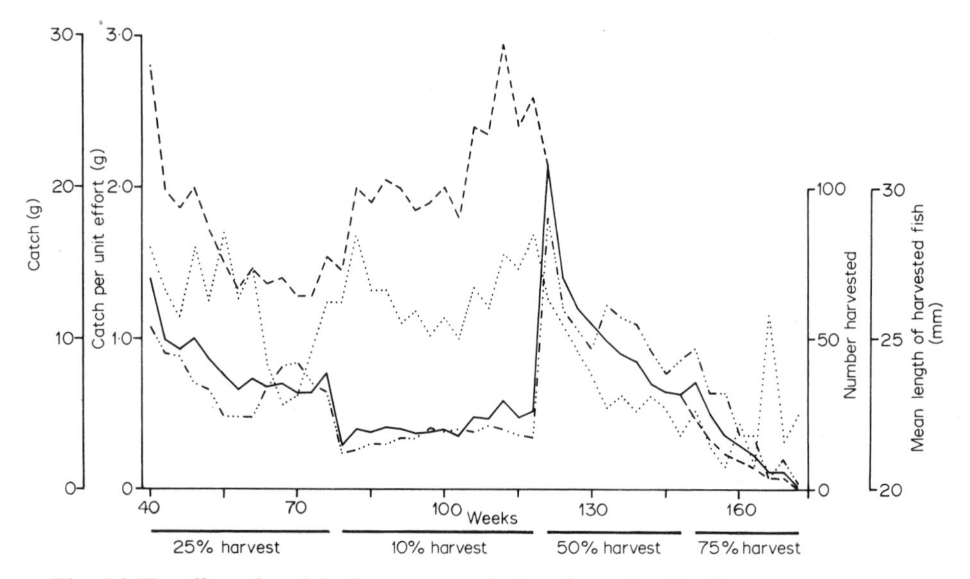

Fig. 5.9 The effect of exploitation on a population of guppies. The lines represent:

————————, biomass of the catch (harvest);
— — — — —, catch per unit of effort;
— · · · — · · ·, numerical harvest;
· · · · · · · · · · · · ·, mean length of a fish in the harvest.

(Drawn from the data of Silliman and Gutsell, 1958)

conclude, in their Fig. 14, that the maximum biomass production is achieved when there is an exploitation rate of between 30 and 40 per cent. Watt (1968) argues that they have not allowed from the decreasing graph when a 25 per cent rate was used, and thus he considers that the maximum production is achieved between 10 and 25 per cent exploitation. Since the graphs for the 25 per cent rate not only flatten out but start to increase (Silliman and Gutsell's Fig. 15), I feel that the fishing rate of 25 per cent is nearer to the rate of maximum biomass production. With a three-weekly harvesting cycle this corresponds to a daily harvest of about two per cent.

Thus we can bring together the various exploitation rates that have been estimated to give maximum biomass production, as does Watt (1962). Table 5.2 shows the data, but it should be remembered that the values for *Lucilia* and *Daphnia* would be substantially less if they were expressed as percentages of the total population.

We can now ask what are the features of populations that are being

TABLE 5.2 The exploitation rate, in percentage of the population per day, giving the maximum biomass harvest

Lucilia	99 per cent of the adults
Daphnia	23 per cent of the juveniles
Tribolium	3 per cent (based on adults, pupae and large larvae)
Folsomia	3 per cent (spread over all the population)
Guppies	2 per cent (spread equally over all animals larger than some determined size)

under-exploited and those being over-exploited. It is clear that the maximum sustained yield is only gained when an optimum exploitation rate has been determined, and both the yield and the rate are functions of the capacity of the animal to replace that part of the population that has been exploited. This capacity is determined both by the fecundity of the animals and by the period of time that elapses before breeding takes place. As a population is successively more heavily exploited the yield increases to its maximum, and then the drop in yield with further exploitation is very steep. The maximum may not be clearly defined, as was the case with the Collembola where numerical yields at 30 and 60 per cent exploitation rates were not significantly different from each other.

Watt (1968) considers the question of maximum yield, and he draws the distinction between heavy and ruinously heavy exploitation, defining the latter as 'that level of exploitation which extends the homeostatic capability of a population beyond the point at which any homeostatic response can occur'.

The first general point to emerge from the laboratory studies is that exploitation of a particular age or size class increases the productivity of that class relative to the productivity of the other classes. Thus, increasing the exploitation up to the maximum sustainable rate increases the abundance of that class relative to the other classes. When the maximum rate is exceeded that age or size class becomes scarcer and eventually extinct. The data in Fig. 5.5 illustrate this general point. Using values of the population size before harvest, the ratio of the number of young to the number in other size classes is listed in Table 5.3. It can be seen that this ratio increases with exploitation rate up to the 90 per cent exploitation which was assumed to be the maximum sustainable rate for the young *Daphnia*. Also in Table 5.3 is the data derived from one of Watt's (1955) experiments where the exploitation was aimed at the adults, pupae and large larvae of *Tribolium*. As the exploitation increased to a maximum

TABLE 5.3 The relation between the abundance of the young class and the abundance of the other classes of *Daphnia* (calculated from the data of Slobodkin and Richman, 1956); and the relation between adult numerical productivity and total numerical productivity of *Tribolium*. (Source: Watt, 1968)

Daphnia		Tribolium	
Exploitation rate	*Ratio of abundance of youngest class to abundance of other classes*	*Exploitation rate*	*Ratio of adult numerical productivity to total numerical productivity*
0%	0·32	63%	0·80
25%	0·32	76%	0·98
50%	0·38	86%	1·28
75%	0·56	100%	0·63
90%	0·83		

sustainable rate so the ratio of the adult to the total numerical productivity also increased. However, when the maximum rate was exceeded the ratio decreased very quickly. It seems then as if such ratios as are listed in Table 5.3 do indicate when exploitation becomes ruinously heavy. When with increased exploitation the ratio decreases further exploitation at that level will lead to the extinction of the population. One point is, however, not clear. When the ratio has shown such a decrease, are there still sufficient homeostatic responses for the population to recover after a respite?

Watt (1968) explains these ratios in the following manner. If we first consider the case of harvesting adults only, then as the exploitation rate increases from zero fewer adults are left behind in the population, and less competition is directed at the sub-adult classes. Thus, a larger proportion of the sub-adult stock survive to become adults, and hence productivity of the adults will increase relative to the other classes. When, however, the maximum sustainable rate is exceeded, the capacity of the adult to replace the exploited animals is impaired, and so there will be a sharp decrease in the adult productivity. If we consider the case of harvesting the young only, an increasing exploitation rate removes an increasingly larger portion of that class of the population, which increases the probability of survival of those remaining. Hence, there is an increase in the productivity of the young relative to the total population. When the maximum sustainable rate has been exceeded there will be too few adults produced to replace the exploited stock, and thus the productivity of the youngest age class will decrease and the population will be faced with extinction.

Other indications of over-exploitation can be gained from the life expectancy and size of the stock, though these are more continuously increasing or decreasing functions, and they cannot be used to determine the exact point when heavy exploitation becomes ruinously heavy. Watt (1968) suggests that once the maximum sustainable rate of exploitation is exceeded the median life expectancy must drop. It can certainly not increase beyond its value for the maximum sustainable rate. It was also seen in Fig. 5.9 that the mean size of the guppies decreased with increasing exploitation. However, such a decrease in size was a feature of heavy exploitation, and is not in itself diagnostic of ruinously heavy exploitation.

The characteristics of balanced harvesting

The laboratory studies that have previously been discussed have differed in one important respect from field studies – they have had a constant environment, and thus the response of the population can be attributed directly to the imposed harvesting rate. In field populations random or nearly random influences are acting to cause changes in the general population size, or in the population parameters such as survival. We need thus to think of density-dependent and density-independent control of population size. Climatic variability is a major cause of density-independent control, and the existence of such presumed control mechanisms influences policy for conservation management.

Watt (1968) discusses the influence of climatic factors on the Pacific sardine, *Sardinops caerulea*, fishery of California. The extreme variability of the sardine catch is shown in Fig. 5.10, where it can be seen that there is a hundredfold variability. It is evident in Fig. 5.11 (*a*) that there is a strong relationship between the size of the year-class and the water temperature when the fish were spawned (with the exception of 1939). There was a considerable drop in water temperature from 1948 to 1949, and from the data shown in Fig. 5.10 it will be seen that four years later there was a dramatic decrease in the catch from over 100,000 tons to only 5000 tons. Most sardines forming the catch are four years old. The highest temperature recorded is in 1940, and this corresponds with the maximum catch in 1944–1945 of 554,000 tons. However, not all of the variability in the size of the year-class can be attributed to water temperature, since the points in Fig. 5.11 (*a*) do not fall exactly on a straight line. Hence we must look for other factors besides temperature. We can see

Fig. 5.10 The annual catch of the Californian sardine. (Drawn from data given by Watt, 1968)

from Fig. 5.11 (*b*) that the survival of the year-class is not related to the size of the spawning stock and hence it is unlikely that there is density-dependent regulation. Another climatic feature is the California current. The larval sardines have a higher probability of survival if they can migrate from the spawning grounds to the coastal nursery grounds. In the years when the average current is strong many of the planktonic sardines will be swept out into the Pacific where their chance of survival would be extremely small. In years when the average current is weak the probability of survival is very much greater. This does, however, somewhat oversimplify the situation since there is considerable variability of the current both in area of ocean and in time.

The size of the year-class is thus a function of temperature and current flow. Both these factors are external to the sardine population and are relatively random in their operation. Radovich (1962) shows that if we plot the year-class size against the spawning stock size we obtain a parabolic relation, with a certain amount of variability on either side of the parabola, Fig. 5.11 (*c*). More generally, Radovich's model is shown in Fig. 5.11 (*d*). The parabola that gives the best fit can be described as the average parabola, and this variability can be enclosed within two other parabolas, as in Fig. 5.11 (*c*), termed the maximum and minimum parabolas. Watt (1968) extends the idea so that this variability, or the maximum and minimum parabolas, are functions of the effect of climate on the population size. Thus, when the effect of climate is as important as in the sardine, the variability defined by the maximum and minimum parabolas is very large. This he calls a 'big weather effect'. Conversely, when there is relatively little climatic effect and the maximum and minimum parabolas are close to the average parabola, Watt refers to a 'small weather effect'.

What conclusions can we draw for conservation management of such

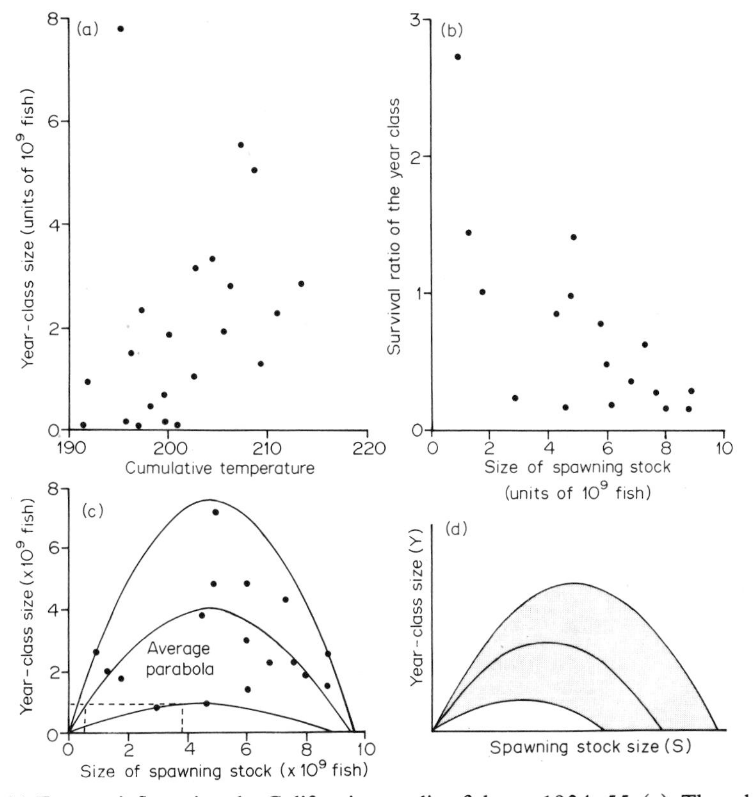

Fig. 5.11 Factors influencing the Californian sardine fishery, 1934–55. (a): The relation between the year-class and a measure of the sea temperature (recorded as the sum of the monthly means at Scripps Pier from April to March). (Redrawn from Watt, 1968); (b): The relation between the survival ratio of the year-class to the size of the spawning stock. (Redrawn from Watt, 1968). (c): The average parabola expressing the relation between the sizes of the year-class and the spawning stock, together with the maximum and minimum parabolas. The horizontal and vertical lines are explained in the text. (Redrawn from Radovich, 1962). (d): A generalized set of three parabolas expressing the relation between year-class size (Y) and spawning stock size (S). (Redrawn from Radovich, 1962)

populations? Naïvely, one can suggest that the most important management practice is to measure the climatic factors from year to year, and from these data predict the amount of harvest that can be taken in due course when the juvenile survival is reflected in the adult population density. Less naïvely, one can calculate what 'safety margin' should exist in the adult population so as to guard against the chance effects of weather. If, for example, we wished to maintain a year-class of 1×10^9 sardines, we can see from Fig. 5.11 (*c*) that under average conditions this could be achieved by a spawning stock of approximately 6×10^8 adults (determined where the horizontal line in Fig. 5.11 (*c*) crosses the average parabola). Under the worst possible weather conditions the data suggest that a year-class of this size would be achieved by a spawning population of $3{\cdot}8 \times 10^9$ adults. The difference between these two estimates of the required spawning stock is considered to be a 'cushion' left behind after exploitation so that there is a safety margin to prevent chance climatic factors destroying the population. It is clear that if the weather effect for the sardines were to be smaller, then the variability around the average parabola would be less, and hence that the 'cushion' that management practice would leave as a safety margin would be very much smaller.

The sardine thus shows the importance of climatic factors in regulating fish populations, and thus determining what the yield can be and how to manage for the conservation of the species. Beverton (1962) examined five species of the North Sea fishery, posing the questions '. . . of whether the populations are in a state of long-term balance and, if they are, of identifying the mechanisms responsible for that balance'. He analyses the trend of the catch over the period from 1906 onwards, with breaks in the records during the two World Wars. He shows that the sole, *Solea vulgaris*, population has an upward trend, having become more common though there are considerable fluctuations about the trend line. The haddock, *Melanogrammus aeglefinus*, has become scarcer, despite the fact that the populations increased in size during the wartime years when there was no fishing. The plaice, *Pleuronectes platessa*, however, shows a very stable population size, and it has the least variability in the annual catches. Thus, in a fishery that has become stabilized (Watt, 1968), it would seem that the population dynamics of the plaice are the most interesting if one wishes to investigate the mechanism responsible for the regulation of the population size.

The data for the plaice are shown in Fig. 5.12. Beverton examines the dynamics of the adult population to determine if population regulation operates at this stage. During the 1939–1945 war there was no fishing, and during this period the size and weight growth of the older fish were retarded by 15 to 20 per cent. During the pre-war conditions mortality of the adults (defined as those animals that are fishable, i.e. older than the three to four year age class) was of the order of 50 per cent due to

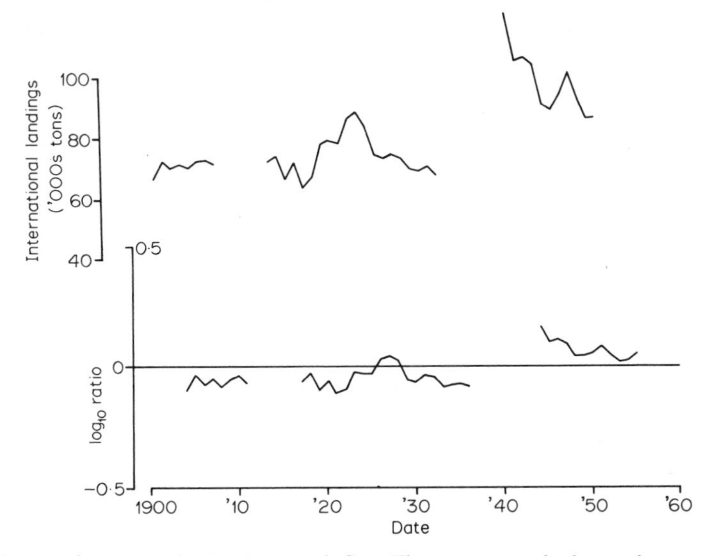

Fig. 5.12 Data for the plaice in the North Sea. The upper graph shows the annual total weight of the international landings during the period 1906–57. The lower graph shows the data for the total landings expressed as the logarithm of the ratio (catch for year/mean catch for 1906–57). (Redrawn from Beverton, 1962)

exploitation, whilst during the war mortality from natural causes was of the order of 10 to 15 per cent. The data thus demonstrate that there is some regulation of the adult population biomass since as the biomass increased the average biomass of an individual fish decreased slightly. However, during the war there was a sixfold increase in the population size, and hence the reduction in the mean weight of individuals is insufficient to compensate for the increase in numbers. Beverton thus concludes that the main regulatory mechanisms must operate in the younger age classes.

Investigation of the spawning area showed that its extent was virtually independent of the density of the eggs. In fish there is a very strong correlation between the adult population biomass and the number of eggs that are laid. Thus it seems that as the density of adults increases so does the density of eggs in the spawning ground, and hence one cannot postulate that survival of eggs or larvae in areas beyond the normal spawning ground will be so much reduced as to regulate the population. Environmental considerations and predators can be eliminated from the initial discussion since their effect will not be directly proportional to the population density. Cannibalism by the adults is discussed, but once again Beverton considers that this one factor will only contribute in a small part to the stability of the population. Finally, Beverton discusses the feeding relationships of the larvae, and he states that they have a strong preference for *Oikopleura*, which is neither abundant in the plaice's spawning grounds nor elsewhere. When *Oikopleura* is present its distribution is very patchy. How then can this feeding factor operate? First, it can operate directly, since if there is insufficient food there will be starvation of a large proportion of the larvae. This effect will be directly related to the population density of the young plaice if the food supply is constant. Secondly, it can operate indirectly, since larvae in poor physical condition will be more adversely affected by climate and predators. It seems therefore that the direct and indirect action of larval feeding is the controlling factor of the plaice population.

However, the actual stability of the population and not just that of the catch should be demonstrated. Figure 5.13 (*a*) shows the relation between

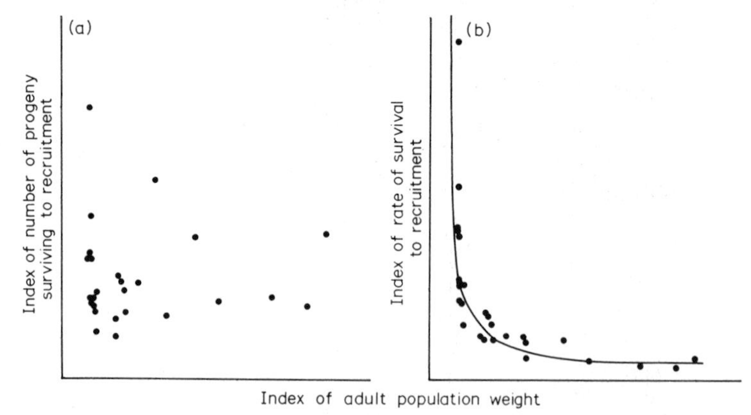

Fig. 5.13 Data for the plaice. (a): The relation between the biomass of the adult population and the number of progeny surviving to recruitment. (b): The relation between the biomass of the adult population and the rate of survival to recruitment. (Redrawn from Beverton, 1962)

the number of eggs that have survived to recruitment (i.e. to fishable age) and the adult population biomass, the latter being a measure of the number of eggs produced. In a statistical sense it can be seen that the points are best fitted by a horizontal straight line, showing that the recruitment is independent of the adult stock. However, if the rate of recruitment (the ratio of the number of recruits to the number of adults) is plotted against the adult population biomass a hyperbolic relation is obtained, Fig. 5.13 (*b*). This led Beverton to investigate the survival from the egg stage to recruitment. There is an average probability of 0·999990 that an egg will fail to develop to a fish of fishable size. The extreme range of this mortality is 0·999970 to 0·999995, though the average range is about 0·999987 to 0·999993. It will be seen from these data that the regulatory mechanism, operating in the first year of the plaice's life, is extremely efficient.

But for conservation management what features do such populations possess? They are perhaps the most simple to manage, since if the regulatory mechanism operates so early in life the number recruiting cannot be altered by management practice, except by over-exploitation. Thus, the rate of recruitment determines the maximum yield that can be taken if the population is to be conserved. If the yield increases above the difference between recruitment and adult natural mortality, then the adult stock will be depleted. Such a small excess may result in heavy exploitation, but a repeated excess will eventually result in a ruinously heavy exploitation.

In conclusion, such fish populations show stability over long periods of fishing, and hence they are not being over-exploited. That they are not under-exploited is shown by the sixfold increase numerically during the respite of the Second World War. But, what models are there for such populations? An extensive treatment of this subject is given by Beverton and Holt (1957). The sigmoid increase, Equation (5.6) has often been applied as a first approximation in model building (Gulland, 1962). If the catch is included in the differential equation for the rate of change of population size, then Equation (5.5) becomes

$$dN/dt = KN(N_E - N) - C \tag{5.16}$$

where C is the catch. It is obvious that when the fishery is in a steady state that $dN/dt = 0$. With a model such as this one can predict that the maximum rate of growth in population size, which corresponds to the steepest part of the growth curve, will occur when $d^2N/dt^2 = 0$ (this is the point of inflection of the sigmoid curve). Using Equation (5.6) the maximum rate of growth is predicted when the population size is half the asymptotic maximum size, i.e. at $0{\cdot}5N_E$, but studies have usually found that the maximum rate of growth occurs when the population size is only a quarter or a third of the maximum. A model such as Equation (5.7) might in these circumstances be of more use. Gulland, however, criticized such models on the basis that they do not include the age structure of the population, and that recruitment is not determined by the present adults but by the adults several years beforehand (three to four years before in the case of the plaice, four years in the case of the sardines). The matrix approach, given generally in Equation (5.8) does not appear to have been tried on fishery populations, though the model would need certain modifications in order to incorporate the time lag between spawning and recruitment into the age classes represented by the matrix.

These models all require expressions for the survival of the fish, and this depends on two factors, natural mortality and mortality caused by fishing. Gulland shows how these can be separated so that equations for predicting populations after t units of time can include parameters for both F (the fishing mortality coefficient) and M (the natural mortality coefficient). Such refinement is essential in models that are to be used for prediction and simulation, since F will be altered as the fishing effort is altered. Since the weight of an individual fish also follows a sigmoid growth curve, the models for the rate of change of the population biomass due to fishing and the rate of change of individual biomass can be combined. Thus, Gulland derives the equation

$$dY/dt = FN_0e^{-(F+M)t_e^{-gt}} \qquad (5.17)$$

which relates the rate of change in the weight of the catch (Y) to the starting population (N_0) as well as to the parameters of fishing and natural mortality: g is a constant. Such a model becomes an even more realistic representation of the actual population by allowing the various parameters to alter in some functional manner.

The characteristics of over-exploitation

Whales have for many years been a subject for conservation discussion since their decrease in numbers has resulted in economic loss of whaling fleets and in the diversification of industry previously dependent upon catching these animals. A history of the efforts at conservation shows the need for international understanding and co-operation, and the almost impossible task of a few enlightened nations in attempting to secure conservation ordinances for a biological resource in international waters. However, such aspects of whaling need not concern us here, since we will use the whale as an example of the results of over-exploitation. Short histories of the whaling operations are given in most contemporary conservation texts, for example those of Ehrenfeld (1970) and Russell-Hunter (1970). A summary of the status of whales and whaling during the three decades 1930–1959 is given by Laws (1962), and Watt (1968) discusses the implications of international disagreements in the 1960s.

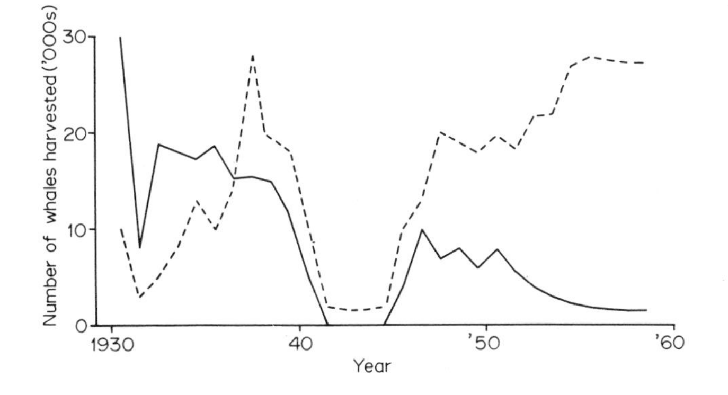

Fig. 5.14 The number of whales harvested during the period 1930–59. The continuous line represents the blue whale and the dashed line the fin whale. (Redrawn from Laws, 1962)

The most important species of whales, which contributed most to the total catch, were the blue whale, *Balaenoptera musculus*, and the fin whale, *B.physalus*. The former species, the largest of the mammals, formed the basis of the industry until the species declined in numbers in the early 1930s. From this time onwards more fin whales were caught, and by the end of the 1950s they accounted for the majority of the Antarctic catch of baleen whales (Fig. 5.14). In order to investigate the reasons for this decline we must look at the population parameters of fecundity and survival.

Ehrenfeld (1970) shows that the female blue whales reach sexual maturity at an age of between four and seven years. The gestation period is approximately a year, and whales produce a single calf which is nursed for seven months during which time the female does not become pregnant again. Coupled with the migratory habit of the blue whale, this indicates that a female will not produce a calf more than once every two years. The maximum age attained by blue whales is in the order of 40 years. No direct statistics for the natural mortality rate are given. However, since fin whales of less than 10 years of age were unlikely to form a significant part of the catch in the 1939–1941 period, an estimate of the natural mortality can be gained from Fig. 8 of Laws' paper. During the first 10 years of life the number of animals surviving had fallen from 1000 to approximately 250 whales. Assuming that the survival of each year-class is similar (justifiable on the ground that the points on the graph form an almost perfect straight line), there is an annual mortality of approximately 13 per cent. Laws also states that the sex ratio in the catches is very close to 1♂ : 1♀, and does not vary with age. Such data can be used for a matrix model of the type in Equations (5.8) and (5.9) thus,

$$\mathbf{M} = \begin{bmatrix} 0 & 0 & 0{\cdot}19 & 0{\cdot}44 & 0{\cdot}50 & 0{\cdot}50 & 0{\cdot}45 \\ 0{\cdot}87 & 0 & 0 & 0 & 0 & 0 & 0 \\ 0 & 0{\cdot}87 & 0 & 0 & 0 & 0 & 0 \\ 0 & 0 & 0{\cdot}87 & 0 & 0 & 0 & 0 \\ 0 & 0 & 0 & 0{\cdot}87 & 0 & 0 & 0 \\ 0 & 0 & 0 & 0 & 0{\cdot}87 & 0 & 0 \\ 0 & 0 & 0 & 0 & 0 & 0{\cdot}87 & 0{\cdot}80 \end{bmatrix}$$

The matrix given above is designed to operate over a two-year period, for the females only, so that the age classes are 0–1 year, 2–3 year, 4–5 year, 6–7 year, 8–9 year, 10–11 year, and 12 years and older. As will be seen the fecundity terms reflect the fact that the females do not breed in the first four years of life, and that breeding is not complete until after the seventh year. The fecundity terms at full breeding are 0·5 since one calf is produced every two years, but this calf has a probability of only 0·5 that it will be a female. The fecundity term for the 12+ age group has been reduced to 0·45 since it is suggested that older animals may not breed as regularly as those at the peak of their breeding life. All the survival terms in the matrix are recorded as 0·87 since 0·13 is the best estimate available of the natural mortality. There is one element in the matrix that is not a feature of the normal matrix model. This is the element of 0·80 in the lower right corner. Since the 12+ age group includes many ages of whales up to 40 years, they cannot all die every two years as the basic model assumes with the oldest age group. Thus, this corner element of the matrix refers to the proportion of the population of that age group that do not die, and it has been assumed to be 80 per cent. This value has been chosen since it gives a mean life expectancy of 7·9 years, for a recruit to the age group implying that there is a total life expectancy of about 20 years for whales that survive to the age of 12 years.

If we estimate the stable population structure as in equation (5.11) we find that

$$\lambda = 1{\cdot}0986$$
$$\mathbf{n} = \{1000, 792, 627, 497, 393, 311, 908\}'$$
$$r = \ln\lambda = 0{\cdot}0940$$

As we can see the value of λ is very close to one, implying that the sustained two-yearly harvest of

$$100(\lambda - 1)/\lambda \text{ per cent}$$

would only be small, being approximately four and one half per cent of the total population per annum. If this harvest rate is exceeded the species would not be able to replace the exploited part of the population unless homeostatic mechanisms acted to alter the parameters of survival and fecundity. It can also be seen that the intrinsic rate of natural increase for the whale population is very small.

A model can be used in this predictive way in order to estimate the catch that can be taken. However, with the whale, far more were taken and hence we can see the responses of the population to this situation. The direct effects of whaling have to be estimated from the whales that are caught, and so they do not necessarily reflect the processes that are going on in the population as a whole. Laws shows that the mean lengths of the blue and the fin whales have decreased rapidly in the pre-war years, with a more gradual decrease in the post-war years, though the data reflect the enforcement of minimum lengths for killing, the selectivity of the catching fleets which became less as the stocks decreased in size, and variations in the opening date of the season. From 1931 to the 1950s the mean length of the blue whale decreased from about 84 ft to about 77 ft, whilst that of the fin whale decreased from about 70 ft to 66 ft. Despite the inaccuracies and the changing conditions of capture Laws considers that the data do show a real change in the size composition of the stock.

A second feature of change has been the proportion of sexually immature animals in the catch. With the blue whale the immature animals, defined as those males and females less than 73 ft and 76 ft respectively, have increased from 10 per cent of the catch in 1931 to over 30 per cent of the catch in the 1950s. There has been an increase of just over 10 per cent in the proportion of immature fin whales, with the percentage being just over 20 in the 1950s. Other data quoted by Laws on the ovaries confirms the general accuracy of taking size as a measure of maturity.

A third feature of change has been the age structure of the populations. Since the proportion of immature animals has increased in the catch there has been speculation that this indicated increased recruitment and possibly even increasing stocks. Laws clearly shows the fallacy of such arguments since the older animals have become much rarer, and as the reproductive rate is so low this will have reduced recruitment, and could not possibly have increased it. By a study of the baleen plate it has been ascertained that the shift towards the younger age groups is greater than is implied in the length statistics or in the proportion of juveniles being caught.

These then are the physical effects of the exploitation, but we can also investigate the homeostatic mechanisms which have been operating. Laws shows that there have been three such responses, First, there has been a slight increase in the pregnancy rate, in both the blue and the fin whales. Secondly, the animals have reached sexual maturity earlier in life. He concludes that the age of puberty had been reduced by about half a baleen group by 1952–53, after which it appears not to have changed. Thirdly, there are suggestions of accelerated growth. Laws speculates that the rate of growth of individual animals will increase since there is less competition for the planktonic food.

Thus, the species of whale do show homeostatic responses to the pressures of exploitation. In order to see if these responses are able to counterbalance the effects of the exploitation it is useful to plot the survival curves. Such curves for the fin whale are shown in Fig. 5.15 (*a*) where there is a comparison of pre-war data and data during the mid-

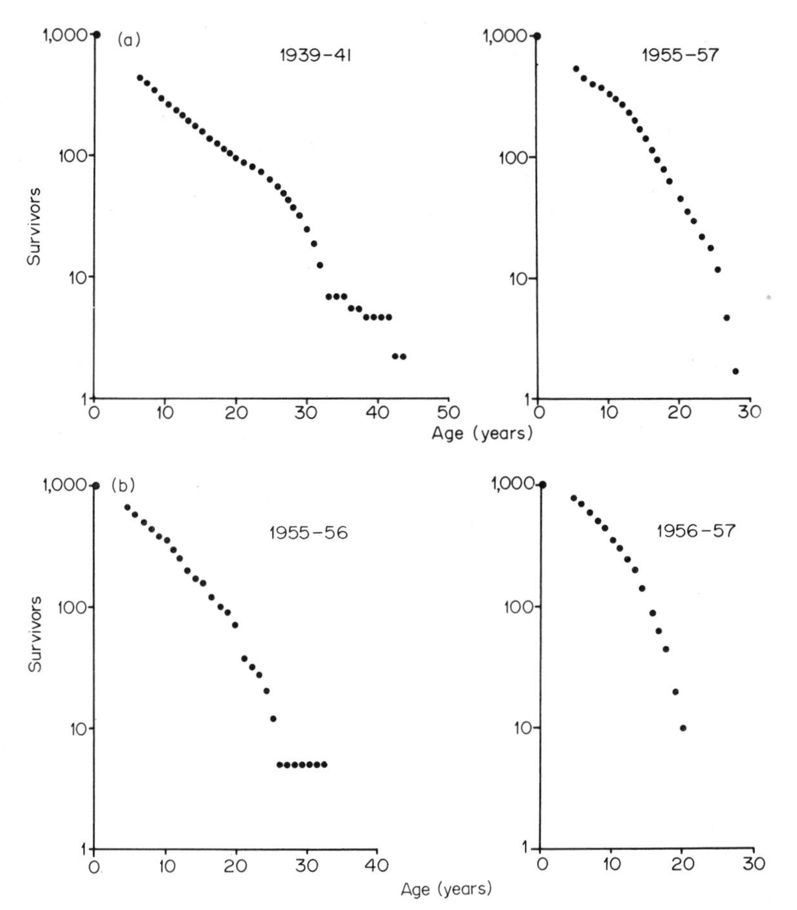

Fig. 5.15 Survival curves for the fin whale, where the number in the first age group is taken as 1000. (a): A comparison of the curves between 1939–41 and 1955–57 in area II. (b): A comparison of the curves between 1955–56 and 1956–57 in area I. (Redrawn from Laws, 1962)

1950s. The curves show clearly that in the region of the older age groups there has been a steepening and truncation, so that by the mid-1950s there were far fewer animals older than 15 to 20 years. Laws shows that this change is not solely due to an increased rate of mortality working through all the age groups, since the process would take over 30 years whereas the two curves are separated by only 16 years. He concludes that the steepening and truncation has been caused by selective fishing mortality with respect to age. In order to underline this rapid change in the survival curves Fig. 5.15 (*b*) shows curves for 1955–56 and 1956–57 for the same area, but different from the area in Fig. 5.15 (*a*). Within only one year the curve has been very much steepened and truncated.

Watt (1968), considering the situation of over-exploitation of the whale stocks, gives six points that are indicative of such a situation. These are:

(1) a decreasing proportion of pregnant females;
(2) a decreasing catch per unit of effort;
(3) a decreasing catch relative to the catch of related species;
(4) a failure to increase in numbers rapidly after a respite from harvesting;
(5) a change in the productivity versus age curve which, when interpreted by using a fecundity versus age curve, shows that the ability of the population to replace the harvested individuals has been destroyed;
(6) a change in the survival versus age curve which, when interpreted by using a fecundity versus age curve, shows that the ability of the population to replace the harvested individuals has been destroyed.

Not all these six symptoms need be present when a species is being over-exploited. Laws' data show that the first is not applicable to the whales, whilst he shows that the catch per unit effort has indeed fallen. The third is more complex to apply to the fin whale since there is now no alternative, though it has been true firstly of the humpback whale in the early part of the twentieth century, and it is true of the blue whale during the 1930s and 1940s. The fourth is difficult to verify, since the length of the respite must be judged according to the breeding cycle of the species. Laws' graphs do, however, suggest that there was little change in the population size during the respite of the Second World War, but even that is a short respite compared with the breeding cycle of the species. The fifth point requires data that are not available, but Watt's sixth point is amply illustrated by the data in Fig. 5.15.

These six points thus allow one to recognize the symptoms of over-exploitation. However, each one of these six needs data that has been measured over a period of time, whereas a conservation manager may want an immediate answer – how can I look at a population and determine if it is being ruinously heavily exploited? Watt produces some survival curves whose shape is indicative of the present state of the population, and these have been modified for whaling by Ehrenfeld (1970). A combination of the curves of these two authors is shown in Fig.

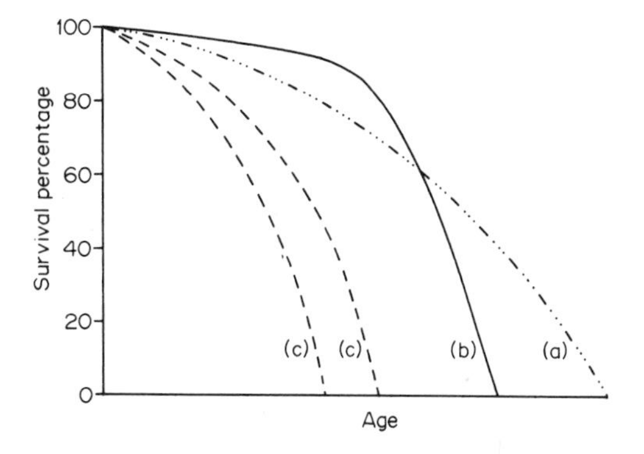

Fig. 5.16 The shape of the survival curve as a function of the exploitation rate. (a) is the curve for an unexploited population, (b) is for a population which has a balanced exploitation, and the curves (c) show over-exploited populations

5.16. Such curves may be useful for a tentative answer to the question posed above, but for a definitive answer it can be seen that long-term studies of data collection and simulation with mathematical models are required.

The characteristics of under-exploitation

In order to demonstrate the effects of over-exploitation or the characteristics of a stable exploitation it is relatively simple to find examples among the natural populations of animals, and from these to estimate the degree of exploitation that has been applied. When man exploits a natural population he usually heavily exploits it, and hence examples of under-exploitation are rather scarce. The subject is not, however, just of academic interest, since preservation of a natural community may result in too little harvest being taken from it, and hence the exhibition of the symptoms of under-exploitation. When an example is found one should ask the question – 'Is there some unusual form of exploitation being aimed at one small part of the population, or is it a genuine example of under-exploitation?'. If the latter is the case why are the predators not completing the exploitation by removing that part of the population untaken by man? Watt (1968) discusses the problems of aiming exploitation at one sex, the males, of the California deer population. The Californian law only allows adult males to be killed, and hence the population is showing the symptoms of under-exploitation since the predators are controlled, though not absent. The red deer, *Cervus elaphus* in Scotland, however, provides an example of genuine under-exploitation.

The situation in Scotland has arisen from three causes. First, there has been control of some predators and extinction of others since these species either were or are incompatible with farming interests, game preservation or the safety of man himself. Secondly, during the nineteenth century there has been progressive deforestation either for timber exploitation or for increasing land available for livestock grazing. Thirdly, in the twentieth century the demand for deer stalking as a sport has decreased, the price of venison has been low since it is unpopular in the United Kingdom, and hence the management of deer forests is no longer economical. Thus the red deer are not being preyed upon and culling has ceased in some areas or been reduced in other areas.

Lowe (1961), in considering the history of the deer in Scotland, states that the species arrived during the Atlantic period (see Introduction), and that evidence of antlers in gravel and peat deposits shows that it has decreased in size by about 25 per cent since that time. During the post-war years interest in the conservation of the species has been aroused since the stags took to marauding farm lands and since a strange loophole in Scottish law allowed these animals to be poached after dark at any time of the year (Lowe, 1961). The essence of the conservation problem was to estimate the population size that could be supported on the minimum amount of grazing land available during the winter period. Surveys carried out at the end of the 1950s indicated that Scotland's population of red deer was between 120,000 and 130,000, with a further 30,000 calves, or a density of about one deer per 9·8 ha on deer forest and one deer per 27·8 ha on marginal land.

This is thus a sketch of the overall position in Scotland, but experimental work on the population dynamics of the species has been continuing on the Island of Rhum since 1957. The population structure on the Island in that year has been estimated by Lowe (1969) and is shown in Fig. 5.17. Lowe constructs a life table for this group of animals, and this is reproduced in Table 5.4. Prior to 1957 the exploitation was very light (approximately 40 stags and 40 hinds per annum), and hence Table 5.4 can be taken as referring to a population that is being under-exploited, with the regulation of numbers being almost entirely due to natural mortality. It will be seen that after the first year of life the natural mortality was most severe in the eight- and nine-year old animals, for

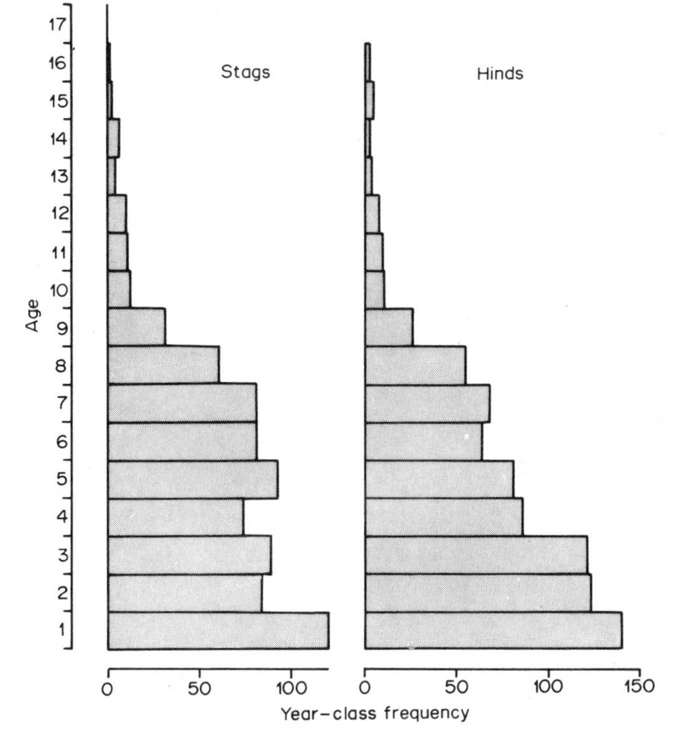

Fig. 5.17 A reconstruction of the year-class frequencies in the red deer population on Rhum on 1 June 1957. (Redrawn from Lowe, 1969)

both stags and hinds. It can also be seen that two- to six-year old stags experience a natural mortality rate of only one per cent, whilst the hinds have a mortality rate of between 10 and 20 per cent. The survival versus age curve is shown in Fig. 5.18 (*a*).

Since the work on Rhum has been continuing annually since 1957, it is possible to follow one cohort of animals from birth. To complete a life table the study would have to continue until the last survivor had died, but after nine years of the study only two per cent of the first year animals in 1957 were still alive. The incomplete life table for this group is given in Table 5.5. This table shows the response of the species to an annual cull of one sixth (16·7 per cent) of the total population of stags and hinds (calves, defined as first year animals, were not used to estimate the cull). It can be seen from Table 5.5 that the natural mortality of the first year animals has been reduced, but that their life expectancy has not been increased. The survival versus age curve is shown in Fig. 5.18 (*b*).

TABLE 5.4 The life tables for the red deer stags and hinds alive on the Island of Rhum in 1957. (Source: Lowe, 1969)

Stags: x (life, years)	l_x (*Survivors at beginning of*	d_x	e_x (*Further expectation of*	$1000q_x$ (*Mortality rate/1000*)	*Hinds:* x	l_x (*Survivors at beginning of*	d_x	e_x (*Further expectation of*	$1000q_x$ (*Mortality*
1	1000	282	5·81	282·0	1	1000	137	5·19	137·0
2	718	7	6·89	9·8	2	863	85	4·94	97·3
3	711	7	5·95	9·8	3	778	84	4·42	107·8
4	704	7	5·01	9·9	4	694	84	3·89	120·8
5	697	7	4·05	10·0	5	610	84	3·36	137·4
6	690	6	3·09	10·1	6	526	84	2·82	159·3
7	684	182	2·11	266·0	7	442	85	2·26	189·5
8	502	253	1·70	504·0	8	357	176	1·67	501·6
9	249	157	1·91	630·6	9	181	122	1·82	672·7
10	92	14	3·31	152·1	10	59	8	3·54	141·2
11	78	14	2·81	179·4	11	51	9	3·00	164·6
12	64	14	2·31	218·7	12	42	8	2·55	197·5
13	50	14	1·82	279·9	13	34	9	2·03	246·8
14	36	14	1·33	388·9	14	25	8	1·56	328·8
15	22	14	0·86	636·3	15	17	8	1·06	492·4
16	8	8	0·50	1000	16	9	9	0·50	1000

TABLE 5.5 The life tables for the red deer stags and hinds that were in their first year of life on Rhum in 1957. (Modified from Lowe, 1969)

Stags: x life, years)	l_x (*Survivors at beginning of rate/1000*)	d_x	e_x (*Further expectation of*)	$1000q_x$ (*Mortality*)	*Hinds:* x	l_x (*Survivors at beginning of*)	d_x	e_x (*Further expectation of*)	$1000q_x$ (*Mortality*)
1	1000	84	4·76	84·0	1	1000	0	4·35	0
2	916	19	4·15	20·7	2	1000	61	3·35	61·0
3	897	0	3·25	0	3	939	185	2·53	197·0
4	897	150	2·23	167·2	4	754	249	2·03	330·2
5	747	321	1·58	430·0	5	505	200	1·79	396·0
6	426	218	1·39	512·0	6	305	119	1·63	390·1
7	208	58	1·31	278·8	7	186	54	1·35	290·3
8	150	130	0·63	866·5	8	132	107	0·70	810·5
9	20	–	–	–	9	25	–	–	–

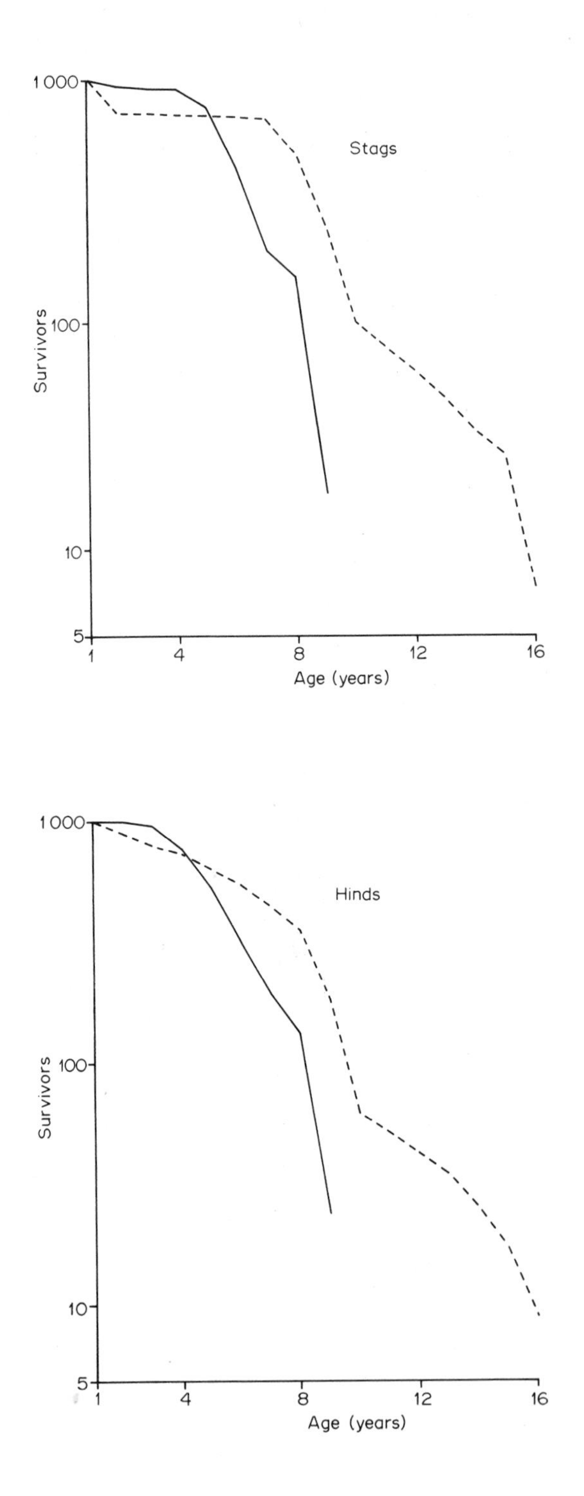
1 000
100
10
5
Survivors
Stags
1
4
8
12
16
Age (years)
1 000
100
10
5
Survivors
Hinds
1
4
8
12
16
Age (years)

We can thus see the response of the red deer population to an annual cull, but it is also useful to develop a model of the population for predictions and for simulation work. Since we need to consider both the age and the sex structure of the population we will need to extend the matrix model given in the introduction to this chapter. The extension is along the lines developed by Williamson (1959). If we consider a simple case where the population is divided into only three age classes, then Equation (5.8) can be written in the form

$$\begin{bmatrix} 0 & f_{m_0} & 0 & f_{m_1} & 0 & f_{m_2} \\ 0 & f_{f_0} & 0 & f_{f_1} & 0 & f_{f_2} \\ p_{m_0} & 0 & 0 & 0 & 0 & 0 \\ 0 & p_{f_0} & 0 & 0 & 0 & 0 \\ 0 & 0 & p_{m_1} & 0 & 0 & 0 \\ 0 & 0 & 0 & p_{f_1} & 0 & 0 \end{bmatrix} \begin{bmatrix} n_{t,m_0} \\ n_{t,f_0} \\ n_{t,m_1} \\ n_{t,f_1} \\ n_{t,m_2} \\ n_{t,f_2} \end{bmatrix} = \begin{bmatrix} n_{t+1,m_0} \\ n_{t+1,f_0} \\ n_{t+1,m_1} \\ n_{t+1,f_1} \\ n_{t+1,m_2} \\ n_{t+1,f_2} \end{bmatrix} \quad (5.18)$$

where f_m and f_f refer to the number of males and females respectively born to a female in the ith age class during the period of time: p_m and p_f refer to the probability that a male and a female respectively alive in the ith age class will be alive in the $i+1$th age class; and $n_{t,m}$ and n_{tf} refer to the numbers of males and females respectively that are in the ith age group at time t.

These data for the population of red deer alive in 1957 can be derived from Lowe's (1969) paper. Thus from his Table 8 the sex- and age-specific survival rates can be calculated, and are listed in Table 5.6. The fecundity terms present more problems. However, Lowe's Table 10 shows the proportion of the population that is breeding, and by taking weighted averages for the age groups two to four years, five to eight years, and nine or more years in Table 15 one can estimate the age and sex specific birth-rates. Southern (1964) states that the red deer usually only have one calf, and this figure has been used in the calculations. The data from Lowe's Tables 10 and 15 with the fecundity estimates are given in Table 5.7.

The data from Tables 5.6 and 5.7 have been used to produce a matrix, similar to that in Equation (5.18), that has 32 rows and columns. Numeri-

Fig. 5.18 Survival curves for red deer stags and hinds on Rhum. The solid line represents deer that were in their first year of life in 1957, and the dashed line represents all the deer that were alive in 1957. The two graphs for each sex thus refer to pre- and post-1957 conditions. (Redrawn from Lowe, 1969)

TABLE 5.6 Data for the p_m and p_f terms of the matrix model. (All probabilities should be divided by 1000)

Age	*1*	*2*	*3*	*4*	*5*	*6*	*7*	*8*	*9*	*10*	*11*	*12*	*13*	*14*	*15*	*16*
Stags	718	990	990	990	990	991	734	496	370	848	821	781	720	611	364	0
Hinds	863	902	882	879	862	840	808	507	326	864	824	810	735	680	529	0

TABLE 5.7 Data for the f_m and f_f terms of the matrix model. (All the fecundity terms should be divided by 1000)

Age	*1*	*2*	*3*	*4*	*5*	*6*	*7*	*8*	*9*	*10*	*11 and older*
Percentage of population breeding	0	41·6	86·3	89·3	95·1	95·2	93·1	98·8	77·5	76·5	85·2
Percentage of calves that are males	–		48·6			38·1				54·5	
f_{mi}	0	202	419	434	362	363	355	376	422	417	464
f_{fi}	0	214	444	459	589	589	576	612	353	348	388

TABLE 5.8 The structure of a stable population of red deer, predicted by a matrix model. The structure has been adjusted so that there are 1000 one-year old stags

Age	*Stags*	*Hinds*
1	1000	1239
2	617	919
3	525	712
4	447	540
5	380	408
6	323	302
7	275	218
8	174	151
9	74	66
10	24	18
11	17	14
12	12	10
13	8	7
14	5	4
15	3	2
16	1	1

cal solution of the equation for the latent roots (see Appendix 1) shows that a stable propulation would have the following characteristics,

$$\lambda = 1{\cdot}1636$$
$$r = \ln\lambda = 0{\cdot}1515$$

and the elements of the latent vector are given in Table 5.8.

Since the data for the matrix were derived from those animals alive in 1957, it can be taken as referring to a virtually unexploited population. Hence, one question might be to ask what harvest could be taken. This is $100(\lambda - 1)/\lambda$ per cent of all age classes, and the analysis would indicate a harvest of 14·06 per cent. It is also interesting to see the predicted abundance of the two sexes when the population is stable. It can be seen that the hinds are overall more abundant than the stags, with a proportion of approximately 100 stags to 119 hinds. However, this abundance is dependent upon age. Table 5.8 shows that the hinds would be the most abundant sex in the younger age classes, but that the males would be more abundant in all classes of animals six years and older.

What then are the features of under-exploitation? First, a study of the survival versus age curves in Fig. 5.18 shows that the smoothly falling curve of the balanced exploitation is less steep and drawn out towards the older age classes in under-exploited populations. These populations thus contain an excess of old animals. These curves can be compared with those in Fig. 5.16 where the effect of over-exploitation was to truncate the curve. Secondly, under-exploitation may not be immediately obvious when mathematical models of the population are developed. The matrix model for the deer population shows that the value of λ indicates a cull of 14 per cent per annum, although it is known that the population can sustain an exploitation rate of about 17 per cent per annum. Although this discrepancy is not very great it stems from the fact that in an under-exploited population the natural mortality factors are regulating the population size, and hence a model with constant parameters will not show the effect of exploitation in reducing the natural mortality. The harvest gained from the population, as well as exploiting the increase in numbers or biomass, can also claim an amount equal to the reduction in natural mortality. Thirdly, the effects of under-exploitation can be reflected in the reproductive activity of the females. The California deer show a reduction in the number of ova being produced; a lowering in the rate of fertilization and implantation of the embryo in the uterus; an increase in the rate of resorption of embryos, abortion and calves that die at birth; an inability of the mother to produce enough milk for the calf; and the commencement of breeding activity at a later age. These are all homeostatic mechanisms on the part of the species concerned, all leading to a reduced rate of recruitment to the population. They tend to lead to an unhealthy stock that is more liable to be affected by climatic extremes and disease, features of a population that are dangerous if a sustained yield is to be gained, and the population conserved.

The exploitation of plant populations

No discussion of the effects of exploitation would be complete if consideration of the plants were to be omitted. Plant populations, however, differ in many features from animal populations, and the mathematics of animal population growth do not necessarily apply to plants, although some sort of sigmoid growth of the standing crop biomass can be detected (Fig. 5.19). As we will see later it is the point of inflection of such curves that will interest us if we are to manage the plant populations for the maximum sustained yield. The essential difference between plant and animal populations is that an animal is a whole unit, and it is either harvested in which case its effect on the population is completely removed, or it is left behind in which case it plays its full role in the population. Things are not so clear-cut with plants, since a harvest may

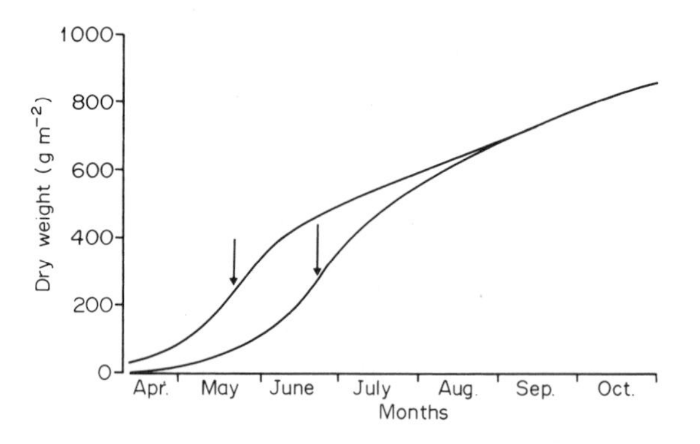

Fig. 5.19 The calculated dry weight increase in undefoliated swards of subterranean clover. The upper graph represents populations that had a density of 2500 plants per m^2, and the lower graph populations with a density of 500 plants per m^2. The arrows indicate the date on which the optimum leaf area index (LAI) was attained. (Redrawn from Black, 1964)

include parts of all or of some of the population, and these harvested plants then have a modified role to play in the community.

In considering plant production one will have to introduce some of the most usual terminology. Blackman (1968) shows that 85 to 90 per cent of the daily gain in dry matter of plants is the accumulation of the products of photosynthesis, and the remaining 10 to 15 per cent is of nitrogen and other minerals. Since photosynthesis accounts for such a large proportion of the whole increase we will concentrate attention on this aspect of production. The daily rate of gain in dry weight, the relative growth rate (RGR), is determined by two sets of factors. First, it is a function of the net assimilation rate (NAR) which is defined as the net diurnal efficiency of assimilation per unit of photosynthetic surface. This net amount is used since it is composed of the photosynthetic gains during the day and the respiratory losses during the night. Secondly, there is the leaf area ratio (LAR) which is the ratio of the total assimilating surface to the total plant weight. For relatively small changes in the leaf area (i.e. provided that the leaf area is not more than doubled in the time t), estimates of RGR and NAR can be made from

$$\left.\begin{aligned} \mathrm{RGR} &= \frac{\ln W_t - \ln W_O}{t} \\ \mathrm{NAR} &= \frac{W_t - W_O}{t} \cdot \frac{\ln A_t - \ln A_O}{A_t - A_O} \end{aligned}\right\} \qquad (5.19)$$

where the measurements are over a period of time t, the initial values of the plant's weight and leaf area are W_O and A_O respectively, and the final values W_t and A_t respectively. A further ratio that is of considerable importance in these studies is the leaf area index (LAI) which is defined as the ratio of the total leaf surface to the horizontal ground surface beneath the leaves.

For considering populations of plants rather than individuals it is the NAR and LAI which are of primary concern, but the complexity of ecological relations can be seen when one relates these two parameters of plant production to the environmental factors, to the stage of plant development, to the population density, and to the structure and arrangement of the leaves. The whole field of these interrelations is reviewed by Blackman (1968).

We will be concerned with management techniques for maximizing the production. Since the NAR is a measure of the efficiency of the plants, which could only be altered by genetic manipulation of the natural population so that it became domesticated, we will more closely examine the relation between productivity and LAI.

The rate of dry matter production is related to LAI in a parabolic manner. There is thus an LAI at which production is maximized, this being due to the balance between photosynthesis and respiration. As the LAI increases so one can conceive of more layers of leaves covering the ground. When there are very few leaves one can postulate that all of the leaves will be in the full sunlight, and that production will be virtually proportional to LAI. As the LAI increases some leaves will be in the shade, and their photosynthetic efficiency will be reduced – in animal concepts we could consider that this is a case of competition for light. At some point as the LAI is increasing the lowest leaves will become marginal producers, since the competition for light is so severe that their photosynthetic activity is exactly balanced by their respiratory activity. They are thus contributing nothing to the net production of the plant population. As the LAI further increases the lowest leaves are submarginal producers, since their respiration is greater than their production, and hence the net productivity of the stand decreases. The curves

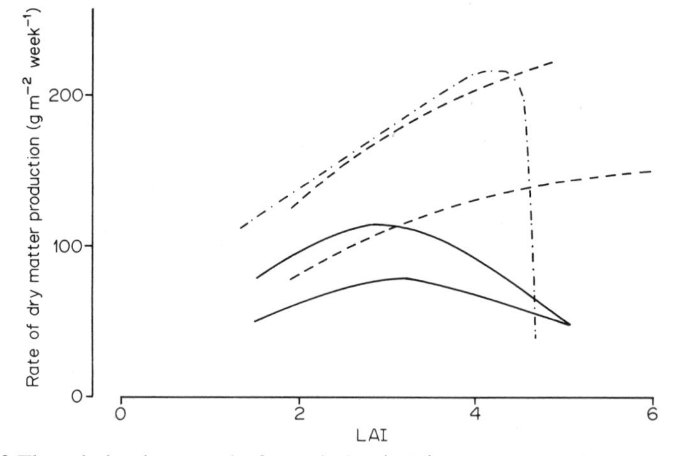

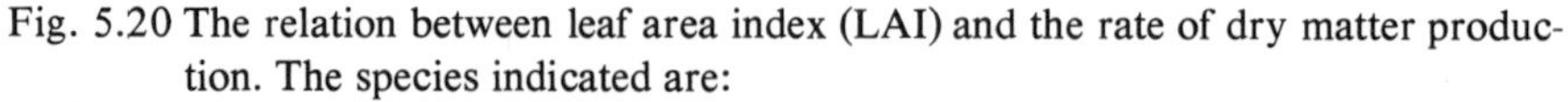

Fig. 5.20 The relation between leaf area index (LAI) and the rate of dry matter production. The species indicated are:
————————, kale;
— — — — — —, sugar-beet;
—·—·—·—·, potato.
(Redrawn from Blackman, 1968)

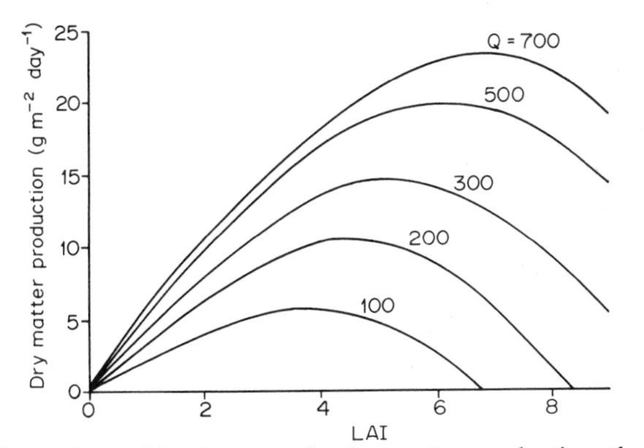

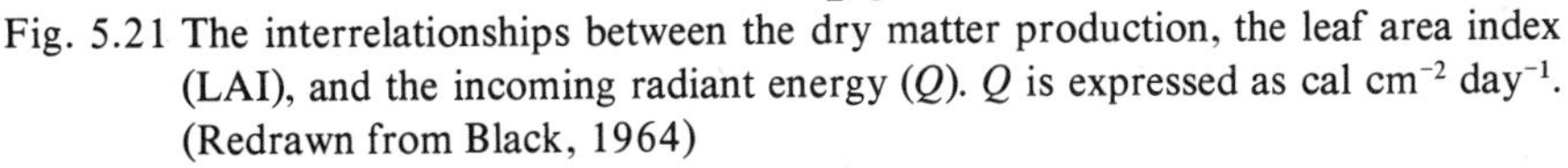
Fig. 5.21 The interrelationships between the dry matter production, the leaf area index (LAI), and the incoming radiant energy (Q). Q is expressed as cal cm^{-2} day^{-1}. (Redrawn from Black, 1964)

obtained in experimental work are shown in Fig. 5.20. There is thus an LAI at which the rate of dry matter production is maximized, and this has been shown by Black (1964) to be a function of the solar energy available to the population. Thus, using the subterranean clover, *Trifolium subterraneum*, Black showed that the optimum LAI was just less than four (the units quoted are $m^2\ m^{-2}$, though as this is a ratio of two areas it is independent of the units in which each area was measured) when there is an energy input of 100 cal cm^{-2} day^{-1}. The optimum LAI occurs at seven when the solar radiation is 700 cal cm^{-2} day^{-1} (Fig. 5.21).

There is an interesting comparison to be made between Figs 5.20, 5.21, 5.11 (*c*) and 5.11 (*d*). Figs 5.20 and 5.21 relate the plant production to the LAI, a measure of the size of population responsible for that production. Fig. 5.11 (*c*) relates the strength of a year-class of sardines, a measure of the potential production, to the size of the fish stock responsible for the production. In other words, both curves show the same thing, production as a parabolic function of the organisms causing that production. In the fish we were able to define a weather effect, and to show how this influenced the yield that could be taken. Plant studies are not sufficiently numerous to be able to draw maximum and minimum parabolas defining the effect of the environmental factors upon the production. However, I do believe that such will be found to be the case. Black's studies of the subterranean clover, Fig. 5.21, have shown that parabolic relationships exist between production and LAI throughout a wide range of values of solar radiation. Whether or not 'big weather effects' and 'small weather effects' can be demonstrated in the plants will have to wait until more information has been gathering. The more generalized curves, Figs. 5.11 (*d*) and 5.21, show very striking similarities.

One can see that for the maximum yield from plant communities one would wish to maintain the LAI at the optimum value. Such a system of management will ensure that the plant population is making the maximum use of the available solar energy, assuming of course that other requirements such as water, nitrogen, minerals and carbon dioxide are not limiting. As can be seen in Fig. 5.19 the optimum LAI coincides with the point of inflection on the sigmoid growth curve. Thus, management of plant populations may well use the same basic idea as management of the North Sea fish populations (pages 172–176) where one attempted to maintain the population at that point on the growth curve for which

$$d^2Y/dt^2 = 0$$

Such studies may determine the optimum yield, but conservation management may wish to preserve some form of structure in the community which harvesting at an optimum level would destroy. Hence, one should see what harvesting does to the structure and composition of the community. Two specific examples will be considered. The first concerns the over-exploitation of a plant community by grazing, and the second a case of under-exploitation following the removal of grazing pressure. Both sorts of exploitation are important in determining the effects of conservation management.

The problems of sheep and particularly goat grazing in the Mediterranean region are well known, and very clearly show the effects of over-exploitation of a plant community not only on the conservation of the plant species themselves, but also on the conservation of the soil, the hydrological system and indeed the whole ecosystem. The possibility of integrating the grazing with forest management in the region is discussed by Margaropoulos (1962). He shows that in Spain there are 0·5 animal units (one head of cattle or eight head of sheep or goats) per ha, whilst the lowest grazing pressure is in Algeria with 0·125 animal units per ha. The potential grazing capacity nowhere exceeds 0·125 animal units per ha. The method of co-existence whereby forestry and grazing are integrated into the same areas is strongly advocated, but the methods require intervention throughout the whole ecosystem. Schemes for achieving these aims require blocks of land for afforestation, protected from grazing during the initial growth period, and separated from each other by tracts of agricultural land. Schemes such as this require considerable capital investment and a wealth of ecological knowledge about the species of trees to be used, but where over-exploitation has destroyed the natural communities it seems that it is possible to re-establish a vegetation cover. Such works, however, move from the sphere of biological conservation to that of soil conservation, agriculture and silviculture.

The second example concerns the effects of rabbit grazing. When in the 1950s a significant proportion of the rabbit population was eliminated by the virus disease myxomatosis, large areas of grasslands became under-exploited. For many years both before and after the epizootic, Watt (1957, 1960a, 1960b) had been experimenting on the influence of rabbit grazing on the East Anglian heaths. On Foxhole Heath, Watt (1957) has demonstrated the effects of grazing on the species composition of the plant communities. There were 14 species of higher plants which had a similar abundance on both grazed and ungrazed areas, nine species that were more abundant on ungrazed areas, and 12 species more abundant on grazed areas. Furthermore, there were 28 species exclusively found on the ungrazed areas and 16 species only on the grazed areas. Thus out of the 79 species 55 per cent were confined to one or other of the areas. The group of species only represented on grazed areas is perhaps the most interesting since they were all annuals with the exception of one species. Only three of the 28 species confined to ungrazed areas were annuals, and seven were grasses.

Furthermore, Watt (1960b) has quantified the changes that have occurred in these heathlands. On grasslands open to grazing, the grass *Festuca ovina* entered the community during the absence of rabbits in 1943, and it increased its cover till 1947 when the rabbits were again present (Fig. 5.22). After the extermination of the rabbits in 1954 the cover of *F.ovina* on the grazed plots again increased. In other areas that were enclosed to eliminate rabbit grazing it can be seen in Fig. 5.22 that the cover of *F.ovina* increased so that it became the dominant grass species.

More generally, it can be said that grazing by rabbits favours the smaller plants and the annuals. In ungrazed swards the grasses tend to become dominant, and ecological succession may gradually alter the community to a woodland (the number of species quoted above include two trees amongst the 28 species exclusively found on the ungrazed area). The aim of conservation management will dictate whether under-

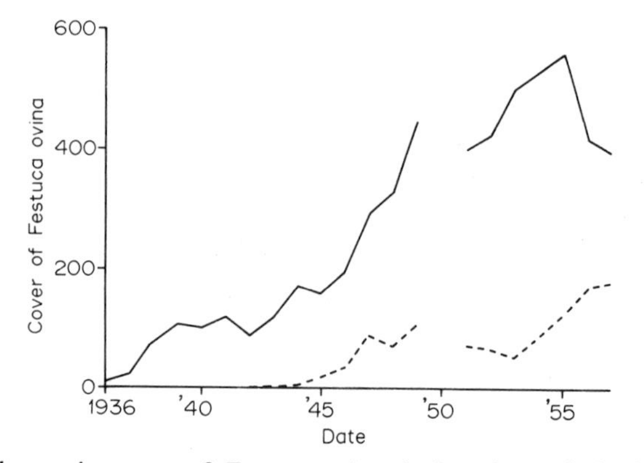

Fig. 5.22 The change in cover of *Festuca ovina* during the period 1936–1957, both within (continuous line) and without (dashed line) the enclosure on a Breckland heath. (Redrawn from Watt, 1960b)

exploitation or a balanced exploitation of the grassland is required. It is known, for example, that, where conservation aims at preserving rare orchid species, the grazing of the grasslands, except at the time of orchid flowering and seeding, is an essential feature of management.

What then are the symptoms of over- and under-exploitation of plant populations? These cannot be drawn up, in the form of survival curves as for animal populations, but they can be expressed as functions of the form of the community. Over-exploitation leads to soil erosion, particularly in parts of the world with a drier climate. In the Mediterranean region it has led to the increased abundance of plants that are relatively non-susceptible to grazing or browsing, such as thorn bushes. The symptoms are so obvious that they need not be stated. When a population is under-exploited the change is towards a taller vegetation community, the elimination of small herbs and annuals, the establishment of bush and tree seedlings, and to an accumulation of grass litter, remaining undecomposed for all or part of the year.

PART TWO
Application

6. Conservation and Biological Management

Definition

> *Conservation*
> The action of conserving; preservation from destructive influences, natural decay, or waste; preserving in being, life, health, perfection, etc.
> *Conserve*
> To keep in safety, or from harm, decay, or loss; to preserve with care; now usually, to preserve in its existing state from destruction or change.
> (Oxford English Dictionary)

As a word in the English language 'conservation' can be defined by the quotations from the Oxford English Dictionary. However, conservation of biological resources is not solely a problem of preservation and protection. It is not hoarding or leaving well alone as this definition may imply, since nature is not static and cannot be maintained in an unchanging state even if this were desirable. Thus, within a biological context any definition of conservation, or search for its meaning, must explore further than a concise form of words. It must review the use of the word in the biological literature and determine the full breadth of its application. It must investigate how the concept of conservation has developed, and how the concept is being put into use. These two features of the word 'conservation' form the basis for the discussion of this chapter. No attempt is made at a complete review, but rather at taking a short series of definitions in the recent ecological literature and at criticizing their bases and implications. No attempt is made at a complete discussion of the application of the concept of conservation, but rather at introducing a restricted application to the management of wildlife and nature reserves.

Definitions and uses of the word 'conservation'

> Any human intervention in nature, even presupposing good intentions, can rarely be reconciled with the idea of strict conservation.

> True, homeostasis in natural systems is always active, and although man tries again and again, he rarely produces truly catastrophic changes in the biosphere. But this is a credit to the efficient organization of ecosystems and not to the wisdom of man. Genuine conservation forbids any interference.
>
> (Margalef, 1968, p. 48)

The view that Margalef is advancing is essentially that man is not part of the natural environment, and that true conservation of the environment is not compatible with any human activity. Since species of the genus *Homo* have evolved as part of the biosphere, man must be a part of that environment and it is false to consider an environment without man. However, the characteristic of the species, the ability of environmental modification, has meant that man has had a greater effect on the environment than any other species. This alone does not justify the view that, wherever and however this effect is felt, there is no longer a natural environment. However, one should try and assess just how much environmental modification there can be before an ecosystem changes from being natural to unnatural or man-made. Hence, there must be a definition of conservation that includes man as part of the ecosystem.

> Conservation is not to be regarded as a negative affair but as a positive, if rather long-term, attempt to set up ecological systems which are viable, stable and productive.
>
> (Macfadyen, 1963, p. 278)

This definition suffers, as did Margalef's, from being too restricted. It is essentially management orientated, being concerned with the setting up of ecosystems – does this imply that there is no ecosystem that is natural or unaffected by human intervention? The pre-requisites of this form of management are that the ecosystem should be stable, a concept discussed by Williamson (1972), and that it should be productive. But what is the purpose and aim of this stability and production? It is, for example, known that some of the ecosystems that are in most need of protection are those that are least stable. Should stability therefore be an aim of conservation management?

> Conservation is the wise use of the country's resources of land and water and wildlife for every purpose including amenity and recreation.
>
> (Countryside in 1970, 1965)

> Biological conservation ... is concerned with the maintenance of natural systems and, where possible, with their utilization – either directly or by way of information obtained from their study – for the long-term benefit of mankind.
>
> (Cragg, 1968)

These two definitions share in common the maintenance or wise use of a biological resource and imply management for the benefit of mankind. Both suffer from the subjectivity of the first premise: what is wise use to one person may be unwise to another, what is maintenance to one may be considered as over-emphasis or despoliation to another. The statement of the utilization of the resources is also open to question. Should this utilization be for the benefit of man, or should there be more of a two-way flow of benefit between man and the remainder of the biosphere?

> A conservationist is a person ... who is concerned that the relationship of Man and Nature evolves in directions that are more beneficial to Man and more acceptable to Nature.
>
> In dealing with the present we must look forward to the future to find better ways of living in and with Nature, bending and shaping Nature to serve our purposes but never breaking her.
>
> Conservation must be for people as well as for the higher plants and mammals, birds, fish, soil, vegetation and water.
>
> (Fisher, 1969)

Fisher's definition of a conservationist and his statements concerning conservation establish his belief in the two-way interaction

$$[\text{Man} \rightleftharpoons \text{Nature}]$$

Fisher has therefore attempted a definition which fits into Black's (1970) category 'I will attempt to manage the resources at my disposal in such a way as to hand them on undiminished to posterity, even though this involves sacrifices on my part, because it matters a great deal to me how they judge my conduct in these matters'. Black argues that such a rationale is a more modern replacement of the older system of stewardship, based on the interactions

$$[\text{God} \rightleftharpoons \text{Man} \rightleftharpoons \text{Nature}]$$

This system implied that Man was accountable to God for his actions and for his management of the biological resources. However, the definition of the type put forward by Fisher fails to explain why there should be any care for the environment that posterity will inherit or for their opinion of this generation. Black concludes

> There is as yet no fully worked-out and satisfying philosophy of conservation, but only a collection of generalities and catch-phrases. These may certainly be useful if not overstretched, but may turn out to be dangerous in that they incorporate their own built-in mechanism for devaluation.
>
> (Black, 1970, p. 117)

It is not the intention of this chapter to attempt to find such a philosophy of conservation, but rather to establish a framework within which the remaining chapters of this book discuss some aspects of conservation.

The definition that will be used incorporates aspects of both Macfadyen's management orientated definition and Fisher's man : nature interaction. One cannot, however, completely ignore Margalef's plea for complete non-intervention by man, but in the Introduction to this book it was seen that the human population is increasing very rapidly, and hence more and more demands for food, living space, recreational space and education will be made upon a rigidly fixed total surface area. In this context it is unlikely that any true wilderness (Fraser Darling, 1969) will remain, although it is evident that such areas are of importance as controls against which managed ecosystems can be compared.

A definition that will describe the use of the word 'conservation' in the context of this book is

> **Biological Conservation is essentially concerned with the interaction between man and the environment.**
>
> **Fulfilment of conservation objectives for a biological resource requires the resource's management in perpetuity on the basis of a sustained production of the resource or biotic component of the environment and a sustained demand on the resource or environment by man.**

This definition therefore broadly equates conservation with biological management, in the full recognition that management can take many forms. It may involve direct human interference in an ecosystem, or the controlled neglect of a wilderness, or any degree of interference between these two extremes. The essential feature, however, of the definition is that there should be sustention in perpetuity, a guarantee as it were that the management practice will not alter an ecosystem to such an extent that biological production is eliminated. In the terms of the previous chapter, conservation is concerned with a balanced harvest of a population or community, attempting to eliminate from the management practice both over-exploitation and under-exploitation.

Application of conservation principles

In the definition of conservation that has been given for use in this book there are two aspects of its application that must be considered. First, there is the aspect of biological management, and one must define the breadth of this subject. Secondly, there is the aspect of sustained production and demand. One must investigate the forces that are acting to hinder conservation management and the methods of overcoming these forces.

The scope of biological conservation is very wide, including as it does all living things. Thus, we can consider that there are six categories of resource which will be managed and may be conserved. These are:

(1) agriculture;
(2) forestry;
(3) water;
(4) fisheries;
(5) recreation;
(6) wildlife.

It is certainly true that the principles of conservation can be applied with any of these resources. Agriculture in Britain, for example, affects approximately 80 per cent of the land area (Cornwallis, 1969), and hence the practices of this one resource can have a great effect upon the whole of the country. Cornwallis, himself a farmer, has described the recent changes in farming as producing a more uniform countryside by abandoning rotational cropping, by replacing livestock by arable crops (particularly barley), by reducing the area of permanent grassland, by the removal of hedges (this was discussed in Chapter 1), and by the drainage of wet areas. These activities are a part of the contemporary system of agriculture. However, if conservation principles are applied there is much that can be done to maintain the diversity of the countryside, particularly by the encouragement of woodland and scrubby species in the uncultivable areas, and by the prevention of soil erosion.

Agriculture, because of the very scale of the operations and because it is supplying the basic human need of food, has not frequently been discussed in a conservation context. The other five forms of land use are, however, discussed in most texts dealing with conservation. Thus Black (1968) considers in separate chapters the soil, water, grasslands, forests, fish and various other animal groups such as ducks and geese, predatory birds, fur-bearing mammals and game mammals. Black relates the changes that have occurred in natural and semi-natural ecosystems to the management or lack of management in order to establish an empirical basis for future management. Dasman (1968) goes further than Black in that he also includes chapters on the demand for recreation, the urban environment and the problem of the increasing human population. By quoting wide series of examples such texts on conservation attempt to demonstrate the principles of management by analogy. There are historically so many instances of the interaction of man and the environment that this empirical approach can be used with a fair degree of success. However, since management is concerned not only with the administration of the present state of the resource but also with predicting the future course of the resource, it would be preferable to incorporate ecological theory into the practices of conservation. A step in this direction has been taken in the text-book written by Ehrenfeld (1970), who analyses the factors that are contributing to the loss of natural communities and species, and incorporates some simple ideas of population dynamics into the discussion of the conservation of some animal populations. Watt (1968) bases his ideas of resource management upon a foundation of ecological theory and mathematical and computer modelling. Watt is not primarily concerned with conservation, although the definition of conservation that has been used for this book would imply that Watt, wishing to sustain the production of his resources, would also be conserving them.

The example of these four text-books on conservation and biological management poses one question. What should be the approach to conservation or biological management? Should it be via 'pure ecology', basing the procedures of conservation and biological management upon ecological theory, mathematical models and statistical analysis? Or alternatively, should it be via 'applied ecology', mostly using the empirical experience of past management as a basis for future management? Text-books

Plate 1 The Oxford ragwort, *Senecio squalidus*, growing beside a railway track on a bridge in Doncaster, Yorkshire

Plate 2 The New Zealand willow herb, *Epilobium nerterioides*, colonising an old airfield runway in Yorkshire

Plate 3 The grey squirrel, *Sciurus carolinensis*

Plate 4 The red squirrel, *Sciurus vulgaris*

Plate 5 The pasque flower, *Pulsatilla vulgaris*, growing to the south of Lincolnshire

Plate 6 The large copper butterfly, *Lycaena dispar*, at Woodwalton Fen National Nature Reserve

Plate 7 Location 12 on the formal nature trail at Skipwith Common in Yorkshire

Plate 8 Part of the semi-formal nature trail in the Gwydyr Forest in North Wales. The mountain in the distance is Moel Siabod

Plate 9 Aberlady Bay Local Nature Reserve. The footbridge at the entrance to the reserve

Plate 10 Aberlady Bay Local Nature Reserve. The Peffer Burn

Plate 11 Aberlady Bay Local Nature Reserve. The sand dune system from the teschenite rock outcrops at Gullane Point

Plate 12 Aberlady Bay Local Nature Reserve. Sea buckthorn bushes providing cover at the Marl Loch

Plate 13 Aberlady Bay Local Nature Reserve. Growth of grasses on the reserve. In the foreground the grasses have been cut, whilst the grasses remain uncut in the distance

Plate 14 Hawthorn, *Crataegus monogyna*

Plate 15 Sea buckthorn, *Hippophae rhamnoides*

Plate 16 Giant hogweed. *Heracleum mantegazzianum*

Plate 17 Aberlady Bay Local Nature Reserve. Erosion of the forward edge of the dune system caused by the north-easterly winds.

Plate 18 Aberlady Bay Local Nature Reserve. Erosion of the dune system caused by public use and mis-use

Plate 19 Aberlady Bay Local Nature Reserve. Small scale erosion of paths on the reserve caused by trampling. The photograph is taken a few yards from the bridge at the entrance to the reserve

on forest management follow the latter approach since there is a long tradition of forest management, particularly on the Continent of Europe. Duffey (1969) shows that the European wildlife conservation movement is still very young, and although there is a basis of past management experience this is not nearly as great as is that of forestry, and there is as yet no general tradition of wildlife management. Partly because of this scarcity of tradition, and because of the greater precision and predictive value of modelling and statistical analysis, the design of this book has been on the former, the 'pure ecology', approach. When ecological theory has been developed and expressed in the form of a model, this model should incorporate sufficient insight to represent the process that it describes (in the sense of a statistically satisfactory fit) and to predict the future course of the process. A model was described in Chapter 2 for the lynx cycle, showing that prediction was possible, but that as prediction was further and further into the future so the accuracy of the prediction became less and less. The model therefore more and more closely approximated to the mean population size as the length of prediction was increased. Thus, in conservation management there should be a prediction of the consequences of a particular action. If the model yields only the expected result then it can be described as being *deterministic*, since the model will always yield the same result if given the same input data. However, this predicted result is not usually sufficient, since one will want to know something about the variance. What is the probability of reaching the predicted result, or what is the probability that some other condition will be reached? Such *stochastic* models are still comparatively rare in the ecological literature, but it may be that the demand for such information from models that are used in conservation management will stimulate more interest in their general application in ecology.

Thus one restriction that is imposed on the breadth of biological management or conservation in this book is that there should be as firm a basis as possible of ecological theory, statistical analysis and mathematical modelling. One other limitation that will be imposed concerns the categories of land use that have already been listed. Each of these six categories involves some form of conservation if the management is not to deplete the resource. However, in this book only the last on the list, namely wildlife, will be considered, and this will often be confined to areas of land which are specifically reserved for the protection of wildlife, types of land that are known as 'nature reserves' when on a small scale or as 'game parks' or 'wildlife parks' or occasionally as 'National parks' when on a larger scale.

Using as a definition 'conservation of a habitat means maintenance or increasing the level of energy flow', Huxley (1961) considers the problems of wildlife and natural habitats in East Africa. He puts forward a programme of 46 points, five of which can be quoted as showing the breadth of application even on the restricted range of wildlife conservation on specially reserved areas:

> [Point 2] 'The ecology of the main wildlife areas, with special reference to the life-histories, habits, and ecological relations of game animals with other larger vertebrates.'

This point raises the issue of establishing an ecological framework on which the conservation operation can be based. It is possible that, if the report had been written a decade later, this point would have made reference to studies of the populations of the animals, since the ecological relations might now be treated as a part of the subject of population dynamics.

> [Point 4] 'The intensive study of wildlife management, including (a) ranching of selected species, and (b) game-cropping, including marketing, in a number of ecologically different areas . . .'

In this point Huxley has incorporated the man : nature relationship, using as a part of the applied ecological approach the need to take a harvest as food. By maintaining or improving the level of energy flow in the definition of conservation, this point leads to the protection of the species whilst at the same time yielding a utilizable harvest for man. However, there can be more immediate effects of man on the system.

> [Point 3] 'The effect of fire on wild habitats, with a view to securing an agreed policy on burning for different types of area.'

Man has an effect through fire but there should be ways of finding some sort of control. Once again if the report had been written a decade later, when there was more public awareness of the effects of toxic chemicals, these might have also formed a control on man's activities in the ecosystem. However, the three points which have been quoted have all been concerned with the ecological characteristics of the site, whereas Huxley goes on to consider other aspects of the use of reserves.

[Point 9] 'Methods for introducing conservation into the educational curriculum . . . as an introduction to natural science.'

[Point 20] 'That the various territories of the region should make a careful estimate of the possible value of their tourist trade . . .'

These two points establish Huxley's belief that conservation of wildlife is not just a matter of protecting the species but that the other forms of land use most closely associated with it are educational and recreational.

A survey of the nature conservation scene in Britain (Stamp, 1969) also shows that any wildlife conservation operations involve a breadth of application. He also discusses education and recreation within a conservation framework, but he adds research as a topic for discussion. Huxley included research as a fundamental in East Africa. Duffey (1969) has shown that in all European countries, with the exception of Great Britain, the research and management operations of wildlife conservation are separated. If the restriction imposed in this book, namely that conservation should be based on ecological theory, is accepted, then it can be seen that management and research are very closely related, and cannot themselves be separated.

Thus, within the present restriction on the area of application of the principles of conservation there must be a broadening in the utilization of the wildlife and the nature reserves. Although it might be safest for the wildlife if it were to be rigidly preserved, yet a more intensive form of land use will have to be adopted since there are human pressures on all land areas, pressures that are likely to increase rather than decrease. Conservation of wildlife on nature reserves must find ways of integrating the safety of the species and the environment with utilization, particularly by the recreational user. The likely effects of this user, and the planning of this integration on a sustained management basis form the topic for Chapter 9. However, a form of utilization that can have built into it the benefit of conservation is education. An educational user causes wear on the environment and may take a harvest (e.g. specimens for laboratory examination). However, it is essential that there should be educational use, if only for public opinion to be changed in favour of the reservation of land for wildlife and its protection. The ideas of the educational use of wildlife and nature reserves form a topic for Chapter 8. However, since much of the conservation effort has been associated with the preservation of environments and species in perpetuity, and ascertaining that there will be a sufficient breeding stock of these species, the preservation aspect of conservation is discussed in Chapter 7.

It will be noticed that these ideas on the application of the principles of conservation, on the breadth of the subject, and on the factors to be considered when planning sustained management, have all avoided the currently topical subject of 'pollution'. A review of the subject is given by Mellanby (1967), and there are numerous papers dealing with specific polluting chemicals or with aspects of air, land and water pollution. There is a number of reasons why this subject has been omitted from a detailed discussion, but the greatest is that it is outside the immediate scope of ecology. In order to understand the implications of pollution one must investigate the toxicological properties of the pollutant, and the physiological responses of all the organisms subjected to the pollution. The ecological ideas only enter at this point, since the effects of pollution within ecosystems may have to be rectified, if this is possible, by the ecologist and conservationist. However, in this book conservation is treated as the application of ecological theory. If pollution were to be discussed then it would be logical to treat it as the application of toxicological and physiological theory, and these theoretical bases would be too wide for a book that deals specifically with conservation.

However, one example is quoted to show the influence of pollution within a conservation framework. The introduction of DDT and the organochlorine pesticides after the Second World War has eliminated insect-borne diseases in many parts of the world. The fight against mosquitoes and tsetse flies has been spectacular in reducing human mortality in many parts of the tropics. Yet there have been some unwanted side effects, and the spatial distribution of the substances has been extremely wide. They have been recorded in Antarctic snows (Peterle, 1969) and Wurster (1969) believes that these chemicals are amongst the world's most widely distributed synthetic substances. In Britain, Tarrant and Tatton (1968) show that organochlorine pesticides are present in all samples of rainwater that they analysed, but they are more concentrated in rural areas due to their use in agriculture (in sheep dips, etc.). Although no effects of their use were noticed at first, yet in time the carnivorous organisms showed a decreased fecundity and an increased mortality. Analysis showed that the pesticides were not being metabolized to harmless compounds, and hence they were being concentrated at each step in the food chain. The dose in the carnivores and top carnivores was sufficient to be lethal. Lockie (1967) shows that the breeding success of the golden eagle, *Aquila chrysaetos*, declined from about 1960. Before this date 79 per cent of pairs nesting in the west of Scotland reared young, but after 1960 the number had fallen to 29 per cent. Many eggs were broken, probably by the parents as a response to infertility, and analysis of these eggs showed large quantities of organochlorine pesticides. Lockie also reports a similar decline in breeding success of the peregrine, *Falco peregrinus*. There have been similar reports regarding birds of prey in many parts of the world (Peakall, 1970).

This example only quotes one kind of pollution, that of a non-biodegradable chemical, but it does demonstrate three of the principles of the subject of pollution. First, many of the conservation problems are the result of accidents. Thus, the Rhine fish disaster in 1969 (Anon, 1969) was the result of the seepage of the insecticide 'endosufan' from a barge into the river water. The menace of oil pollution round the coasts is usually the result of an accident at sea, as with the *Torrey Canyon* (Smith, 1968). Secondly, chemicals causing pollution in one case may be a direct benefit in another case. In the insecticide example although the application of these chemicals has resulted in damage to the predatory organisms, it has resulted in the saving of a large number of human lives and in an increased production from some of the domestic livestock that form the basis of food and textile industries. Is the use of these insecticides therefore to be judged as definitely harmful? Thirdly, pollution is essentially only controllable by legislation since the effects of pollution are felt beyond the area of origin. If legislation is to be passed then there must be both a public request for action and the scientific advice on which to establish any necessary controls. The bans and partial bans that have been enforced on some chlorinated hydrocarbon insecticides are evidence that such a feed-back mechanism can be made to work, and it must be a part of conservation education to show that there must be a control on the waste products and chemical contaminants in the environment.

Pollution is thus peripheral to the main ideas of conservation or biological management. It has become one of the factors acting on the environment, the control of which is usually beyond the immediate reach of the manager. A long-term aim of conservationists will, however, be to educate the general public to accept that accidents with toxic chemicals must become rarer by more stringent safety measures, and that the general exposure of wildlife and man to these substances should be reduced by finding safer substances or by more efficient disposal of the present range of substances. Biological conservation is itself concerned with finding how to organize the relation between man and the environment, on the basis of the sustained production of both, given the set of influences that act and will act on both sides of this relation.

7. Conservation and Preservation

Introduction

From the definitions of the previous chapter it can be seen that preservation is a central theme of conservation, since any form of management on a sustained basis requires a nucleus of breeding stock, be this of either plants or animals. In Chapter 1 the genetical implications were also considered, and hence this breeding nucleus must be large enough not only to replace any mortality (natural or a harvest) but also to act as a gene bank for the species. In Chapter 5 there was a discussion of large and small weather effects, and the theory showed that a buffer was necessary, this being a portion left behind in the population in case there was a combination of adverse climatic conditions. Conservation is therefore centrally concerned with ways of preserving species, in terms of both numbers and genes and also in terms of a form of insurance against the worst possible conditions for the species survival.

The methods of achieving these conservation aims are diverse, but they can be broadly grouped into field and non-field research and projects. In the former category we shall be concerned with the preservation of one or more species *in situ*, and of the management of the habitat so as to ensure the survival of those species. In the latter category the work of zoological and botanical gardens will be considered. The distinction between 'field' and 'non-field' can at times be false, since supplementation of field populations by laboratory or cage reared stock may be undertaken. An example of this sort of management concerns the large copper butterfly, *Lycaena dispar batavus* in Britain. The example, which will be discussed more fully later in this chapter, clearly shows the dual role of 'field' and 'non-field' operations in attempting to preserve a species in a natural habitat. It is obviously beneficial for a species to be able to survive with a minimum of management intervention, but the laboratory or the cage has the advantage over the field in that there is a rigorously controlled environment and predation and parasitism can be eliminated or reduced to a minimum.

Besides this dichotomy as to the place where conservation operations are carried out, there is another dichotomy between the organizations responsible for the conservation management. In many examples quoted in this chapter the person or persons responsible for the management are trained biologists or ecologists. However, there is a strong voluntary movement of enthusiastic amateurs in Britain (see Chapter 8), and similar enthusiasm is seen in many other countries (Duffey, 1969). The studies and operations undertaken by these people are often vital to the well-being of some of the rarest species, and the intensity of amateur concern has on many occasions led to official concern and the injection of national or local finance and trained expertise into a conservation project. Two examples of amateur work, the first relating to one of Britain's rarest orchids and a second relating to a bird of prey, will be discussed at the end of this chapter. The very success of these operations lies in the ability of the amateur organization to spend a lot of time in wardening or carrying out tedious operations.

In the literature there are many examples of the conservation management of particular species of animals, but there are far fewer examples relating to the preservation of plant species. The aim of this chapter is not to attempt to catalogue the preserved species, but to take a series of examples that draw the distinction between scientific research-based operations and the success of amateur studies, and between the field-based operation and the operation centred on the establishment of populations away from the field. A full summary of conservation operations around the world is given in *The Red Book* of wildlife in danger (Fisher, Simon and Vincent, 1969).

Research and management in the field

Conservation management for plants

Conservation is often associated with the management of plants that are either rare or are very uncommon in a particular part of their range. It is thus of particular importance to understand something of the history of the species, and to understand the particular sequence of events that have led to the scarcity of the species at the present time. Besides the historical phase there are two other phases of study that have to be undertaken. These three phases can best be listed by asking the kind of questions that a manager will need to ask.

(1) Why is the species so rare today? What is known of the species historically, and what trends have there been in the species' abundance during the last 200 to 300 years?

(2) What are the environmental conditions required for the optimal growth of the species? Are these conditions available at the present time, or how can management establish optimal or sub-optimal conditions?

(3) What are the means of dispersal of the species? Does this species spread vegetatively, or does it set fertile seed? What are the characteristics of pollination – is it self sterile or can it be self-pollinated readily? What conditions are required for the germination of the seed?

It is obvious that, to some extent, these three questions overlap, but in them is the information that is needed in planning a management strategy for the plant species. An example of the application of these principles is in the management of the pasque flower, *Pulsatilla vulgaris* (Plate 5), in England by the Nature Conservancy.

A general description of the biology of *P.vulgaris* is given by Wells and Barling (1971), who show that the species is almost completely confined to steep south and south-west facing slopes where insolation is high, soils are shallow and competition from other species is restricted. The British distribution (Fig. 7.1) remains, however, a phytogeographical problem. The species is confined to soils derived from chalk and oolitic and magnesian limestone, but it is absent from the North and South Downs, from Hampshire, Dorset and Wiltshire, and from the Yorkshire Wolds. All these habitats have a similar climate to those habitats where the species occurs today. Wells and Barling discount an earlier hypothesis of Perring's that humidity may be an important factor in determining the species distribution, but they do suggest that the history of the chalk grasslands is more likely to be the factor determining the geographical distribution of the species in Britain.

Historical investigations have been carried out by Wells (1967, 1968). He records the extinction of the species from 39 sites, the dates of the extinctions being recorded in Table 7.1. From these studies Wells concludes that the most important contributory factor in the extinction has been the destruction of the habitat, particularly by ploughing following Parliamentary enclosure. Jones (1969) disagreed with the last statement, claiming that although ploughing was responsible for the disappearance of *P.vulgaris* from 64 per cent of Wells's sites, there was no evidence to relate this to enclosure. Wells (1969a), however, shows that the last

TABLE 7.1 The extinction of *Pulsatilla vulgaris* from sites in England. (Source: Wells, 1968)

Date	*Before 1800*	*1800–1825*	*1825–1850*	*1850–1875*	*1875–1900*	*Unknown date before 1900*	*Post-1900*	*Total*
Number of recorded extinctions	10	3	9	7	4	4	2	39

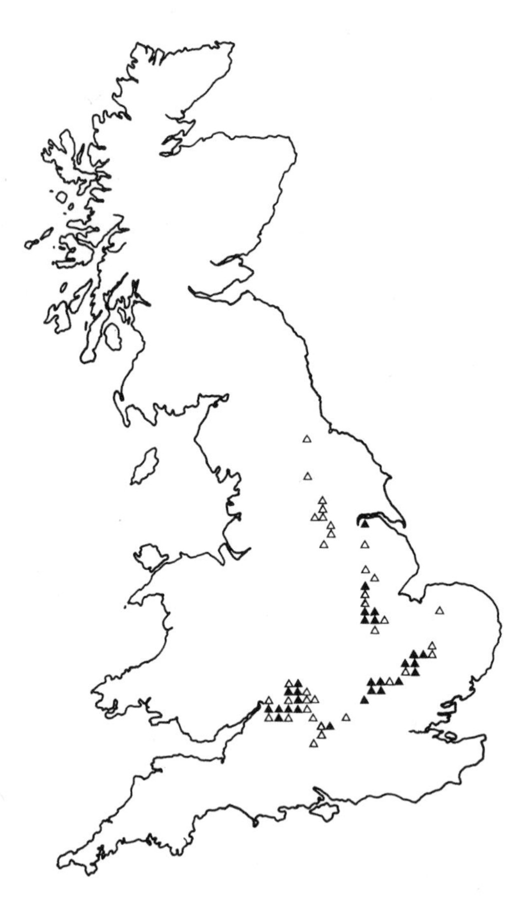

Fig. 7.1 The distribution of *Pulsatilla vulgaris* in Britain. The closed triangles relate to records made in 1930 or later years, and open triangles relate to records before 1930. (Source: Wells, 1967)

record of *P.vulgaris* from a site on which it has become extinct usually post-dated the enclosure of the habitat.

The first of the questions asked about plant species can now be answered. During the last 200 years there has been an underlying trend of reduction in the number of habitats occupied by *P.vulgaris*. Fig. 7.1 shows that the species no longer occurs in Yorkshire (the last record for the County was 1879). The loss of habitats has meant that the northern limit to this species distribution, originally in Yorkshire, has been pushed substantially further south into Lincolnshire. This contraction in range has resulted from the destruction of habitats, either by ploughing or by the preparation of land for ploughing. Other contributory factors have been the destruction of habitats for quarries and building land.

The second question related to the optimum conditions for the growth of the species. As a research project this is extremely difficult to determine, since the present distribution, so dependent upon the land use history, may not reflect the growth of *P.vulgaris* in its optimal habitats. However, the site conditions relating to the present distribution of the species in England have been assessed by Wells and Barling (1971). In these habitats *P.vulgaris* requires an adequate supply of calcium ions and would appear to be a calcicole species. On all of the sites investigated, except for one, there was a pH between 7·1 and 7·8. The environmental conditions also affected the growth and flowering performance of *P.vulgaris*. Data from an enclosed trial for the period 1964–1969 is shown in Table 7.2, from which it can be seen that enclosure, resulting in a reduction in the grazing pressure, has favoured the spread of the species. *P.vulgaris* became more abundant by the vegetative growth of the old plants, but the number of these plants has subsequently decreased slightly. However, the number of flowering plants and also the proportion of plants in flower have decreased as the height of the grasses has increased. Wells

TABLE 7.2 Percentage of plants of *Pulsatilla vulgaris* flowering in an enclosure in Bedfordshire. The plot was enclosed in May 1963 and has an area of 552 m^2. (Source: Wells, 1967; Wells and Barling, 1971)

Year	*1963*	*1964*	*1965*	*1966*	*1967*	*1968*	*1969*
Number of plants	6	138	469	612	849	760	654
Number of flowers	6	89	291	57	92	52	39
Percentage flowering	100	65	62	9	11	7	6
Height of grasses (cm)	1	7	10	15	20	18	22

(1967) postulates that *P.vulgaris* is able to survive in both grazed and ungrazed grasslands, but that its ability to flower decreases when the height of the sward exceeds about 20 cm. In these tall swards the plant survives in the vegetative state and flowers when the competition is removed. Nothing is known, however, about the length of time that the plants will survive in the vegetative state.

An instance of the beneficial effects of the removal of competition is at Therfield Heath (Wells and Barling, 1971). In its 'natural' state *P.vulgaris* was rare on a hill top on this Heath, but now the species is abundant in an area that is intensively used by the public for recreation. The heavy trampling (see Chapter 9) stimulates the development of deep-seated adventitious root buds which vegetatively produce small rosettes near the parent plants.

The third of the questions was concerned with the production of seed and the spread of the species. In discussing the optimum environment for the species it has been seen that vegetative reproduction is a frequent feature of this species. However, seed is freely set in wild populations of *P.vulgaris* in Britain, and Wells and Barling (1971) record a germination of 92 per cent. However, seedlings of *P.vulgaris* are only rarely found amongst wild populations. Experimental planting of seed has shown that the growth of seedlings is very slow. Two-year old seedlings had a rosette of only 0·5 cm, and some seedlings of this age had only two or three leaves, each 0·2 cm long. Under conditions of competition from other species it is likely that seedlings with such a slow growth rate would have little chance of survival.

Thus it can be seen that *P.vulgaris* is a species that is unable to compete with the taller plant species, its own leaves being only 2 to 9 cm long. The scarcity of *P.vulgaris* at the present time in Britain is due to the land use history, and the species is known to have relatively specific habitat requirements. Young plants are rare, and most new plants that are seen are due to the vegetative reproduction of old plants. However, conservation management will need to balance the interest of this species with the interests of the taller species of chalk and limestone habitats. The techniques of chalk grassland management are discussed by Wells (1969b), where it can be seen that grazing, or artificial grazing by mowing, are the most useful management tools. The timing of the mowing operations are important, but in general this is undertaken once per year at the time when the dominant grasses are growing most rapidly.

Conservation management for animals

In discussing rare plants three sets of questions were posed, dealing with the history, environment and reproduction of the species. Similarly, these three sets of questions will have to be asked about an animal population. There will be, however, one difference. Plants, except for the rust fungi, are confined to one habitat and conservation management is directed towards the conditions of this habitat. Animals are mobile, and can move from habitat to habitat randomly, or there can be a directional movement from habitat to habitat, a migration. Hence the questions that must be answered about a rare animal species are:

(1) Why is the species so rare today? What is known of the species historically, and what trends have there been in the species abundance during the last 200 or 300 years?

(2) What is the biology of the species? Is it confined to one ecosystem, or does it utilize several adjacent ecosystems for food, shelter, etc.? Is there any migration?

(3) What are the environmental conditions most suited to large populations of the species? Are these conditions available at the present time, or how can management establish optimal or sub-optimal conditions?

(4) What are the means of dispersal of the species, and does it use them? What are the breeding characteristics of the species and what is the intrinsic rate of natural increase? What factors limit the increase in the population size?

Two examples of animal species will be considered. First, the polar bear, *Ursus* (*Thalarctos*) *maritimus*, will be used as an example of a species whose numbers are declining. Secondly, the trumpeter swan, *Olor buccinator*, will be used as an example of a species that was almost extinct but which is now sufficiently numerous for there to be little or no concern for its survival.

The decline in number of polar bears started in the seventeenth century (Fisher, Simon and Vincent, 1969) when shipping entered the Arctic waters and there was vigorous hunting. During the eighteenth and nineteenth centuries the species was hunted and heavily exploited in Spitsbergen, Hudson Bay, Novaya Zemlya and any other Arctic areas that were accessible. Towards the end of the nineteenth century there was a transfer of ships from whaling to sealing, resulting in even further pressures on the polar bear particularly in the Canadian Eastern Arctic.

The development of the fur trade has resulted in still further exploitation of polar bears. The present situation is that about 6000 to 7000 of these animals live in the Canadian Arctic, and this accounts for probably more than half of the world population. Polar bears also live in Alaska (U.S.A.), Greenland (Denmark), Norway and the Soviet Union, although, being a marine animal for much of the year, many of the bears occur outside territorial boundaries. Historically, then, the species has declined in numbers, and fears have been expressed that the species may become extinct if steps are not taken for its preservation.

The biology of the species is well known. The polar bear spends most of the year away from land during which time it is carnivorous, mainly eating seals. During the summer break-up of the ice the species comes to land where there is relatively little food for it. On land polar bears are omnivorous and carrion feeders. In the autumn the bears, except for pregnant females, return to the ice and sea. Mating takes place in April, denning begins in October and the young are born in early December. The female feeds the small cubs, and she leaves with them for the sea in March or April. The group remains as a family unit until the next spring. Sexual maturity is reached in the females at three years of age, and young females usually only produce one cub. Older females usually produce twin cubs, though three or four cubs are known to have been born at the same time. Since females do not become pregnant again whilst they are tending their young cubs, pregnancy only occurs every third year unless the previous cubs have died. This slow rate of breeding implies that the intrinsic rate of natural increase would be very small, and hence that only a small rate of exploitation could be sustained by this species.

The optimum environment for the species has not been assessed, though as evidence suggests that polar bears are becoming slightly more numerous in the Hudson Bay area (Jonkel, 1970) this environment may be taken to be optimal or sub-optimal. The Canadian authorities have established a reserve of 7000 square miles (18,130 km^2) on the southern margin of Hudson Bay (Fig. 7.2), providing the summer land requirements and denning ground for polar bears. The reserve can be divided into four sections of sub-arctic tundra, an interior zone of almost treeless bog, a transition zone of mixed forest and fen, and a coastal zone of boulder-strewn mud-flats. Although this area is now a Provincial Park, access by the public is restricted so that there is a minimum of disturbance for the polar bears and the other animal species (Brunelle, 1970).

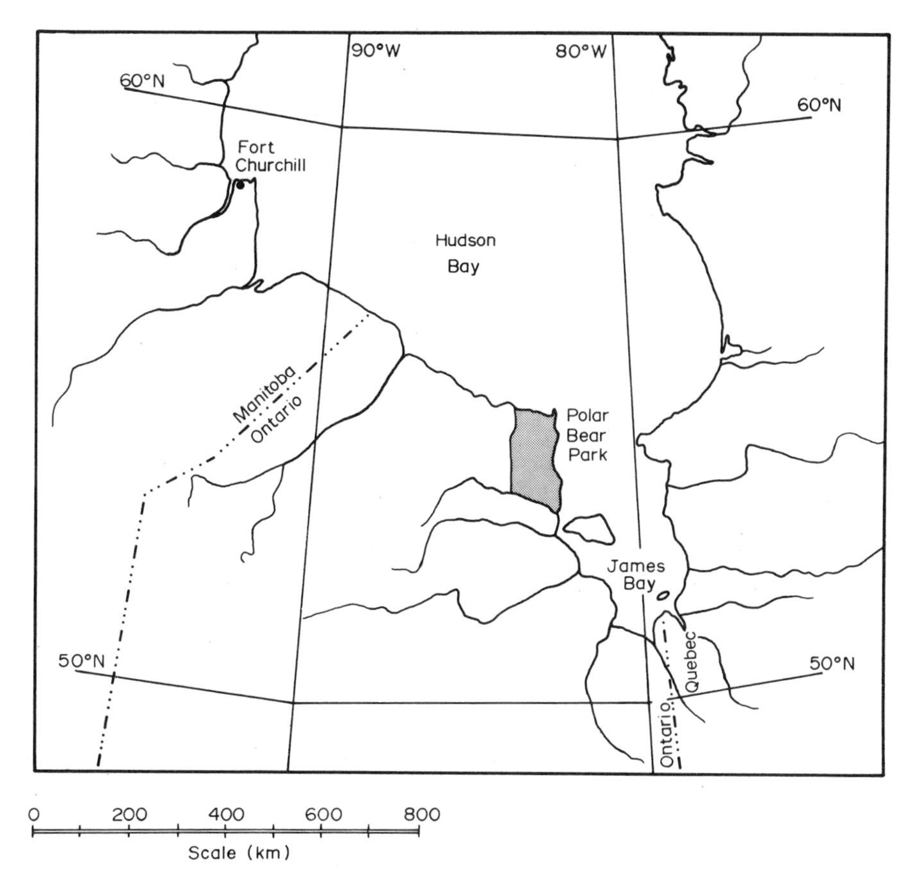

Fig. 7.2 A map of part of Canada showing the locations of the Polar Bear Park and Fort Churchill

Further west along the southern shore of Hudson Bay at Fort Churchill the polar bear is a relatively common animal, feeding on rubbish dumps near to the centres of civilization. Canadian authorities have restricted the killing of the species, there being very few white hunters, but the Eskimos, further to the north, are able to take a catch. A few polar bears are killed legally in the Churchill area (those animals that are considered to be dangerous to man), and about 20 to 40 per year are killed illegally or quasi-legally (Jonkel, 1970).

From this summary some of the features of over-harvesting (Chapter 5) can be detected. The ability of the species to replace harvested animals is very limited, since during the most productive years of life a female will only produce two cubs every three years. Assuming that the sex ratio at birth is 1♂ : 1♀ it can be seen that this capacity to replace harvested individuals is only slightly greater than that of the blue whale. However, Jonkel (1970) states that there is an inadequate research background to polar bear studies. Although survival curves and data on the size structure of the blue whale populations are known (see Chapter 5), there is no similar data available for the polar bear. Conservation management is therefore based on empirical ideas about the populations of this species, and at the present time management generally aims at restricting the level of exploitation. Thus in Canada there are restrictions on the hunting of the species by non-Eskimos, and restrictions on the equipment that can be used for approaching, hunting and trapping the animals. The Soviet Union has completely banned the trapping and killing of polar bears.

It seems therefore that as research yields more information on the population dynamics of the species a form of international control on the harvest must be worked out, determining maximum allowable quotas that can be taken each year. This is important not only to the long term survival of the polar bear, but also to the economy of the Arctic peoples who rely on the native animal species for food, dog food and fur.

The second example of the conservation of an animal species is the trumpeter swan, the largest of the water birds of North America. Unlike the polar bear that is decreasing in numbers, this bird is currently increasing after a long period of decrease. In 1932 there were an estimated 69 birds in Montana, Idaho and Wyoming (Anon. 1970b). A few other birds also occurred in Alaska and the neighbouring Provinces of Canada. During an aerial survey in 1968 a total of 3641 birds were counted in the United States, and it is estimated that the population was between 4000 and 5000 birds with an unknown number north of the boundary in Canada.

The trumpeter swan has been used as food by humans for thousands of years (Fisher, Simon and Vincent, 1969), but in the mid-nineteenth century the species was heavily exploited due to the demands of hunting and the trade in feathers and swan-skin. In 1935 the Red Rock Lakes National Wildlife Refuge was established in Montana, and this refuge has played a major role in the conservation success since about half of the population of these swans outside Canada and Alaska have nested here. By 1938 the total American population was 98 birds, and this rose to 207 birds in 1944, 417 birds in 1951 and 560 birds in 1954. During the 1950s and 1960s there was an average of 476 adult swans. In these breeding areas hunting and shooting of the trumpeter swan was prohibited, and a technique that was used to further protect the species was to restrict the hunting of any similar species. Thus, restrictions were placed on the shooting of the snow goose, *Chen hyperborea*, since swans were being accidentally killed when they were mistaken for geese.

Having gained a numerical increase in this small area of the United States, the next management decision was to restore the trumpeter swan

TABLE 7.3 The fecundity of the trumpeter swan as a function of the population density in Red Rock Lakes Refuge (Source: Denson, 1970)

Population density	*Number of cygnets produced per mated pair*
46	1·50
64	1·12
76	1·12
103	1·00
123	0·73

to suitable areas within their former breeding range (Denson, 1970). Ten birds were transported to the nearby National Elk Refuge in Wyoming, and the first young bird was fledged there in 1944. In 1939 birds were transported to fenced ponds in Oregon, but there was no breeding. It is now known that the transplanted birds failed to reproduce due to the crowded conditions of this habitat. The programme of transportation has continued until by 1970 free-flying flocks of trumpeter swans have been established in eight of the States in the U.S.A. However, the birds seem unable to colonize new areas by themselves, since marked birds seem to remain in the refuge where they were hatched or to which they have been transported. Some idea of the population dynamics of these swans has been gained from field studies by Banko (1960). These show that the breeding success of the swans is inversely proportional to their population density (Table 7.3), and this is probably the cause of the stable population size in Montana, Idaho and Wyoming during the 1950s and 1960s. The average population size of 476 adult swans during this period is a measure of the carrying capacity of the environment, and accords with Banko's finding that nesting trumpeter swans had large territorial requirements.

Based on the success of the trumpeter swan operation, Denson (1970) draws up a model for the conservation of 30 other species of wildlife threatened with extinction in the United States. The six stages of his model are:

(1) immediate physical protection from man and from changes in the environment;

(2) educational efforts to awaken the public to the need of protection and so to gain acceptance of protective measures;

(3) life history studies of the species so as to determine their habitat requirements and the causes of their population decline;

(4) dispersion of the stock so as to prevent loss of the species by disease or by a chance event such as fire;

(5) captive breeding of the species so as to assure higher survival of young, to aid research, and to reduce the chances of catastrophic loss;

(6) habitat restoration or rehabilitation when this is necessary before reintroducing the species.

Thus it can be seen that such a model is similar to the four points listed at the start of this section on the conservation management of animals. Denson's model goes further, however, in three respects. First, he advocates immediate action to stop any further exploitation of the species or its environment (point 1). Secondly, he acknowledges the need for work away from the field locality, in zoological gardens or research centres (point 5), a subject that is for discussion later in this chapter. Thirdly, he stresses the need for an educational or propaganda effort so as to establish public support for conservation operations (point 2). This subject is further discussed in the next chapter of this book.

Supplementation of field populations

In the two previous sections we have seen some of the techniques that have been applied to the conservation of wild plants and animals in the field, and have asked the questions that research will have to answer in order to establish a scientific basis for management. Later in this chapter we shall investigate the role of zoological and botanical gardens in preserving rare species. But between these two extremes, management in the field and management away from the field, lie a whole series of management techniques in which field populations are periodically strengthened by the release of captive individuals. The logic of such operations has been mainly developed in relation to species of commercial interest because conservation management has been unable to apply such a strategy in many instances due to the large and continuing cost of the operation. However, before discussing a conservation example, it will be as well to look at the logic of supplementing populations of animals of commercial interest.

Such a logic has been developed by Watt (1968), who asks three questions relating to a species and its management. These are:

(1) Are we dealing with density dependent or density independent regulation of recruitment?

(2) How great is the carrying capacity of the environment?

(3) How long will it take to recapture the animals that have been reared in hatcheries?

Such a series of questions is relevant when we are dealing with the stocking of a river or lake with trout or a woodland with pheasant, since these animals will be harvested. In relation to conservation management of a rare species it is unlikely that the third of these questions will have any meaning, since it is unlikely that any harvest will be taken. However, the first two questions are relevant, and the answers to them cannot be separated one from the other since the carrying capacity of the environment will be related to the regulation of the population.

Watt discusses the small mouth bass, and he shows that there is considerable year to year variation in the size of the year-class, similar to the variation that was seen in the sardine population in Chapter 5. An investigation of the environmental effects showed that 94 per cent of the variance in recruitment of four-year old fish from year to year could be accounted for by the June to October temperature in the year when the fish were spawned. The release of marked bass showed such a low recapture rate that it was evident that the mortality of young, hatchery-reared bass was also affected by the temperature. What, then, is the advantage of supplementing a population where the regulation of recruitment is independent of the population density? It is evident that mortality affects not only the natural population but also the introduced population, though the intensity of effect may vary between the two populations (it is likely, however, that the introduced population will suffer more than the natural population from climatic extremes). If a large part of the mortality is operating at a definite point in the development of the organism, then there is no reason to supplement the natural population at any developmental point prior to the point at which the mortality operates. If, however, the hatchery can grow an animal in a suitable environment to one point beyond the point where mortality factors are operating in the natural population, then a natural population may be successfully supplemented. It is obvious that the supplementation operation is more feasible for species where the operation of the density independent mortality factor is early in the development, and it is especially useful when the hatching of the eggs is the point of greatest mortality.

Watt also discusses density dependent regulation and uses as an example some ponds stocked with fish. Two similar ponds were stocked in May, one with 37 oz. of bluegill fry per acre and the other with 180lb of year-old bluegill per acre. Analyses indicated that the food availability throughout the year was similar in each of the ponds. When the ponds were drained in November the first pond had 105lb of fish per acre and the second pond had 92lb per acre. Evidently the ponds had a carrying capacity of about 100 ± 10lb per acre, and the population was therefore regulated by the biomass density of the fish in the pond. In these experiments it is also clear that two processes are at work – one regulates the numerical population density and the other regulates the growth of individual animals in the population. These two compensatory mechanisms are also compensatory with respect to each other, since a pond can either have a few large fish or many small fish, either adding up to approximately the same biomass. What, then, can be derived from these experiments in relation to supplementation? If the natural population density is less than the carrying capacity of the environment, then it is possible that the introduced animals will contribute to an increase in the population density to approach the environment's carrying capacity. If the natural population density is already close to or exceeding the carrying capacity of the environment, then supplementation will have no effect other than to either increase the natural mortality or to decrease the size of individual animals in the population.

It is thus clear that, when supplementation is to be considered as a means of conservation management, the regulation of the population must be considered. If the regulation is found to be density independent, then one must ask if a large part of that mortality operates sufficiently early in the life of the organism for the laboratory or hatchery reared animal to be released after the operation of this mortality factor. If the regulation is found to be density dependent, then one must ask if the natural population density is significantly less than the carrying capacity of the environment.

If the answer to either of these questions is 'yes' then it may be appropriate to consider a management strategy based on supplementation. If the answer is 'no' then it is more than likely that supplementation will have no use as a management tool. An example of management planning with supplementation concerns the large copper butterfly, *Lycaena dispar batavus* (Plate 6), at Woodwalton Fen National Nature Reserve in Britain.

A history of this species in Britain is given by Frohawk (1924) and Ford (1945). The British subspecies, *L.d.dispar*, became extinct in the middle of the nineteenth century, with the last recorded capture being of five specimens in Holme Fen in 1847 or 1848, although Duffey (1968) has found a record of the subspecies being taken in Bottisham Fen in 1851. Frohawk concludes that 'there is but little doubt that the extinction of this beautiful insect was brought about by the draining and burning of the fens when they were reclaimed', though Ford says 'the cupidity of collectors does, however, appear to have been responsible for the extinction of one of our butterflies, and this the finest of them all, the large copper'. It seems likely that a combination of both of these factors, over-collecting and a reduction in the area of the habitat, were responsible for the extinction of *L.d.dispar*.

In Britain the species is single brooded, flying in July and August (Frohawk, 1924). The eggs are laid on both surfaces of the leaves of the great water dock, *Rumex hydrolapathum*. The larvae feed on the under surface of the dock leaves, and hibernate in their second instar in the folds of damp, dead leaves. Following hibernation the larvae once again feed on the dock, and they pupate on a stem or under a leaf. The feeding larvae are green and the pupae are brown in colour, darkening prior to emergence to show the colouring of the adult.

Ford (1945) documents the attempts to reintroduce *L.dispar* into Britain. The first of these attempts was in 1909 when the Continental subspecies *L.d.rutilus* was released in Wicken Fen, Cambridgeshire. Because the food plant was uncommon in this locality the attempt met with little success and the subspecies did not survive. A few years later, in 1913 and 1914, the same subspecies was released in a snipe-bog in Southern Ireland. The habitat had been prepared and planted with the greater water dock, and the introduction was successful with the subspecies surviving till 1936. *L.d.rutilus* was introduced from Ireland into Norfolk in 1926, but it only persisted for two years. It is said (Duffey,

1968) that this introduction failed because the food plant grew along the fen drains and not in the open marsh where the insects prefer to lay their eggs.

In 1915 a new subspecies, *L.d.batavus*, was discovered in the Province of Friesland, Holland. This subspecies closely resembled the original *L.d.dispar* and was confined to a few fens in Friesland. It now represented the most western form of *L.dispar*, a position previously held by the British subspecies. Although *L.d.batavus* was scarce, a stock of 25 males and 13 females were obtained and liberated in Woodwalton Fen in 1927 in an area that had previously been cleared of bushes and planted with great water dock. Despite prolonged flooding during the 1927–28 winter the survival of the hibernating larvae was good and over 1000 adults were said to have flown in the summer of 1928. Since that time the Woodwalton colony has been maintained, though on occasions the numbers have fallen to such an extent that supplementation of the stock was essential. It is this colony, then, that can be used as an example of the supplementation of a field population as a part of conservation management.

Population counts of the Woodwalton colony have been kept since the Nature Conservancy took possession of the reserve in 1956. In 1960 additional clearings had been made, great water dock planted, and a

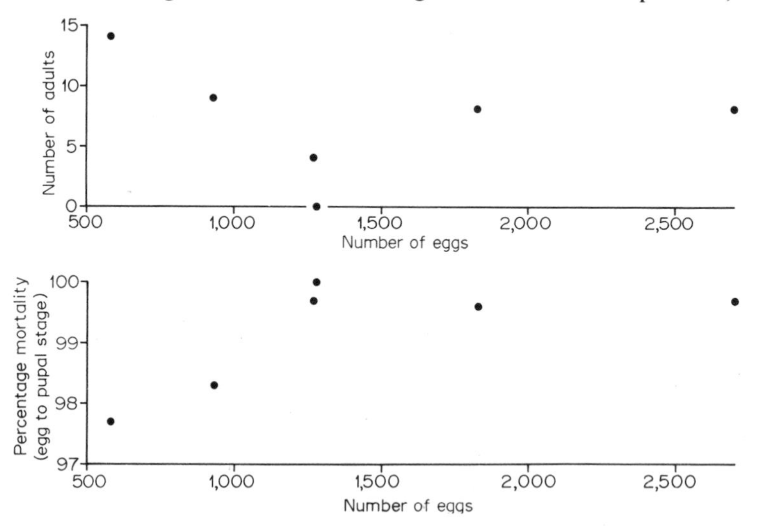

Fig. 7.3 Data for the large copper butterfly, *Lycaena dispar batavus*. (a). The relation between the number of eggs in one year and the number of adults produced from those eggs in the next year. (b). The percentage mortality from the egg to the pupal stage as a function of egg density. (Source: Duffey, 1968)

colony known as the 'wild colony' was established with 13 females and 11 males from another part of the fen. The history of this colony is recorded by Duffey (1968) who shows that, although it theoretically became extinct in 1966, it is being maintained by individuals from other parts of the fen. Fig. 7.3(*a*) shows that there is no relation between the number of adults and the number of eggs from which these adults were derived. However, in Fig. 7.3(*b*) it would seem that the percentage mortality from egg to adult is less when the density of eggs is small, though again there is no clear relation between survival and density. The scatter in these diagrams would tend to indicate that the population is not being regulated by density dependent mechanisms, and that some independent mechanism is operating.

Duffey also considers the stage of development at which mortality is occurring, and the details of the analysis are given in Table 7.4. The data can be used for a *k*-factor analysis, a technique that is described by Williamson (1972). Thus, if the number of eggs laid in 1960 is denoted by N_e and the number of spring larvae in 1961 by N_s, the proportionate loss during the first instar and during hibernation (winter 1960–61) can be measured by

$$k_1 = \log N_e - \log N_s$$

Similar expressions give the shortfall in egg production (k_0) and the loss from spring emergence to pupation (k_2). Summation of k_0, k_1 and k_2 or the subtraction of the logarithm of the number of pupae from the logarithm of

TABLE 7.4 Counts of eggs, spring larvae and pupae (assumed to be equal to the number of adults) in the 'wild colony' of large copper butterflies. The *k* and *K* values are explained in the text. (Modified from Duffey, 1968)

	Dates					
	1960–61	*1961–62*	*1962–63*	*1963–64*	*1964–65*	*1965–66*
Maximum potential number of eggs (number of females ×300)	19,800	7,500	17,400	32,100	13,200	51,600
Eggs laid	579	934	1,272	1,833	2,701	1,278
Spring larvae	37	94	44	32	85	19
Pupae	14	9	8	8	8	0
k_0	1·534	0·905	1·136	1·244	0·689	1·606
k_1	1·195	0·997	1·462	1·758	1·503	1·828
k_2	0·422	1·019	1·041	0·602	1·026	–
K	3·151	2·921	3·639	3·604	3·218	–

the maximum potential number of eggs, gives the total generation loss, denoted by K. The values derived in Table 7.4 are plotted in Fig. 7.4. Since not one of the graphs of k_0, k_1 or k_2 is appreciably nearer to the graph of K than the others, it can be concluded that the mortality factors do not operate at any specific developmental stage. If there had been a finer division of the developmental stages, say into larval instars, then more information might have been gained. However, it appears from Fig. 7.4 that k_1 is most closely correlated with K, and that k_0 and k_2 are

Fig. 7.4 An analysis of the mortality factors acting upon populations of *Lycaena dispar batavus* in Woodwalton Fen. The derivation of the four lines is discussed in the text, and the data are given in Table 7.4. (Source: Duffey, 1968). The lines represent:

——————— K
— — — — — k_0
— · · · — · · · k_1
- - - - - - - - - - k_2

compensatory with respect to each other. The values for K and k_2 cannot be calculated for the 1965–1966 season since the population became extinct, and the logarithm of zero (no pupae) is negative and infinite.

Thus, for the supplementation operations, it does not appear as if mortality is affecting any one of these developmental stages more than any other stage. Duffey does not give an answer as to why there is a shortfall in eggs (k_0), though he does relate egg laying to air temperature. It is unlikely that any caging operation could significantly increase the number of eggs laid per female, and hence supplementation cannot overcome the shortfall in eggs. Duffey also considers mortality during the egg, first larval feeding and hibernation phases (k_1), but the causes of mortality at this stage are uncertain. Experiments with flooded and unflooded plants have failed to demonstrate if flooding inhibits hibernation and thus increases the mortality. Until more of the causes of mortality at this stage are known it is unlikely that supplementation will be a useful management technique. However, the mortality in the post-hibernation and pre-pupal stage (k_2) is understood. A number of larvae are parasitized by the fly *Phryxe vulgaris*, and others are eaten by birds and small mammals. Hence, supplementation of the wild population has been concentrated on collecting larvae as soon as they emerge from hibernation. These larvae are kept in muslin cages where they are protected from predators and parasites (experimental work has indicated that six times as many larvae are killed by vertebrate predators as by parasites and invertebrate predators), and on emergence the adult butterflies are released. Mean values for survival of populations under various types of caging are shown in Table 7.5.

TABLE 7.5 The survival of post-hibernation larvae to adults in caged and uncaged conditions. (Source: Duffey, 1968)

Experiment	*Percentage survival*
Control (no cage)	4
Cage of 6 mm plastic mesh (this excluded vertebrate predators but observation indicated that it did not exclude invertebrate predators or parasites)	71
Muslin cage (this excluded both vertebrate and invertebrate predators and also parasites)	83

Supplementation of a wild population in this example is unusual in that it is not concerned with rearing eggs to some stage when the young can be released. Rather, the mortality factors at one stage (k_2) have been identified, and supplementation takes the form of protecting the species against these mortality factors. However, the example does demonstrate the principles behind any supplementation operation – first identify the method of population regulation and then quantify and identify the mortality factors acting on that population.

Planning the management of a whole environment

In the examples that we have been considering the aim of the conservation management has been directed at a single species. However, it is often the case that several rare species, both of plants and animals, occur in the same locality and hence conservation management will need to consider the whole environment. The questions asked of individual species will also have to be asked in respect of each species in this environment, but wider planning is also needed. An example will be drawn from a nature reserve managed by the Yorkshire Naturalists' Trust.

Wharram Quarry is a nature reserve of 6 ha situated in the Yorkshire Wolds, about 200 m above sea level. Quarrying operation began in 1919 and ceased in 1930, although operations were recommenced on a small scale between 1938 and the mid-1940s. The material removed was chalk of the Upper Cretaceous period, and because of its high calcium carbonate content it was largely used for making cement and as a source of lime for agricultural purposes. A plan of the nature reserve is shown in Fig. 7.5.

When the quarry was finally abandoned it had a flat floor that was being colonized by calcicolous plants. The diversity of these species indicated that the site would have scientific interest, whilst the presence of rare species, particularly the bee orchid, *Ophrys apifera*, and the

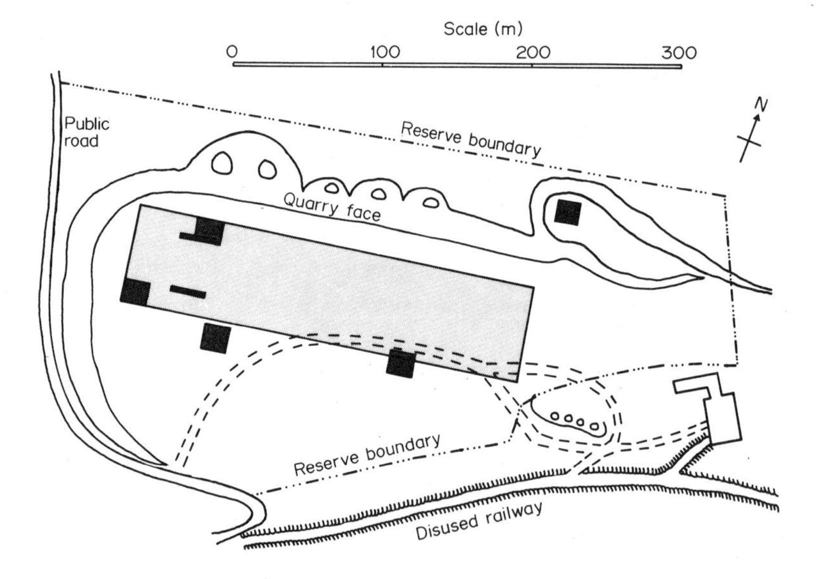

Fig. 7.5 A plan of the Wharram Quarry reserve. The large oblong pale shaded area on the quarry floor was used in the classificatory studies, and the five darkly shaded areas were used in soil and environmental analyses. Pattern analyses were carried out on the two transects, coloured black, at the western end of the pale shaded area

clustered bellflower, *Campanula glomerata*, made the vegetation in the quarry worth protecting. Nature reserve status was declared in July 1966.

Many amateur naturalists considered that by the time the nature reserve was declared the quarry 'wasn't as good as it had been', since visual estimates of the bee orchid indicated that this species had become scarcer. Since the quarry had been successively abandoned since about 1930, various stages in the succession from bare chalk towards some form of edaphic climax community could be expected to be present. No scientific studies or survey work had been carried out prior to the declaration of a nature reserve, and hence management decisions could not be based on definitive data about the site. It was thus decided that management of the reserve should instigate long-term studies, but that the initial management decisions should be based on short-term or on-the-spot studies on the reserve. In this way a first approximation to the dynamic ecological processes of the reserve could be determined, whilst in the years to come, as more accurate data became available, the management prescriptions could be modified. Such an approach has one obvious disadvantage. These initial management decisions might affect the course of development of the reserve in such a way that the long-term studies did not represent the natural ecological processes. It was, however, felt better to take this risk and plan on the basis of short-term studies rather than carry out no management practices until the long-term studies yielded reliable data.

The first short-term study was a survey of the species present on the reserve. This survey gave a basis for further studies, since it indicated the approximate abundance of many of the species and drew attention to those species that are particularly scarce and in need of protection. Such species lists are also useful for drawing comparisons between sites. The survey at Wharram Quarry listed 135 species of higher plants, of which the species of most interest for protection was the bee orchid. The list of vegetation was typical of Wold habitats, with the grasses *Dactylis glomerata* and *Arrhenatherum elatius*, or less frequently *Brachypodium pinnatum*, being dominant where there was a closed cover of vegetation.

Having carried out such a survey a complex question needs to be asked – what is the succession from a chalk quarry floor towards an edaphic climax? Is there just one edaphic climax for the whole of the reserve, or are there several? Intuitively, one could consider that the succession would be:

$$\text{Bare chalk} \rightarrow \text{Pioneer plants} \rightarrow \text{Chalk grassland} \rightarrow \text{Scrub} \rightarrow \text{Woodland}$$

It was obvious that the grassland stage of such a succession was associated with dominance by *D.glomerata* and *A.elatius*, and in some areas of grassland there were small scattered bushes of hawthorn, *Crateagus monogyna*. The speed of succession was probably dependent upon the size of the rabbit population, since the small, flat areas on the quarry face (see northern boundary in Fig. 7.5), besides having scrub, also had small ash, *Fraxinus excelsior*, and crab apple, *Malus sylvestris*, trees. Rabbits were naturally excluded from these quarry face areas since the ground was too steep for more than the occasional rabbit to ascend or descend. Trees are absent from the quarry floor.

TABLE 7.6 A list of species recorded on the area of Wharram Quarry used for a numerical classification. The maximum number of occurrences is 495. Nomenclature is after Clapham, Tutin and Warburg, 1952

Species	*Number of occurrences*	*Percentage frequency*
Ononis repens L.	94	19·0
Medicago lupulina L.	181	36·6
Lotus corniculatus L.	466	94·1
Crataegus monogyna Jacq.	51	10·3
Chamaenerion angustifolium (L.) Scop.	125	25·3
Heracleum sphondylium L.	293	59·2
Gentianella amarella (L.) H.Sm.	266	53·7
Euphrasia nemorosa (Pers.) H.Mart	68	13·7
Thymus drucei Ronn.	78	15·8
Prunella vulgaris L.	88	17·8
Plantago lanceolata L.	250	50·5
Campanula rotundifolia L.	68	13·7
Galium verum L.	31	6·3
Tussilago farfara L.	95	19·2
Achillea millefolium L.	348	70·3
Carlina vulgaris L.	46	9·3
Centaurea nigra L.	135	27·3
Leontodon sp.	435	87·9
Hieracium pilosella L.	397	80·2
Dactylis glomerata L.	430	86·9
Arrhenatherum elatius (L.) J. & C. Presl	386	78·0
Holcus lanatus L.	311	62·8

In order to investigate this problem of succession and plant communities an objective method of classification, essentially that described by Williams and Lambert (1959) (see Chapter 4) was employed. The data were collected from a rectangular area comprised of 495 regularly spaced 1 m square quadrats forming nine rows each of 55 quadrats. Quadrats were spaced such that the centres of adjacent quadrats were 5 m apart. The approximate location of the sampling area is shown in Fig. 7.5, and a list of species with a frequency of more than 5 per cent is given in Table 7.6. No species had frequencies greater than 95 per cent. Full details of the collection and analysis of the data are given by Bloomfield (1971).

The first analysis to be performed followed exactly the procedure outlined by Williams and Lambert (1959). The χ^2 value with Yate's correction was calculated for every pair of species, and for each species the significant values of χ^2 (i.e. $\chi^2 \geqslant 3{\cdot}841$, $p \leqslant 0{\cdot}05$) were summed. The presence or absence of the species with the largest sum was used to divide the population of quadrats. The process of division was continued until there were no significant values of χ^2, or until there were so few quadrats

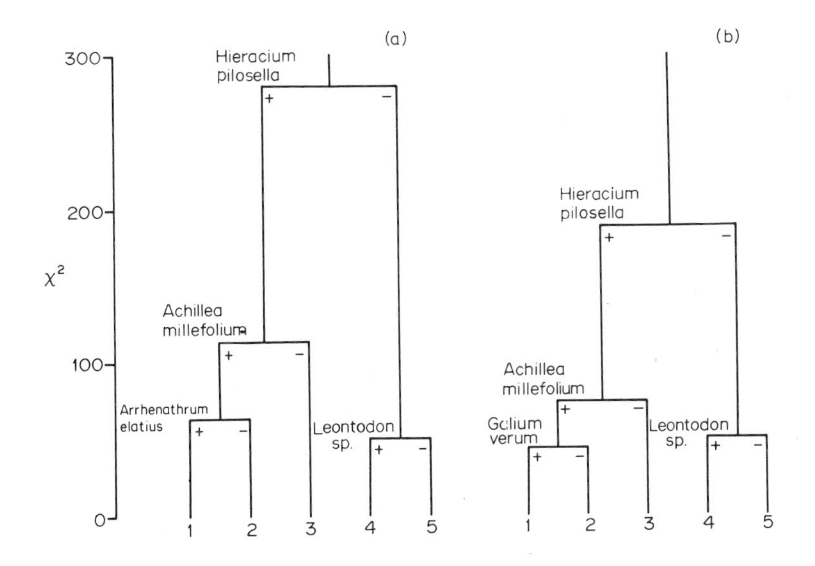

Fig. 7.6 The numerical separation of the vegetation on part of the floor of Wharram Quarry reserve into five groups by two methods (a and b). The methods and the selection of the plant species are described in the text

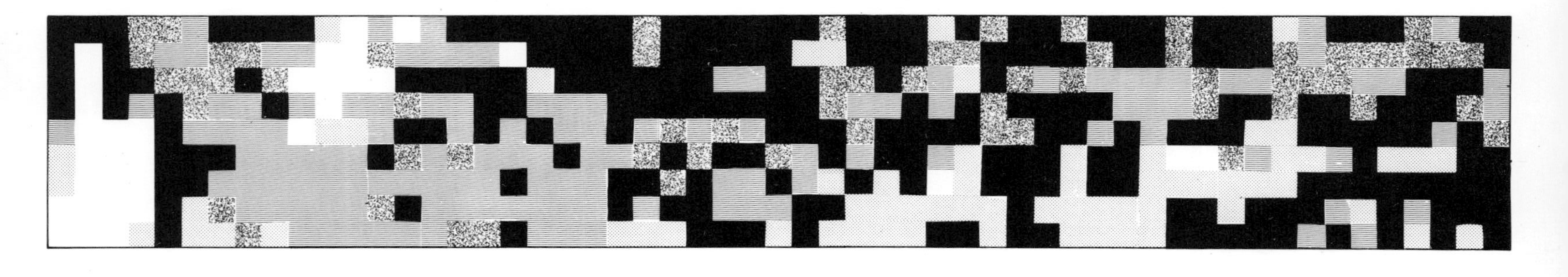

Fig. 7.7 A plan of part of the floor in Wharram Quarry reserve showing the distribution of the five groups of vegetation. The groups are shown in Fig. 7.6 (*a*), and their derivation is discussed in the text

forming a group that the χ^2 criterion could not be calculated. Seven groups were produced. However, if we follow the convention of Williams and Lambert (1959, 1960) and terminate the process of division when the value of the sum of the χ^2s fall below an arbitrary value, then we need only consider the implications of the higher order division. Using an arbitrary summation value of 50, five groups were formed which are shown in dendrogram form in Fig. 7.6(*a*) and as

TABLE 7.7 The frequency of occurrence of species in the five groups distinguished in the association analysis. All figures are given as a percentage, and nomenclature follows that in Table 7.6

Species	*Groups* 1	2	3	4	5
O.repens	20	11	10	42	18
M.lupulina	35	25	52	22	41
L.corniculatus	98	100	94	92	62
C.monogyna	13	5	5	17	9
C.angustifolium	20	19	42	14	29
H.sphondylium	61	46	65	66	38
G.amarella	53	87	67	20	12
E.nemorosa	18	13	11	11	3
T. drucei	15	33	17	5	3
P.vulgaris	22	33	8	14	3
P.lanceolata	34	25	11	67	32
C.rotundifolia	16	16	11	9	12
G. verum	8	0	1	17	6
T.farfara	19	22	28	9	3
A.millefolium	100	100	0	86	53
C.vulgaris	7	14	19	0	0
C.nigra	32	17	14	42	38
Leontodon sp.	95	95	89	100	0
H.pilosella	100	100	100	0	0
D.glomerata	93	78	80	94	71
A.elatius	100	0	66	94	97
H.lanatus	69	49	48	74	79

a map in Fig. 7.7. In order to investigate the nature of these groups the frequencies of plant species in each of the groups is given in Table 7.7.

The χ^2 criterion in the association analysis has been the subject of criticism, and other measures such as χ^2/N, $\sqrt{(\chi^2/N)}$ or the summation of χ^2 irrespective of significance have been advocated (see, for example

Williams and Dale, 1965). Goodall (1953) has argued that negative associations have no particular meaning, since they can be artifacts of the quadrat size. An association analysis summing only the χ^2 values for positive associations had been carried out on the Wharram Quarry data. A dendrogram dividing the sample into five groups is shown in Fig. 7.6(*b*). The frequency of the other species in these groups is similar to that shown in Table 7.7 for the first analysis.

In both of these analyses the first division of the 495 quadrats is based on the presence (397 quadrats) or absence (98 quadrats) of *Hieracium pilosella.* This species has numerous stolons terminating in overwintering rosettes (Clapham, Tutin and Warburg, 1952), a form of life more adapted to growth in open conditions. The first division of the analysis thus indicates the height of the vegetation; those quadrats without *H.pilosella* (groups 4 and 5) being in areas where there is tall grass and a closed vegetation stand.

In both of the analyses the division of quadrats containing *H.pilosella* was according to the presence or absence of *Achillea millefolium.* Looking at the map of these vegetation groups, Fig. 7.7, it becomes clear that those quadrats without *A.millefolium* are distributed in an area where road grit has been dumped, and on the quarry floor where larger pieces of chalk and flints were dumped prior to the end of quarrying operations. This group, group 3 in both analyses, is indicative of very open conditions that are sparsely vegetated. It will be seen from Table 7.7 that this group is associated with the greatest frequency of *Chamaenerion angustifolium, Tussilago farfara, Medicago lupulina* and *Carlina vulgaris.* The first two of these species are commonly found on disturbed sites in Britain.

Those quadrats in which both *H.pilosella* and *A.millefolium* were present were further divided by the presence or absence of either *A.elatius* or *Galium verum.* Although the two analyses gave these different divisions, it is likely that they are very similar. Table 7.7 shows that all of the quadrats with *G.verum* also contained *A.elatius.* It can also be seen that group 1 had a large frequency of occurrence of *D.glomerata,* whilst in group 2 there is the maximum frequency of *Gentianella amarella, Prunella vulgaris* and *Thymus drucei.* These last three species all have brightly coloured flowers that add to the visual attractiveness of the reserve.

Those quadrats without *H.pilosella* were divided in both analyses by the presence or absence of *Leontodon* (both *L. autumnalis* and *L.hispidus* occur on the reserve). Those stands with *Leontodon* (group 4) tend to have a smaller frequency of group 3 species, such as *C.angustifolium* and *M.lupulina*, and they tend to have a larger frequency of taller plants, as for example *Heracleum sphondylium*. The data for the only shrubby species, *C.monogyna*, show that this species is most frequent in group 4.

Can these five groups be considered as communities or as successional stages, or as both? It would seem that they give an idea of the succession, and from the evidence of this analysis a schematic representation of the

TABLE 7.8 A schematic representation of the stages of the succession from bare chalk towards a shrubby community in Wharram Quarry

Group in analysis	*Characteristics*	*Transition process*
3	*Hieracium pilosella* with bare ground	
		Achillea millefolium enters
2	Many flowering plants such as *Gentianella amarella*	
		Arrhenatherum elatius enters
1	Vegetation stand closes and grasses become dominant	
		Hieracium pilosella dies out
5	*Arrhenatherum–Dactylis* grassland. Community becoming taller	
		Shrubs begin to appear
4	A transition type between grassland and a true scrub	

succession is given in Table 7.8. However, to be certain of the actual course of the succession long term studies are needed, but this analysis does give, from measurements at one point of time, an indication of what might be happening over a long time scale.

It is an easy matter to compare the rare flora with the groups listed in Table 7.8. The bee orchid is most frequent in group 2 areas, whilst the clustered bellflower is most frequent in group 1 areas. The protectionist role of the nature reserve would therefore be satisfied if groups 3, 2, 1, and 5 were maintained since no further species are added to the reserve when group 4 becomes dominant, and hawthorn scrub tends to eliminate many of the more interesting (from the conservation point of view) plant species.

In the first stages of the colonization of the chalk quarry floor it is important to understand the influences acting upon the establishment and

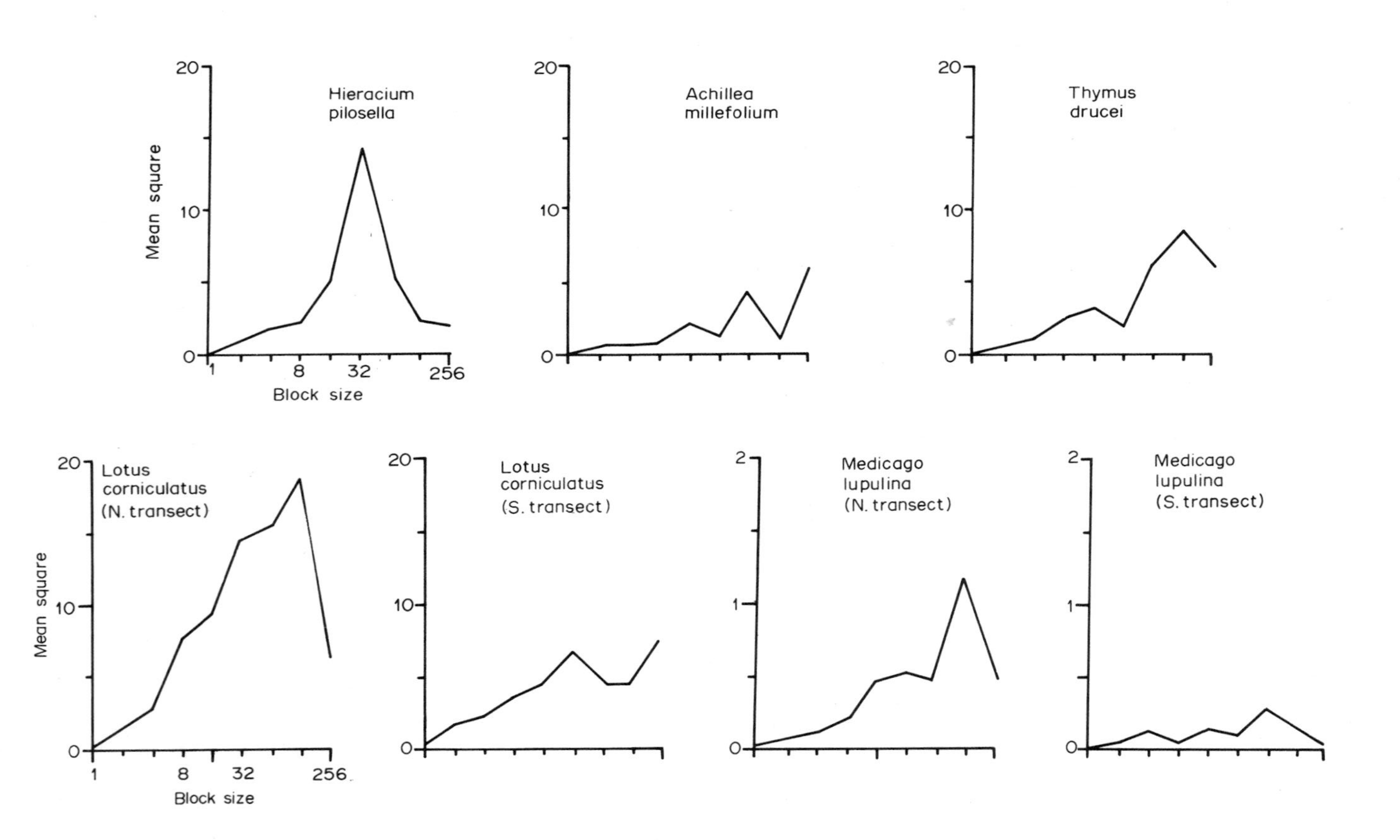

Hieracium pilosella
Achillea millefolium
Thymus drucei
Lotus corniculatus (N. transect)
Lotus corniculatus (S. transect)
Medicago lupulina (N. transect)
Medicago lupulina (S. transect)
Mean square
Block size
0
10
20
1
2
8
32
256

TABLE 7.9 The percentage cover of the species used in the pattern analyses. The two groups refer to the type of successional stage as determined by the association analysis

Species	*Group 3*	*Group 2*
L.corniculatus	23·48	22·10
H.pilosella	–	14·57
A.millefolium	–	3·88
T.drucei	–	7·88
M.lupulina	12·72	1·09

distribution of the plant species. Although to determine this information it would be necessary to carry out long term studies with repeated observation of a small area, similar to those of Watt (1960b) discussed in Chapter 1, an indication of the processes can be gained from an analysis of the pattern of these communities. In order to do this Burden (1970) laid out transects in areas corresponding to groups 3 and 2. The transects consisted of contiguous quadrats, each 2·5 cm square. Both transects had a length of 600 quadrats (15 m), and a width of 4 quadrats (10 cm) on the group 3 site and 8 quadrats (20 cm) on the group 2 site. In each of these quadrats the plant species were scored according to a visual estimate of their cover, the scores ranging from 0 (species absent), 1 (present with a cover of less than 20 per cent) to 5 (cover over 80 per cent). A total of thirty species were encountered on the transects, but only the five species listed in Table 7.9 had a sufficient mean cover for analyses to be performed.

Pattern analyses of the type described in Chapter 1 were used, and the results of the analyses for the first 512 quadrats in both of the transects are shown in Fig. 7.8. However, since the transects were each of 600 quadrats it was possible to carry out the stepping procedure described by Usher (1969b). The numbers of peaks at each block size are given in Table 7.10. The analyses showed that there was no systematic cycle of shifts in the peak mean square, but they also showed that the results of the analyses were dependent upon the starting position in the pattern. One other criticism of pattern analysis is that peaks can be an artifact of changes in the relative abundance of a species along a transect (Goodall,

Fig. 7.8 Pattern analyses of various plants in Wharram Quarry reserve. In all cases, except for the two indicated, the pattern was measured on the northern transect. The locations of the northern and southern transects are shown in Fig. 7.5

TABLE 7.10 The results of the pattern analyses showing the percentage occurrence of peak mean squares at each of the block sizes

		Block size								
Species	*Group*	*1*	*2*	*4*	*8*	*16*	*32*	*64*	*128*	*256*
L.corniculatus	3	0	0	0	18	56	28	0	58	42
L.corniculatus	2	0	0	0	1	39	47	34	0	100
M.lupulina	3	0	0	1	10	26	16	31	42	58
M.lupulina	2	0	12	36	25	33	35	52	32	44
H.pilosella	2	0	0	0	0	7	60	17	42	28
A.millefolium	2	0	0	0	17	35	20	58	0	100
T.drucei	2	0	0	0	25	36	33	32	42	58

1961). It was suspected that, for example, this may have caused the occurrence of peaks at block size 256 for *L.corniculatus* in the transect on the group 2 site. If a correction is made for trend, as in the manner outlined for time series in Chapter 2, then the occurrence of peak mean squares at the various block sizes is indicated in Table 7.11. The data for *A.millefolium* is also given. It will be seen that the results of the two sets of analyses are very similar. What, then, is the interpretation of the results of the pattern analyses?

Burden (1970) removed turfs from the nature reserve so as to determine the morphological pattern, based on the sizes of rosettes and clumps of the plants. The average diameter of a clump of *L.corniculatus* was 40 cm, which corresponds to block size 16, which was the most frequent peak mean square for the species in the earliest successional stage. *L.corniculatus* is usually cross-pollinated and the seeds are dehiscent. This results in a number of plants aggregating around an individual colonizer, accounting for the most frequent peak mean square being at block size 32

TABLE 7.11 A comparison of the percentage frequency of the peak mean squares when the pattern analysis is performed on the original data and on trend corrected data

	Block sizes					
	8	*16*	*32*	*64*	*128*	*256*
Lotus corniculatus						
original data	1	39	47	34	0	100
trend corrected	2	35	51	10	32	68
Achillea millefolium						
original data	17	35	20	58	0	100
trend corrected	17	34	23	53	0	82

(80 cm) in the analysis on the group 2 site. These peaks can therefore be attributed to a morphological scale of pattern. Using an approximate measure of intensity (Greig-Smith, 1961a), the pattern on the group 3 site was more intense than the pattern on the group 2 site. This accords with Greig-Smith's suggestion (Greig-Smith, 1964) that patterns are more intense in the earlier stages of the succession. The cause of the heterogeneity associated with block sizes 128 and 256 (320 and 640 cm respectively) is unknown, but since it is not an artifact of trend it is probably environmentally determined.

M.lupulina is a small species having a diameter of 5 to 10 cm, and hence morphological pattern can only manifest itself at the small block sizes. It can be seen in Table 7.10 that this species does in fact have peak mean squares at smaller block sizes than any of the other species.

L.corniculatus and *M.lupulina* are probably the first species to colonize the quarry floor, and it has already been shown that the scale of heterogeneity of the former is of the order of 40 to 80 cm, the scale increasing as the succession moves from the group 3 to the group 2 stage. *M.lupulina*, although present throughout these stages in the succession, becomes less important since its cover decreases to only about 1 per cent in the group 2 stage. Both of these species are leguminous plants capable of fixing atmospheric nitrogen. It will be noted from Table 7.10 that the pattern of *H.pilosella*, *T.drucei* and *A.millefolium* has the same scale as that of *L.corniculatus*. It seems likely that the build up of humus and nitrogen compounds in the areas where *L.corniculatus* grows is a factor determining the establishment of plants associated with the later stages of the succession.

It is, however, essential to remember that these studies in Wharram Quarry were carried out at one instant in time, and the results of the analyses can be taken as no more than hypotheses which can be tested by long-term studies. Although conservation management is concerned with processes that continue over long periods of time, such on-the-spot studies have indicated that when *H.pilosella* becomes absent from the vegetation and when *Leontodon* and *C.monogyna* become established, the succession has developed too far for the protection of the rarer species. Management of the reserve would therefore prescribe that small areas of this group 4 vegetation be returned to the conditions of a bare quarry floor, thus allowing the process of succession to be restarted. By clearing areas of a relatively small size, each not more than about 250 m^2 in the form of a rectangular strip, and as a mosaic over the quarry floor, the species complement of the quarry should not be reduced and there should be sufficient seed production for the colonization of the newly cleared areas.

Permanent quadrats would be demarcated so that long term quantitative assessments of the successional changes can be made. The methods of recording these quadrats would be specified in the management plan, so that data is gained not only on the species changes during the succession but also on the speed of succession. Since the initial abandonment of the quarry took place about 1930, it probably takes in the region of 40 years for the succession from bare chalk to grassland with scattered hawthorn scrub. It would therefore be planned to treat between one eighth and one tenth of the total reserve area every five years, allowing for this proportion to be altered as data from long-term studies becomes available. Since these studies have only started in 1969, it is likely to be many years before long-term studies yield data that is directly useful, or can give an insight equivalent to the quantitative handling of these on-the-spot studies.

Research and management away from the field

Botanical gardens

There are obvious advantages to conservation operations taking place in the field, preserving a species or several species in their natural environment. However, it is possible that a species threatened with extinction cannot be protected by field management alone, and that a part of the breeding stock should be moved into the protection of a zoological or botanical garden. When such a transplantation of a plant species is envisaged, there are two prerequisites (Polunin, 1969), namely:

(1) The species should be brought into cultivation in good time, i.e. before there only remain one or two plants in the natural population.

(2) There should be sufficient knowledge of the habitat requirements and conditions for propagation so that the species will not only live in the botanical garden but that it can be propagated.

The pressures on wild populations come from two directions. If the species has become rare due to changes in the environment one can question whether the botanical garden can be anything more than a living museum. In these conditions it is unlikely that the environment will ever revert back to its previous condition when the species could be reintroduced, or the depleted natural population supplemented by the cultivated plants. If, however, the species has become rare due to over-collecting or some other form of mismanagement, then, provided that there is a control on further collecting or mismanagement, the species can be reintroduced by using cultivated plants.

Three examples can be quoted to demonstrate the role of botanical gardens. First, *Sagina boydii* was discovered in 1878 (Britten, 1892), when it is thought to have been collected on Ben A'an, a mountain near Braemar in Scotland. It was not noticed as being a distinct plant until the finder, Mr. W. B. Boyd, was planting out his collection, and he neither remembered gathering it nor did he remember the exact locality where it was growing. The species has remained in cultivation ever since. Elliston Wright (1938) again records the characteristics of the plant, hoping thereby to stimulate enough interest amongst botanists to search for the species in the mountain country around Braemar. Elliston Wright discards the possibility that Boyd's plant had originated in Switzerland, and he also discounts the possibility that it is a hybrid between two other British *Sagina* species. However, despite the interval of nearly a century since the species was first found it is still only known from botanical gardens. Living plants have never again been found in the wild state in either Scotland or Switzerland, though it is possible that it may yet be rediscovered. The alpine plant *Diapensia lapponica* was unsuspected in Britain until it was discovered on an isolated hilltop, in a district with no botanical rarities, in the west of Scotland in 1952. Thus *S.boydii* remains a species known only in cultivation, and its preservation must lie in the activity of botanical gardens.

A second example is the dawn cypress, *Metasequoia glyptostroboides*. The genus was known from fossil specimens, but this species, the first living specimen of the genus, was found in China in 1941 (Fisher, Simon and Vincent, 1969). The population of trees surviving in Central China is very small and usually confined to the vicinity of religious establishments. The species is thought to have decreased in abundance so that the Chinese populations are now the relics of a previously more extensive and abundant distribution. However, *M.glyptostroboides* has been established throughout the world and the species is no longer in any danger of becoming extinct. It is widely grown in the northern temperate regions, and groves of the trees have been planted. If the natural population ever needed supplementation it seems likely that there are sufficient young trees around the world for such an operation to be feasible. It must, however, be remembered that the initial cultivation was made from a very few wild plants, and hence the cultivated stock of this species has a very depleted gene complement.

The third example does not concern a single species, but the whole family Orchidaceae. Orchids have had a special fascination for a long time, and they have been over collected for the herbarium, for garden cultivation and for commercial cultivation. The lady's slipper orchid, *Cypripedium calceolus*, in Britain has now become virtually extinct as a result of collectors and gardeners digging up plants, and extinction faces many of the more attractive tropical species of this family. Botanical gardens do have a role to play in propagating large numbers of these plants for the eventual supplementation of the wild populations. Legislative action in Europe has protected many of the orchid species, and with a changing public opinion as to the conservation of these species it becomes more possible to reintroduce orchids to localities in which they have become extinct, or to supplement natural populations that had been reduced to very small numbers of plants. The botanical garden has at its disposal two techniques for the propagation of the orchids. First, the flowers can be fertilized, the seeds ripened and then germinated under optimum environmental conditions. Secondly, there is the technique of meristem culture, which is a relatively recent development that can be used to produce many plants of a similar genotype. Once again care should be taken that the gene complement of cultivated stocks, or of the reintroduced wild populations, is not depleted to too great an extent.

Just as with management in the field, the effectiveness of botanical gardens can be assessed by asking the question – 'Why has the species become rare?' If the rarity is due to environmental changes, then the natural population may be doomed to extinction and its supplementation, or the reintroduction of the species, will not succeed. If, however, the answer to the question is that there has been over-exploitation of the natural population, then supplementation by cultivated material may be the best means of preserving the species in its natural environment.

Zoological gardens

The recurrent theme running through this chapter is that of the history of the species, of the community or of the environment. It is this basic information that determined the form of the management that will be required if the species, community or environment is to have a maximum probability of being preserved. Thus, we must consider the factors that have led to the rarity of the species concerned when we consider the role of zoological gardens in preserving animal species. This role will depend on whether the species has become rare because its habitat has been destroyed (and this includes the introduction of other species that compete with the native species for the habitat) or on whether rarity is the result of mismanagement, over-collecting or hunting, or a succession of adverse and random influences. Very general introductions to the conservation role of zoological gardens are given by Stamp (1969) and Jenkins (1970). Rather than repeating the generalities, the conservation operations of zoological gardens will be demonstrated by three examples, each selected as a step in the progression from a totally wild population to one that is being wholly maintained in captivity.

The first example concerns the preservation of the tamaraw, *Anoa mindorensis*, in the Philippines. This species, resembling a small water buffalo, is threatened with extinction since its total population is about 100 animals living on the island of Mindoro. The pressure on this species has come from two directions. First, the habitat on Mindoro has been fragmented by cultivation and commercial activity. Whereas a century ago the tamaraw occurred throughout the island it is now confined to three widely separated mountain areas (Harrisson, 1969). Secondly, there is considerable hunting of the species, both by the use of firearms and by night hunting with spotlights. These twin pressures have resulted in the extermination of the tamaraw everywhere except in the rugged mountain country. In 1961 the Mount Iglit Game Refuge was established (Fisher, Simon and Vincent, 1969) though in early 1969 there were only about 20 tamaraw surviving in the reserve.

Although there is much hope that the species can be preserved in the wild state (Harrisson, 1969), a project away from the field is being implemented. Four tamaraw (one bull, two cows and a bull calf) are in captivity in the world, but there are no plans to capture further wild animals. In 1969 negotiations were proceeding in order to bring these four captive animals together in a large enclosure so as to breed a larger number of captive stock. However, it is known that the species is exceptionally difficult to keep under ordinary captive conditions and it has only once previously bred in a zoological garden.

The tamaraw example thus demonstrates the incipient stage of conservation operations in a zoological garden. Although the habitat of the species has been partially destroyed, yet the population had been further over-exploited by hunting. The establishment of the Mount Iglit reserve together with the creation of a wildlife conservation association has made it possible for the greater protection of the species in its wild habitats, and perhaps the supplementation of this wild population with animals bred in a zoological garden will encourage a numerical increase in wild-living tamaraw.

The second example is that of the Swinhoe's pheasant, *Lophura swinhoei*, in Taiwan (Wayre, 1969a,b). This species was discovered in 1862 in the hill forests of Taiwan, and a century later very little still seems to be known of the extent of its natural range or the actual population size of wild pheasants. Some estimates suggest that the numerical strength of the captive population, nearly 600 birds in at least 65 zoological gardens, might exceed the number of native birds in Taiwan. The captive birds have largely been bred from a pair of birds received in France in 1866 and by a further pair received in Norfolk, England, in 1958. In 1967 the Ornamental Pheasant Trust sent 15 pairs of Norfolk-bred birds out to Taiwan. Six of these pairs were released into a forest reserve, and the remaining nine pairs were used to form a breeding nucleus so that further releases could be made.

The Government of Taiwan has passed a law protecting the Swinhoe's pheasant, as well as other bird and mammal species, and hence the conservation management of this species can be aimed at increasing the wild population by suitable acclimatization of bred birds followed by their subsequent release. In the past it is considered that the increasing hunting pressure from the rapidly expanding population of Taiwan has been responsible for the numerical decline of the wild populations of Swinhoe's pheasant. However, if a policy of control, prohibiting the killing of this species, can be enforced, and supplementation of the natural population with birds bred in a zoological garden can be maintained, then it seems possible that Swinhoe's pheasant can be preserved in its natural environment. These management steps have been taken at a time before the wild population has become extinct, or before it has become numerically so small that none can be removed to captivity, and they have also taken place when there is a legislative framework for enforcement of the management aims.

However, extinction of a wild or semi-wild population has not always been avoided. The example of Pere David's deer, *Elaphurus davidianus*, demonstrates the role of zoological gardens in this situation. Pere David's deer ceased to exist in the wild about 3000 years ago when the swamps of the Chihli Plains were cultivated. The species was maintained in parks, but when it was first seen by an European in 1865 the only surviving herd was contained in the Imperial Hunting Park south of Peking. In 1866 the first two skins were sent to Europe, and from these the species was described. Several live specimens were later shipped to Europe, and their progeny were distributed to a number of zoological gardens.

In 1894 flood waters breached the wall of the Imperial Hunting Park and the deer escaped into the surrounding countryside, where they were killed and eaten by the peasants. A few animals survived this disaster, and the majority of these were killed by a rebellion in 1900. By 1911 only two Pere David's deer remained alive in China, and the species became extinct in that country ten years later.

As a result of these events in China, the Duke of Bedford established a herd of sixteen animals in Britain, from various European zoological gardens. By 1922 this herd had increased to 64 animals, and by 1963 there were over 400 Pere David's deer. In 1964 four specimens were sent back to China. There has as yet been no attempt to reintroduce the deer back to their original habitat, and thus to complete the conservation of this species by establishing a semi-natural population again. However, it is one example where the work of zoological gardens has prevented a species from becoming extinct. Another example of the prevention of extinction is the European bison or wisent. The conservation of this species has been complete, since although it became extinct as a wild animal a population of captive animals has been established in the Bialowieza Forest and it is now essentially managed as a wild population.

From these examples it can be seen that the role of the zoological garden is to maintain a breeding stock of animals that can be used to supplement wild populations if they are faced with extinction. It is, perhaps, important to realize that some species have become extinct or nearly extinct very rapidly or unexpectedly following some chance event. If there is no captive population before this rapid numerical decline of the wild population, it may be impossible to contemplate a subsequent supplementation operation. It is therefore important that a large breeding collection of animal species should be kept in zoological gardens just as an insurance against the chance events and the danger of extinction in any of the wild populations.

Amateur successes

In the previous sections of this chapter we have been concerned with the scientific approach to conservation, in general involving projects supervised by trained ecologists and biologists. However, this approach has probably presented conservation as a subject for the scientist, an efficient form of management dependent upon national and international finance. Such a picture is far from the truth, since many of the successes are directly the result of the amateur effort, either by amateur leadership or by amateur participation. Two examples will be considered in detail, and in both there is a balance between amateurism and professionalism in the planning and execution of the conservation operation.

Monkey orchid in England

The monkey orchid (*Orchis simia*) had been recorded periodically in Kent from 1777 till 1923, after which date a period of 32 years elapsed during which no plants were recorded in the county. In May 1955 one plant was discovered growing in open downland. The efforts to protect this plant and the locality in which it was growing are recorded by Wilks (1960). During the summer of 1955 this first plant was eaten, probably by a slug, before the seed capsules matured. The following year the same plant flowered again, and four other plants on the same downland appeared and flowered. However, the spring of that year was unusually dry, and all of the flowering plants withered before any seed was set. In 1957 a total of 35 plants were found, seven of which flowered (these included the five plants that flowered in 1956). Of these seven, three were eaten or accidentally destroyed, and the remainder set seed although, since the plants withered early in the summer, it is not known if the seed was viable. A further 28 plants, three of which flowered, were discovered in a nearby wood.

These first three years had shown the need to protect the orchid from three threats (Wilks, 1966). First, many of the flowering spikes had been eaten by herbivores, the most important of which were the rabbit and the slug. In 1955 myxomatosis had reduced the rabbit population, but in 1958 the rabbits were again so numerous that the orchids had to be protected during the flowering and seeding season by conical cages made of rabbit netting. Moth balls were found to be a useful rabbit repellent. Slug killer was used in the vicinity of the orchids to reduce the danger from the slugs. Secondly, there was a threat from man and his domestic animals. The fencing off or the area of downland with these orchids reduced the public access, and the rabbit netting cages prevented accidental damage due to trampling either by man or horses. Horses had been responsible for eating two of the flower spikes in 1957. Thirdly, there was the problem of the fertilization of the orchid flowers themselves. The evidence of the previous years suggested that little seed was being set, and hence the Kent Trust for Nature Conservation set about artificial pollination. A finely pointed pencil proved to be too clumsy, and a needle was too damaging to the flowers. Finally, broken stems of the brome grass (*Zerna erecta*), which grew on the site, were used, and a 25 per cent successful fertilization was achieved. When mature, the flowering stems were cut into polythene bags, dried and the seed rubbed into the downland soil where the parent orchids were growing.

Protection from herbivores and from man yielded results, since the population size of both flowering (Fig. 7.9) and non-flowering plants

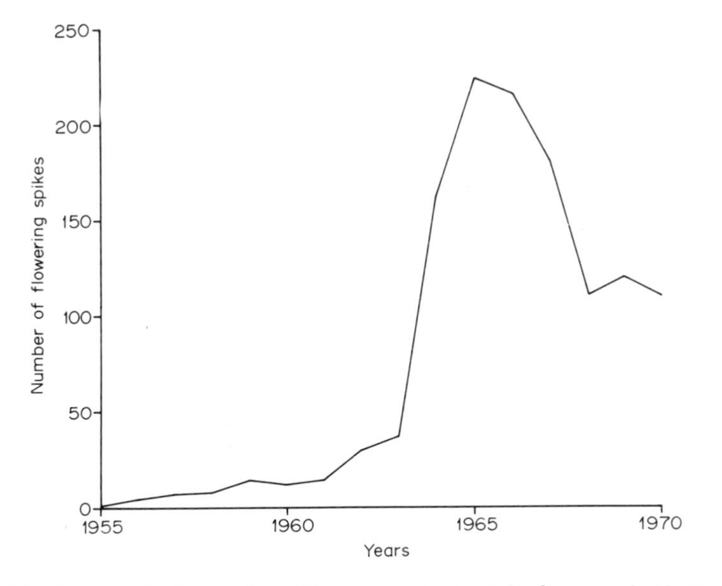

Fig. 7.9 The increase in the number of flowering plants of *Orchis simia* in Kent after the first plant had been found in 1955

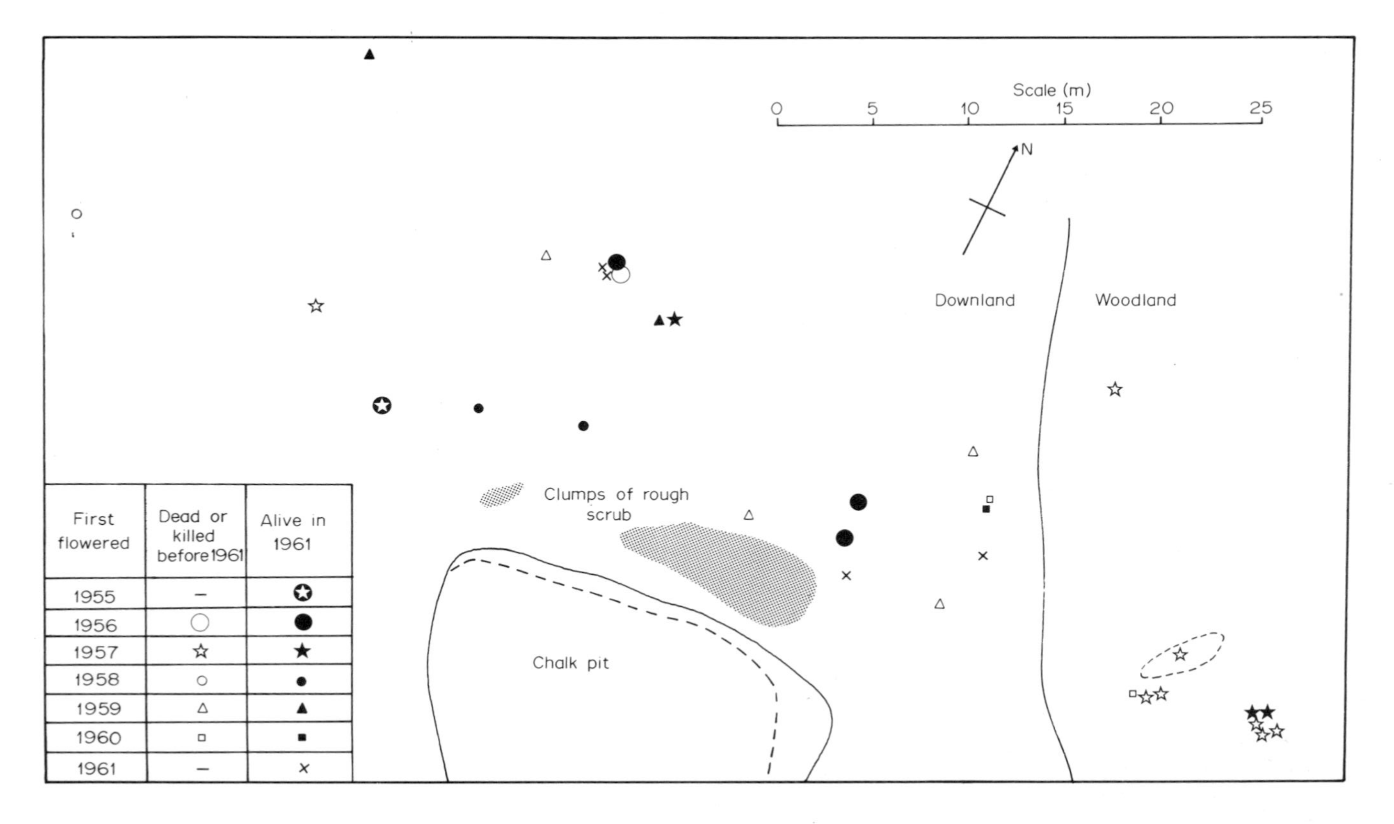
Scale (m)
0
5
10
15
20
25
N
Downland
Woodland
Clumps of rough scrub
Chalk pit
First flowered
Dead or killed before 1961
Alive in 1961
1955
1956
1957
1958
1959
1960
1961

TABLE 7.12 The longevity of the flowering period of those *Orchis simia* that were alive in 1961 and those that had died before 1961

Number of years known to have been flowering	*Number of plants flowering in 1961*	*Number of plants that had died before 1961*
1	4	5
2	1	0
3	2	2
4	2	1
5	1	0
6	3	0
7	1	0

increased, till in 1963 and 1964 there were 37 and 162 flowering plants respectively. The fertilization campaign also yielded results. Seed sown in 1958 produced plants with flowering spikes in 1965, a developmental period of seven years from seed to flowering. The graph in Fig. 7.9 records the increase in population size to over 200 flowering plants in 1965 and 1966, and the subsequent decrease to what is thought might be a stable size of flowering population of slightly over 100 spikes in 1969, 1970 and 1971.

The development of the monkey orchid in the field during the first seven years of the conservation operation, 1955–1961, is shown by the map in Fig. 7.10. Although Fig. 7.9 shows that the maximum number of flowering plants during this period was 14, the maps records a total of 22 plants, and the characteristics of the mortality of the individual plants can be investigated. Data for the longevity of the flowering period of these 22 plants is recorded in Table 7.12 where it can be seen that five of the eight plants that had died had flowered for only one year, whereas nine of the fourteen plants still flowering were in flower for at least their third season. It would therefore be seen that there is no definite number of years over which a plant flowers (unless this is more than seven years). It would also seem that if a plant has succeeded as far as its second flowering season it has then passed through the period of greatest mortality.

The conservation of the monkey orchid in England has thus resulted from an amateur effort, overcoming by empirical means the various

Fig. 7.10 A plan of the Kent habitat of *Orchis simia*. The symbols show when plants first flowered, as well as indicating whether the plants were alive in 1961 (solid symbols) or whether they had died (open symbols)

problems of protecting the plants from herbivores and humans and the problem of fertilization. Without a dedicated effort to overcome these problems it is unlikely that the monkey orchid would be so abundant, or even present, in Kent in the 1970s.

Ospreys in Scotland

The second example concerns a bird of prey that is now regularly nesting in the Highlands of Scotland. The history of the osprey, *Pandion haliaetus*, shows that it nested over a wide area of the Highlands in the eighteenth and nineteenth centuries, but the demands of collectors resulted in many birds being shot and many clutches of eggs being taken. By the end of the nineteenth century the osprey had virtually ceased to be a breeding bird in Britain. However, occasional birds were seen in Scotland and in the 1940s the number of birds seen began to increase and single birds started to turn up in the breeding season.

The story of the efforts of the Royal Society for the Protection of Birds (R.S.P.B.) during the 1950s is told by Brown and Waterston (1962). There were unconfirmed reports that a pair nested in the Cairngorms region of Scotland in 1952, and as a result of these rumours and other birds being seen in the area the R.S.P.B. realized that it was possible that the osprey may, with protection, become once again a regularly breeding species in Britain. Discussions between R.S.P.B. representatives focussed upon the possible dangers to nesting birds – the actions of gamekeepers in reducing populations of birds of prey; the possibility that birdwatchers who are frequent in the Cairngorms would continually disturb the nesting birds; and the possibility that bird egg collectors would attempt to take the clutch. Less immediate problems concerned the possibility of forest fires, disturbance due to forestry operations and crows taking the eggs when the ospreys were away from the nest. During 1953 and 1954 it is possible that ospreys either nested or attempted to nest in the area. In 1955 a pair was found building a nest in the Sluggan Pass, but in later years it was realized that this nest was a repeat nest, being built after the first clutch of eggs had been destroyed by some unknown means. No breeding took place in the Sluggan Pass. In 1956 a nest was apparently built in Rothiemurchus forest, but only the repeat nest near Loch Morlich was found. Once again the first clutch of eggs must have been destroyed, and no breeding was successful in 1956.

It was evident that protection operations must begin as soon as the ospreys returned to Scotland after the winter, and it was important to locate the eyrie before any eggs were laid. Volunteer watchers in 1957 started their survey of the lochs of the area a few weeks before the first osprey was seen. However, although one osprey was seen and although it took an interest in the old eyrie it was never joined by a partner, and there were rumours that an osprey had been shot in Speyside that year. However, in 1958 two ospreys were on the eyrie site near Loch Garten, and by mid-May there was one egg in the nest. On this day when the watchers were starting to plan the operations two incidents occurred which disturbed them. First, a birdwatcher, unaware of the operation about to be mounted, disturbed the birds by walking out on to the bog towards the eyrie tree. Secondly, another man approached and climbed the tree. It was obvious that a 24-hour watch would have to be mounted. However, just at dawn after the very dark night of 2–3 June, a man climbed the tree, taking or breaking the eggs, and leaving two painted chicken eggs in their place in the eyrie. The ospreys built a repeat nest in the vicinity, but there was no breeding.

Success, however, came in 1959. After the experiences of the previous year it was obvious that more watchers were needed so that two people could be on watch during the night, and better communications were established between the watchers and the base camp. In 1959 the male osprey arrived on 18 April and the female arrived on 22 April. They nested in the repeat eyrie of the previous year, and the chicks were known to have hatched by 9 June. With a public interest in this conservation operation, the watchers had to organize the visitors, operating a special hide with high-powered binoculars for them to observe the eyrie. The full account of the breeding in 1959 and the public response is given by Brown and Waterston.

The success of re-establishing the osprey as a breeding species has been maintained during the following decade, during which time a few other pairs have nested in Scotland. The species now appears to be a regular breeder in Britain, and it is hoped that this position will continue and that more eyries will become established. Although much of the operation has been directed by professional ornithologists, yet the success with the ospreys has depended upon sufficient man-power being available to watch the birds and to warden the visitors. It is here that the amateur effort has stimulated one of the most widely publicized conservation operations in Britain.

8. Conservation and Education

Introduction

In the previous chapter the real core of the subject of conservation was discussed, since conservation is associated with the preservation of the quality of the environment within which man lives. In the next chapter, 'Conservation and Recreation', there will be an analysis of the effects of the human pressures upon the environment, and an indication of how integration of both the preservation and utilization of land can be planned. However, there is a wide gap between preservation and exploitation in the form of recreation. The most important means of bridging this gap is the education of the recreational user. His actions and interests are immediate or only short term, but he must be induced to think of the long term preservation of the resources that he is using today and that he will wish to use in the future. The idea of sustained utilization of resources is an essential ingredient of education at all levels.

In the introduction to this book an analysis was made of legislative programmes introduced to foster conservation. But, can legislation and its associated penalties achieve anything until it is supported by an educated body of opinion? If such opinion is to be expressed, then it must be based upon some ecological understanding, whether this be superficial or a sound scientific training. Broad (1969), generally discussing the relations between education and conservation, said:

> '. . . in my opinion, in education lies our only hope of solving this vast problem of ensuring that the natural biological resources of the earth are used with wisdom and restraint so that it will continue to provide a suitable habitat for man.'

Broad has thus stressed the importance of education, but what truth is there in his assertions? It is true that conservation as a subject can have a co-ordinating role, but this is because it is an applied form of ecology. Ecology as a science cuts right across the familiar divisions of 'botany' and 'zoology', since the ecosystem is composed of both plants and animals as well as the abiotic environment. Conservation as a science can therefore only be taught when there is also some knowledge of the underlying ecological principles.

Education at the school level is thus an important aspect of conservation management, but several questions must be asked. First, what sort of demand is there in schools for the field teaching of biology, and how is this demand likely to develop during the forthcoming years? Secondly, how can an area of land be planned so as to maximize its usefulness in educational activities? Thirdly, what sort of projects can be carried out? Should they have a practical application, or is field teaching an extension of laboratory teaching where, if it is wished, the same experiment can be repeated every year? And, fourthly, how can educational and conservational uses of an area of land be integrated so that neither adversely affects the other?

These questions form topics for later sections of this chapter. However, before attempting to answer them the scope of education must be expanded to embrace not only children but adults as well. Broad (1969), so encouraging about education in schools, seems despondent about adult education. He says:

> '... that only by a continuous process of forcing information home can anything be made to penetrate very far, and ... the effect is slight and transitory.'

Within areas of particular conservation interest, such as nature reserves and national parks, there is greater opportunity for adult education. The recreational user of the countryside is the main target for the educational drive, since he has come from his home into the area and is generally keen to find out about his surroundings. Surveys that have been carried out, for example those of Mutch (1968) in forests and Usher, Taylor and Darlington (1970) on nature reserves, have indicated that information is one of the facilities in greatest demand by the recreational user of the countryside. Conservation management should attempt to satisfy this demand by supplying information that contains the essential facts about the environment and its conservation. In both the United States and Europe this information has often been given in the form of 'nature trails', which can be defined as self-guiding walks. Self-guidance is achieved by some sort of demarcation on the ground and by a leaflet or booklet that contains details of the route to be followed and the plants, animals or other objects that can be seen. The planning and implementation of nature trails, designed to interest and educate adult users, is considered within the following section.

Adults and conservation education

Nature trails

The nature trail has arisen as a result of demand, and hence it is not surprising that there is no *pro forma*, or recipe book, for the preparation of a successful trail. Indeed the bodies and organizations setting up trails are so numerous that a diversity of trail types exists. However, amongst this diversity, there are three patterns of trails, which can be defined as:

(1) *Formal nature trails*, which provide a rigidly guided tour from location to location, such that at each location the neighbouring fauna, flora and physiography are demonstrated (Plate 7).

(2) *Semi-formal nature trails*, which also provide a guided tour from location to location, but the emphasis is placed on the habitat, and its fauna, flora and physiography, that occupies the space between the locations (Plate 8).

(3) *Informal nature trails*, which do not provide a guided tour, but which provide information that allows the user to walk where he will within the area seeking out those features that interest him most.

The concepts behind these trails differ, largely because formal and semi-formal trails impose a route which has to be followed from start to finish, whereas an informal trail user decides both the distance and where he wishes to walk. In surveying 20 nature trails, more or less randomly selected from areas all over Great Britain, it transpired that five trails were informal, one was semi-formal and the remaining 14 trails were formal. Various data on these last two types of trail are given in Table 8.1. Of the nature trails that formed the sample only two were less than one mile in length, two were greater than three miles and one third had a length of $1{\cdot}5 \pm 0{\cdot}2$ miles. The average British trail would thus appear to be considerably longer than the average trail in the United States (about 0·5 mile).

Table 8.1 indicates some data to be considered in planning nature trails. The number of locations was never less than seven, and over half of

TABLE 8.1 Data from a sample of 15 formal and semi-formal nature trails in England, Scotland and Wales

Character	*Mean*	*Mode*
Number of locations	13·6	7
Length (miles)	1·67	1·5
Length of guide (words)	3300	2500
Number of maps in guide	1.13	1
Number of illustrations in guide	9·92	0, 2 or more than 20

the trails had 12 or fewer locations. The large discrepancy between the mean and the mode is indicative of a skewed distribution, the mean being inflated by one trail that had 36 locations. The number of locations should be decided so that, on the one hand, there are sufficient locations to make the educational points that should be put across, whilst on the other hand there are few enough locations for the user to maintain interest throughout. The length of the guide is important since a large amount of printed matter is expensive, whilst something that is too small is considered to be of too little value. One method of establishing the price of a trail guide will be discussed in Chapter 9. None of the formal trail guides used in compiling the data for Table 8.1 contained more then 4000 words, whilst all but one of the informal and semi-formal trail guides contained more than this number of words. The provision of illustrations provides another problem. All trail guides provide a map of the area, showing the route to be taken, or, in the case of informal trail guides, the paths and areas of particular interest. There are, however, two approaches to further illustration, as indicated by the modal value in Table 8.1. Either one considers that the inclusion of illustrations is useful to the trail user, as it is part of the educational process and leads to immediate awareness of the environment and species. Or one considers that the inclusion of illustrations unnecessarily increases the cost of the guide, and that a popular handbook on birds or flowers would provide more information. But, how many trail users possess such handbooks? If illustration is to be carried out, then obviously it must be done thoroughly.

Having defined such physical characteristics of trails and their guides one must investigate how these are interpreted in practice. A set of three short examples will be taken, each referring to one or more plants and explaining both where they can be found and something about the habitat in which they grow.

First, as an example of an informal trail, we can consider Barns Ness in East Lothian, Scotland (the booklet entitled *Barns Ness* is published by the East Lothian County Council):

> 'In the drier areas growing in short limestone turf one of the most conspicuous and widespread plants is Eyebright (*Euphrasia officinalis*). This small (2–9 inches) annual is a member of the foxglove family and is semi-parasitic, drawing most of its food from other plants to whose roots its own are attached by suckers. The small flowers (June–September) are usually white but are sometimes tinged with lilac or streaked with purple.'

This extract from Section 7 of the Barns Ness guide clearly shows the educational impact of the information contained in the guide. Although no route has been given it is quite clear where *Euphrasia* is to be found. Some idea of the biology of the plant is given, and an attempt is made (not included in the extract) to summarize the ecological factors affecting the whole area.

An example of a semi-formal trail is taken from Gwydyr Forest in Caernarvonshire, Wales (the booklet entitled *Gwydyr Forest Trail* is published by the Forestry Commission). On the stage 1, going from an elevation of 150 to 210 m above Ordnance datum, the guide states:

> 'Observe the ground vegetation. Among the mosses are: *Polytrichum*, *Thuidium*, *Hylocomium*, *Dicranum*, and *Plagiothecium* spp. Generally we can classify mosses in their relation to the type of soil. Thus at one extreme we should distinguish *Mnium* and *Thuidium* as indicative of good humus conditions, *Hypnum* and *Dicranum* on medium humus conditions, and *Rhacomitrium* and *Sphagnum* on bad humus conditions.'

From the guide we know some of the mosses that can be found on this stage of the trail, a length of approximately 300 m, but the locations of the species are not exactly demarcated. The trail user has been taken to the correct area and is then left to discover the exact habitat of the plants for himself.

A third example, one of a formal nature trail, is taken from Skipwith Common in Yorkshire, England (the booklet entitled *Skipwith Common*

Nature Trail No. 1 is published by the Yorkshire Naturalists' Trust Ltd.). At location 14 on this trail the guide says:

> 'On the left hand edge of the path, there is the little prostrate bog pimpernel (*Anagallis tenella*), . . . Beyond, in an area that looks as though it was once a small peat cutting, there is an insectivorous plant, sundew (*Drosera rotundifolia*). It forms small rosettes of oval reddish leaves covered with long sticky hairs. Insects adhere to these hairs which secrete enzymes to digest them. Sundews usually grow on poor acidic soils, and obtain from the insects the element nitrogen which is not easily available in such soils. If you leave the path to look for the sundew, please be careful where you step because the plants are only small and will not survive trampling. In July, a flower stalk grows up about six inches and bears white flowers, unusual in their six petals, which only open in hot sun.'

In this example from Skipwith Common there is no doubt that the user of the trail will see a specimen of the sundew, provided that the whole area around the stake marking location 14 has not been trampled. One of the locations on the Skipwith trail is shown in Plate7.

What then are the advantages and disadvantages of the three trail types? It can be seen that with formality comes a degree of certainty of seeing the organisms described in the trail guide. Thus, for visitors who know very little of the ecology of the area, the formal trail allows for very specific education, and for most of the essential points concerning the environment and its conservation to be made. In the example of the informal trail at Barns Ness, if one was unfamiliar with limestone turf it would be impossible to find the eyebright, except by chance, whereas the sundew on Skipwith Common should be found by anyone walking to location 14.

It can also be seen that as the formality of the trail increases, so the user is channelled into a smaller area where trampling can become of major importance. The effects of recreational trampling are discussed in Chapter 9, but in planning formal trails it is essential to consider these effects upon the environment. Locations on the formal trail may have to be periodically changed so that the wear on the ecosystem is kept within acceptable limits.

Thus, the overall aim of the nature trail needs to be defined. Are the users to be channelled to some parts of the reserve, or are they to be allowed to walk and search over the whole area? The latter can result in the effects of trampling being spread over as wide an area as possible, but the whole of the area is subjected to some kind of disturbance. The former results in an integration of the conservation and education interests by zonation, since the route of the trail can be planned to avoid particularly rare species, experimental areas or parts of the habitat that would be susceptible to disturbance and trampling. Thus, the planning of the nature trail for adult use depends upon the conservation objects of management, as well as upon the demand for a facility of this sort. However, the requirements for school use will be found to be more specific, and are discussed in a later section of this chapter.

Organizations

A lot of emphasis has been placed upon the role of the nature trail in adult education, and for the manager of wildlife or nature reserves this will be his main link with the public. In such areas of land the nature trail can be manipulated so as to emphasize the points of particular conservation interest, and the responsibility for the planning and design of such an educational effort lies with the manager. It is therefore important in a text-book on biological management to stress this aspect of adult education. But it is only one aspect of the subject.

Many people are more aware of the educational effort made within their homes or within the urban or residential area within which they live. Effective means of adult education within the home are radio and television, and, although few conservationists are trained to appear on these media, it is their responsibility to assist in the preparation of these programmes. Newspapers are read by millions of people each day, and, though few conservationists are journalists, the journalist relies upon the information of the conservationist for writing articles and features. Secrecy has often surrounded the conservation operation for fear that advertisement will attract vandalism. It is, however, more certain that advertisement attracts sympathy for the cause, and in this respect Broad's gloomy conclusions about adult education, quoted in the introduction to this chapter, may be partially unfounded.

Away from the home, but based in the urban or residential surroundings, are film shows, Societies and Trusts. Films about wildlife and particular conservation operations are extremely useful educational aids. The Royal Society for the Protection of Birds have made many films, dealing not only with the birds of a particular geographical region but, more importantly, with special projects. Films have been made about the work involved in re-establishing species that previously nested in Britain (see Chapter 7), or of the results of accidents such as that of the *Torrey Canyon*. The activities of Societies and Trusts have changed during the last half century. In Britain today we see the gradual decline of many of the 'Natural History' Societies whose main influence was felt towards the end of the last century and early this century. Although not all of these Societies have declining memberships there are many that have come to an end during the last 10 or 20 years. However, the reverse situation is characteristic of the Conservation Trusts. First Norfolk was founded in 1922 and then Yorkshire in 1946, and during the subsequent 20 years all of the counties of England and Wales have been included in a national network of Trusts. Membership of these bodies has been increasing rapidly, and many of them are showing membership increases of over 20 per cent per annum. These Trusts have a prime responsibility in adult education not only to their members but also in making known their conservation objectives to a wider population. These activities, besides the management of nature reserves, have been in the showing of films, the arranging of talks, lectures and exhibitions, and in the dispersal of their literature about the local conservation efforts. Other developments more recently in Britain have been the founding of the Council for Environmental Education, and the coming together of many organizations interested in the countryside to form the Committee for Environmental Conservation (CoEnCo) in 1970.

The considerable and growing interest in the activity of Trusts and similar organizations indicates that adult education can work, and that there is a developing climate of opinion favourable to the conservation of species and habitats in Britain.

Schools and the teaching of conservation

In the introduction to this chapter three questions were asked in relation to schools and the teaching of conservation. However, before discussing these questions, we shall investigate some of the recent trends in school teaching ideas, based upon the biology teaching projects of the Nuffield Foundation.

If we begin by looking at the teaching of junior science, the teachers' guide (Anon., 1967) contains a section on conservation. This introduction stresses the relation between man and the environment, showing how man's actions can affect other species. The basis for the 'out-of-doors' approach to teaching is given in the extract (p. 297):

> 'There is a great need in Britain to bring the children in contact with the natural environment ... This is not only because they need experience of natural habitats but because out of doors there is a rich variety of objects and stimuli to provide starting points for enquiries.'

The advice to teachers is to care for and respect the environment in which their children are being educated, instilling, one hopes, into these children not only the ecological basis of the subject but also the idea of responsibility for what might happen if the area is maltreated.

At the next stage of the educational system in Britain, the O-level syllabus, there is less direct concern for application but more concern with the scientific principles (Anon., 1966). Thus, amongst the aims of the course we see:

> '(*a*) to develop and encourage an attitude of curiosity and enquiry;
> (*f*) to teach the art of planning scientific investigations, the formulation of questions, and the design of experiments.'

However, besides these aims that establish a basis for scientific ideas there are other aims that relate directly to the man : nature relation (see Chapter 6). These are:

> 'c. to develop an understanding of man as a living organism and to place him in nature;
> e. to encourage a respect and feeling for all living things.'

These points are further developed in the aims of the A-level syllabus (Anon., 1965), where the emphasis is largely placed upon the scientific foundation. The man : nature relation no longer appears amongst the aims of education at this level, but the sentiment is still expressed in the seventh listed aim:

> 'The ability to evaluate the implications of biological knowledge to human beings. The applications and implications of biology will be stressed whenever possible.'

Thus the contemporary thought in biological education is to encourage discovery at an early age, and to build up a critical ability during the subsequent stages of education. The relation of man to the biological systems is stressed throughout, balancing the teaching of a pure biological science with its application for human populations.

The demand for facilities

The role of field teaching in school curricula has been relatively small (Anon., 1963; Broad, 1969). Broad states that in primary education six to eight hours per year were spent in out-of-doors study, whilst in secondary education this figure was only two to two and a half hours per year. Such survey data indicate that the demand from schools for field studies in the 1960s has been very small.

However, if the supply of educational facilites is investigated, a contrary situation would be indicated. The provision of field centres has developed in Britain since the Second World War largely through the efforts of the Field Studies Council and Local Education Authorities. The creation of these centres has not kept pace with the demand from schools for the facilities that are offered. It may therefore be that the small amount of field teaching in the school curricula is more a reflection of the supply of rather than the demand for field facilities.

An example of a habitat that has been used by schools for day excursions is Bishop Wood near Selby in Yorkshire. The wood, approximately 340 ha, was largely felled during the First World War and has been managed by the Forestry Commission since 1921, who have replanted much of the wood with both conifers and deciduous trees, but some of the wood has been allowed to regenerate naturally. Such areas now have a mixture of tree species with a scrub layer and a very diverse ground flora. Although the Forestry Commission's primary object of management is to produce timber economically, it is also policy to provide rural facilities. The educational effort in Bishop Wood is a part of this policy, and the organization is jointly carried out by the Forestry Commission and the West Riding of Yorkshire County Education Department.

A survey of schools using Bishop Wood was carried out by Taylor (1969). The distribution of the schools, Fig. 8.1, shows that most of the schools are situated on the eastern fringe of the industrial zone of Yorkshire. An analysis of the frequency with which schools visit the wood is given in Table 8.2. Of the 24 schools that provided data, 20 had

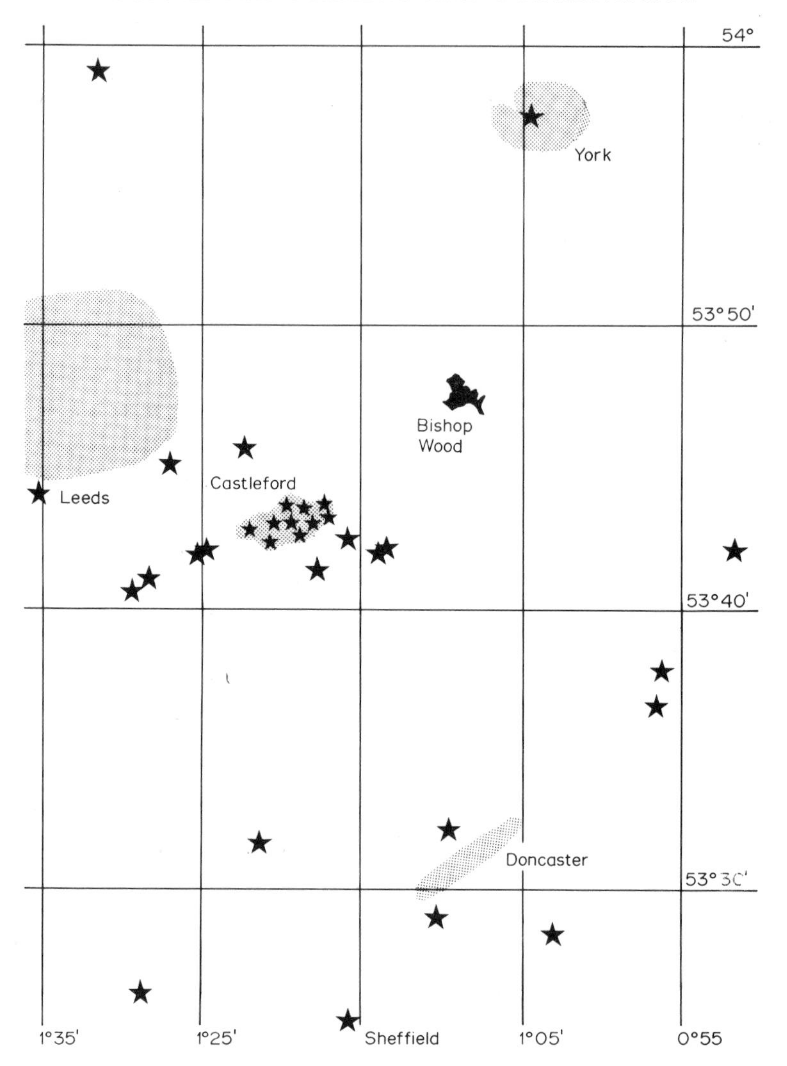

Fig. 8.1 A map of southern Yorkshire showing the location of schools that have used Bishop Wood for field teaching

used the wood in summer, and 15 had used it in winter. It is known that a total of 32 schools were using or had used the wood at the time that the survey was carried out. Further data from Taylor's survey indicated that an average visit was approximately 30 children spending half a day in the wood. Using the 'full year' data in Table 8.2, the educational pressure on the wood in 1968 was approximately 1040 child days. The survey also investigated the planned use of the wood by these schools in 1969, from

TABLE 8.2 The frequency of visits made by schools to Bishop Wood, Yorkshire (Compiled from Taylor, 1969)

Frequency (visits per period)	*Periods*		
	Summer	*Winter*	*Full year*
1	11	12	10
2	6	3	7
3	1	–	4
4	2	–	1
5	–	–	0
6	–	–	2

TABLE 8.3 The potential use of Bishop Wood in 1969 as a function of the use in 1968 (Compiled from Taylor, 1969)

Proposed visits in 1969	*Number of schools*
More than 1968	7
Same as 1968	10
Less than 1968	5
Don't know	4

which it was found that there was a potential increase in the usage of the wood (see Table 8.3). To be added to these data are the number of schools that will first visit the wood in 1969. Thus, Bishop Wood, which was planned to accommodate day visitors only, has an increasing educational use.

But what studies are being conducted by the schools? The Bishop Wood survey asked schools to what curriculum studies the field teaching related, and a supplementary question asked if there were reasons, other than formal learning, for the visit (Table 8.4). From the replies it can be seen that the environmental studies are strongly represented on the curricular subjects, though field classes do, particularly in primary education, have a wider appeal in subject matter. The fact that seven of the schools were concerned with social questions and the preservation of the habitat indicates that some basic ideas about conservation are forming a part of the field education, though this aspect could obviously be expanded.

Some of the features of the demand for field educational facilities are clearly demonstrated in Taylor's study. First, although there is a scarcity of field teaching in the curricula of many schools, the whole concept of

TABLE 8.4 Curriculum subjects to which a visit to Bishop Wood relates, and other reasons advanced by schools for their visit

Curriculum subject	*Number of schools*
Nature study	23
Art, Craft, Music	18
English	18
Mathematics	12
History, Geography	11
Science	5
Extra-curricular activities	
Preservation and Social studies	7
Aesthetic appreciation	13
Recreation	6

field education is increasing the demand for facilities. The Bishop Wood survey has shown a potential increase in the demand from schools in the order of 20 per cent per annum. Secondly, although the teaching in the field is not necessarily concerned with conservation, it is apparent that biological studies form the focus for such teaching. However, having established that a demand exists for such facilities, how are they to be planned within a framework of biological conservation?

Planning for school use

Education within the conservation framework has two aims. First, there is the more factual content of demonstrating the structure and functioning of the whole ecosystem, of establishing the concepts of energy flow and nutrient cycling that have been discussed in Chapter 3. Secondly, there is the need for freedom of action so that there can be a discovery of this structure and function. These two aspects of education, together with providing a measure of protection to the environment, must be carefully planned.

Planning has often taken the form of establishing special nature reserves or nature trails for the use of schools. A study of such facilities was carried out by a Study Group set up by the Nature Conservancy in the United Kingdom in 1960. The report (Anon., 1963), after examining the role of biology and field studies in schools, training colleges, technical colleges and universities, looks at the facilities and planning of educational nature reserves. The report concludes (p. 135):

'Nature reserves for intensive educational use should ideally:

(1) contain the diversity required for the demonstration of a wide range of habitats, communities and species, and the operative ecological factors;

(2) not be liable to destruction or irreparable damage by controlled but heavy use; e.g., certain types of bogs are easily damaged by excessive trampling and are unsuitable for large parties;

(3) not contain rare features or species which ought to be permanently conserved for research and similar activities, in the national interest;

(4) be reasonably accessible to users; educational reserves near to cities and towns will serve a larger number of schools than those in remote areas;

(5) have rides, paths and other access routes for safe and rapid movement of parties within the area;

(6) contain a field museum or similar centre; and

(7) have a qualified warden.'.

What then are the essential features of planning for this kind of educational use? On the ground it is essential that there is something to follow in the form of a nature trail along a demarcated path, guiding the user to the habitats and avoiding areas which are particularly susceptible to trampling. The most appropriate form of nature trail is one that is a hybrid between the formal and the semi-formal types, allowing both for specific instruction at several locations and for application and experiment within one or two larger areas where the children can work more freely. The provision of information on these trails has also to be considered. Taylor's (1969) survey of Bishop Wood, that has two nature trails described in printed guides, showed that the most frequent answer to the question 'Can you suggest any improvements . . .?' was to suggest more information, not only by an enlarged handbook but also demarcation on the ground and the possibility of an information office acting partly as a museum. The information that is needed is not a long list of species, but rather a summary of the structure and functioning of the main habitats or ecosystems. Very often appeals for lists of species may indicate that the teacher has an inadequate training or experience of field conditions. This of course poses one question: 'How much field instruction should be given to student teachers?' If, as the developments of the

Nuffield schemes suggest, field projects are to become more important in the education of the future, teachers will need to have more knowledge and experience of field teaching techniques.

School projects

In discussing the educational effort in schools we have looked at the demand for facilities, and the planning of reserves for school use. Schools also require areas where their teaching is essentially an extension of the school laboratory, where the same things can, if it is wished, be seen year after year and where similar experiments and exercises can, if it is wished, be performed year after year. An introduction to this kind of use of nature reserves and other rural areas is given in a booklet prepared by the Berkshire, Buckinghamshire and Oxfordshire Naturalists' Trust (1970). This booklet is aimed at the teachers, and its content falls into two parts. First, it gives a very simple introduction to ecology and conservation, stressing the trophic–dynamic structure of the ecosystem and the multiple use concept of nature reserves. Secondly, it describes possible projects that could be included in field teaching programmes. These projects not only involve the children in collecting field data but they also have an elementary form of analysis, introducing simple concepts of statistics.

However, nature reserves have a wider educational use, involving the application of ecological principles. In particular the work of the Conservation Corps in Britain is such an activity (Stamp, 1969). The Corps has been organized as enthusiastic bands of young people working for a week or two on a specific project in a specific area. These projects have concentrated upon nature reserves, creating and maintaining footpaths, clearing scrub and unwanted species, planting and tending introduced plants or regenerating species, and generally carrying out the management programme of the reserve. A history of the movement and a list of all projects carried out has been published (Anon., 1970a). Such projects require careful organization and supervision since the young people carrying out the tasks are, at least initially, unaware of the full implications of conservation management. However, these projects provide a benefit in both directions – the education of the helpers is benefited by the acquisition and application of environmental and conservational principles, whilst at the same time the conservation of the habitat is benefited by the work being undertaken. Whereas conservation and education are generally integrated within a habitat by zonation (i.e. nature trails), possibly moving the zones in response to wear on the environment, conservation projects take place throughout the reserve, integrating fully the educational and conservational land use policies.

There are, however, opportunities for more extensive surveys involving large numbers of schools and young people. Although this potential has not been used directly in conservation operations, it is starting to be utilized in environmental studies. One such study concerns the broom, *Sarothamnus scoparius*, which is widely distributed in Scotland and a very easily recognized species (McKelvie, 1970). The aim of the survey was to determine some of the features of the growing season in northern Scotland. It was decided that measurements of climatic factors, although related to the growing season, were inappropriate for two reasons. First, they require considerable man-power if accurate climatic records are to be kept in sufficient places for a sufficient length of time. Secondly, one or two measurements, such as temperature, are unlikely to be really representative of a complex ecological factor such as the growing season. Perhaps the most important aspect of the growing season is when growth can start in the spring since this determines when a plant can first use the available solar radiation. Thus, by taking a plant that has a distinct flowering period, an attempt can be made to assess the combined effect of the factors influencing the length and initiation of the growing season. A spring flowering plant was preferable to one flowering later since the spring flowering one would tend to show when the growing season actually started and would have only a small component that reflected the climate during that growing season.

McKelvie's survey used 525 primary schools, with a request for 25 volunteers from each school. Each volunteer was asked to record the date and location when any flower buds of broom first showed a tinge of yellow. Schoolteachers were asked to keep a piece of broom in water in the classroom so that the children could recognize the species. Of the schools initially approached, 390 returned data, 75 replied to say that there was no broom growing locally and 60 schools did not reply.

The data from this survey are shown in Fig. 8.2. McKelvie gives details of his methods of assessing the validity of the data, and the map has been prepared from data that was considered to be valid. Since some children forgot to record when the flowers first showed yellow, and recorded a date when the plant was fully in flower, the date for the district in which the school is situated was assessed as being the average of the 10 earliest

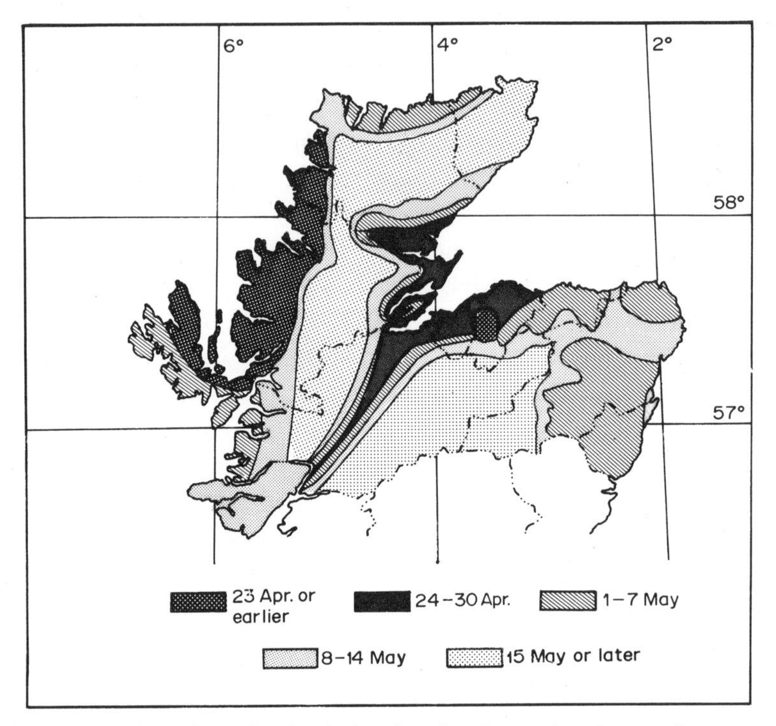

Fig. 8.2 A map of northern Scotland showing the dates when broom first came into flower. (Redrawn from McKelvie, 1970)

dates recorded by children from that school. The map, Fig. 8.2, confirms that the commencement of the flowering season is related to known areas of early production. Thus, the earliest areas are on the north west coast and on the Moray Firth, whilst the central areas and Aberdeenshire were the latest.

This survey was aimed at gaining data useful in predicting the potential of an area for agricultural production, and hence a second part of the survey was concerned with an intensive assessment of the growing season in a smaller area, that of Aberdeenshire. The results were also plotted on a map, with isophenes joining points on the map for which the date of commencement of broom flowering was the same.

The fact that about 10,000 records were gathered from 390 schools is indicative of the interest generated by such a project. These projects can be characterized by a two-way flow of benefits. Firstly, McKelvie would have been unable to draw his maps without the help of thousands of recorders watching for the day when the broom flowers first opened.

Secondly, the schools were able to introduce an element of biological science into their curricula by encouraging the survey, and by keeping broom in their classrooms. After the data had been analysed the participants received coloured maps enabling them to compare their results with those of the rest of northern Scotland.

Conclusion

Throughout the discussion of conservation and education it has been assumed that some educational activity was taking place on the nature reserve, and the concept of controlling it, or of integrating it in the overall conservation management programme, has been used in such a way as to indicate, at least at times, that these two forms of land use are essentially competing for the same resources. The educational activity has had one of two aims, either to teach ecology or to teach the application of this science in a practical form of conservation. Education is, however, not competing for the same resources in the same way that recreation is, since education is far more deeply involved with conservation. Education remains the basis for the conservation of the world's natural resources in perpetuity (Pritchard, 1968). Pritchard has discussed the training of the people who will in the future have an impact on the environment, and it is clear that only if they have an understanding of the processes that have created and are maintaining that environment will they manage it on a sustained basis. Thus, Broad's comments, quoted in the introduction to this chapter, are once again important – that the aim of education should be directed towards the younger generation. School projects and tasks, linking the sustained management of nature reserves with education in ecology and biology, are thus an essential feature of conservation management, and should not in general be considered as competing for the same biological resources.

9. Conservation and Recreation

Introduction

The term 'recreation' is used to describe a large number of pursuits followed by people as leisure-time activities. Before the relation between these pursuits and biological 'conservation' is investigated it is essential to derive a recreation classification (Burton, 1967). Discussing the demand for recreation, Burton postulates three types of pursuits, as detailed in Table 9.1. It is clear that, whilst the third category is essentially associated with biological conservation, there may be competition between the conservation interests and the recreational users of the second category. Burton further divides the classification according to the timing of the pursuit, whether it be seasonal or all year or whether it be overnight or daily. Such divisions are of minor interest in conservation studies, since the very nature of conservation implies that it will be an all year activity, and hence the land cannot be handed over to the exclusive use of the recreation interest for part of the year. Thirdly, Burton divides the location of the pursuits into three locational categories, listed in Table 9.2, where it will be seen that these demands on location are important in determining the level of action required in conservation management. We can thus see that an integrated form of land use can only be developed between a few recreational activities and biological conservation. These

TABLE 9.1 The three categories for the types of recreation pursuits given by Burton (1967)

Pursuit	*Examples*
Cultural	Attendances at theatres, museums and art galleries, and participation in amateur plays, concerts and exhibitions.
Sports and Physical	Participation in such traditional games as football, cricket and hockey and, also, in the (relatively) newer pursuits such as golf and water skiing.
Informal	Walking, driving for pleasure, camping, taking picnics and informal nature studies.

TABLE 9.2 The three categories for the location of recreation pursuits given by Burton (1967)

Location	*Definition*
Local demands	Demands which are concentrated upon an area within a radius of, say, 5–10 miles of the participant's place of residence.
Regional demands	Demands which fall within an area of radius of, say, about 25 miles or an hour's travel or some time-distance factor equivalent to these.
National demands	Demands upon which there appears to be no time-distance limitation.

recreational activities can be defined as being all year, informal pursuits, either with a local, regional or national demand.

The classification of recreation supply has two aspects, namely the facilities available and the source of supply of these facilities. The conservation interests are themselves the source of supply, and hence discussion will centre on the facilities. Burton classifies the facilities into two categories, 'user-orientated' and 'resource-based', the latter being defined as those facilities 'whose location is determined primarily by the fixed location of the major resources included in them. Thus, for example, National Parks and Areas of Outstanding Natural Beauty are based upon areas of unique landscape which are fixed in location'. Each of these classes of facilities is further divided into 'single use' and 'multipurpose' facilities. The latter is defined as those facilities which 'can be used for recreation pursuits and other purposes, such as nature reserves, or for a number of different recreation pursuits, such as sports halls and flooded gravel pits'. Thus, the supply of facilities by the conservation interests can be classified as resource-based and multipurpose.

If we imagine an area of land whose primary use is biological conservation, then the supply of recreation facilities is not only a measure of the recreational potential of the site, but this supply is regulated by the nature of the conservation management. The supply is thus controlled by the conservation interest, but the demand for recreation facilities, also a measure of the recreational potential of the site, is generated by forces beyond the control of the conservation interest. If there is to be no direct confrontation of these two competing forms of land use, then some method of integrating the recreational and conservational uses needs to be found.

Forestry has, perhaps, been the form of land use that has been integrated most with recreation and with the provision of recreational facilities (Douglass, 1969). However, in searching for integration between recreation and other land uses there are three questions that will need to be asked and answered, namely:

(1) What will be the biotic effects of the recreational pursuits on the environment?

(2) How can the recreational demands be assessed and perhaps compared with the conservation usage of the environment?

(3) What planning is required so that the two forms of land use are integrated one with the other, and not competing with each other?

These three questions will be discussed in detail in the three remaining sections of this chapter.

The effects of recreation pressure on the ecosystem

Conservationists, thinking of preserving a habitat, might be tempted to say that any recreational pressure was bound to be adverse to the environment. However, the discussion in Chapter 5 showed that almost all populations can be exploited, and that a population that is being under-exploited is being mismanaged. In natural communities the situation may not be so clear-cut, since an under-exploited population may be experiencing a higher natural mortality rate (perhaps due to inter-specific competition) than an exploited population. Such would be an example of a homeostatic mechanism, where the level of mortality is approximately the same whether it is made up completely of natural mortality or of a mixture of natural mortality and a harvest.

In considering the recreational pressures on populations, a harvest is being taken. This harvest can be direct, as with wildfowling, or indirect, as with the wearing of vegetation, a function either of trampling or of mowing or of both. Inevitably some populations are more productive than others, and hence for each population a balance will have to be determined where the production of the population approximately equates to the recreational harvest plus the natural mortality. Environments will also differ in the potential harvest that they can produce if properly managed. An analysis of environmental stability will therefore be necessary in order to determine how this component of the ecosystem will react when the pattern of land use is changed or modified.

This implies that there are three features of the recreational use of land which merit consideration. The following sub-sections deal with these each in the form of an example, namely:

(1) direct recreational harvest: example the controlled wildfowling on nature reserves;
(2) indirect recreational harvest: example the effect of horses trampling plant communities;
(3) the response of the abiotic environment to recreational use: example the stability of the sand dune ecosystem.

Wildfowling

Wildfowling has been incorporated into the general land use policy of several nature reserves in Britain, as, for example, at Caerlaverock in Dumfriesshire (Huxley, 1964) and at Aberlady Bay in East Lothian (Usher, 1967; reproduced in the latter part of Chapter 10). Most wildfowl are migratory species that winter in Britain, and nest in remote, widely dispersed habitats within the Arctic Circle. Hence, although a harvest may be taken on a nature reserve in Britain, the conditions at the birds' breeding grounds at a different season of the year may be critical for the species conservation. Policies that are applied at one habitat may not be of much importance if the policy at the other habitat is inappropriate for the species. We shall, however, assume that there are no influences acting to drastically reduce the populations at the breeding grounds, and thus concentrate upon the management of the wintering grounds where shooting can be controlled.

During the winter there is also much movement of birds between areas, not only between sites within Britain but also between sites in Britain and on Continental Europe. General data on some species of wildfowl in Britain is given in Table 9.3, where there is no indication that there is excessive mortality, or imposed mortality, at the breeding grounds. Overall, the populations of most species are stable or are increasing slightly in Britain. Unfortunately there is no similar data available for other European populations.

Huxley (1967) discusses the control of wildfowling on reserves, working from the premise that where possible, subject to certain human needs, 'wildfowl stocks should be maintained in enough places around the country to give reasonable opportunity for some people to watch wildfowl and also for some to shoot them'. The issuing of permits for shooting forms

TABLE 9.3 Changes in the population of some species of wildfowl and the estimated over-wintering population size, in Britain in 1963. The data are extracted from Huxley (1967)

Species	*1963 Population*	*Notes*
Mute swan	18,000	Population has increased, now stable
Bean goose	–	Population has decreased in the past years
Greylag goose	36,000	In 1962, 36,000 were counted, probably the largest number to occur in Britain in the twentieth century
Shelduck	50,000	Population has increased
Teal	150,000	Population has increased; since 1963 some evidence has suggested that this may be due to teal coming to Britain from the Continent, where it is decreasing possibly due to the Dutch Delta Scheme
Mallard	500,000	A small population increase from 1948 to 1962
Wigeon	250,000	No sustained trends in numbers between 1948 and 1962
Pochard	13,000	The population has doubled between 1948–51 and 1960–62

the best method of control (Huxley, 1964), since this limits the number of wildfowlers, and conditions can be applied to the permit, such as the maximum number of birds or a limitation on the species that can be shot. Permit renewals can be made as provisional upon a return being made on a previous permit, and hence statistics can be kept on the harvest that has been taken. Thus, although there is a certain loss of freedom of action for the individual wildfowler, such data and control systems should ensure that there are sufficient stocks of birds and that disturbance, especially at the roost, is minimized. A system of permits can lead to an improved recreational experience, whilst at the same time ensuring the stability of the birds' population on the reserve.

Huxley (1967) concludes with two main reasons why wildfowling is compatible with conservation. First, he could find no evidence to show that mortality due to wildfowling is reducing the wintering wildfowl populations. Secondly, wildfowling is a resource-based recreational activity, and hence any over-exploitation will imply that there is a decrease in the recreational experience, and it is in the interests of the recreational user to conserve the populations. Thus the aims of a wildfowler and a biological manager wishing to preserve a population of wildfowl are very similar, and the two forms of land use can be integrated provided that a control system on the shooting can be established and implemented. Feed-back from this control system, in the form of data on the numbers shot, provides the data on which forward planning and management can be based.

Trampling

In discussing wildfowling there was a very quick feed-back from the population of wildfowl to the wildfowler and thence back to the manager. Unfortunately, the general public using the countryside is not so intimately aware of the conditions of the population that it is exploiting, and hence any feed-back is very much slower. If the trampling of vegetation is to be considered, it is unlikely that this vegetation is the main resource upon which the recreational user relies. For example, in Watson's (1967) studies on public pressure in the Cairngorms in Scotland, the resources from which the public derive the recreational experience are the ski lift and the scenery, and certainly not the soil, plants and animals in the vicinity. Trampling by human feet and, particularly, damage during the construction of the ski lifts have resulted in areas that are almost bare of vegetation. Watson's study has indicated that several species are relatively resistant to trampling, amongst which are the herb *Rumex acetosella*, the grasses *Poa annua* and *Aira praecox*, and the mosses *Polytrichum juniperinum* and *P.piliferum*. These species, together with the natural colonizers of gravel and peat outwashes, are being used to assess the best performance in rehabilitation experiments. Whilst these experiments are being carried out the public has to be excluded either by fences or by the diversion of footpaths.

A more intensive study of the effects of trampling upon a vegetation sward has been carried out in relation to galloping on race courses (Perring, 1967). The study area of approximately 810 ha was situated at Newmarket Heath on the border of Cambridgeshire and Suffolk, an area of Middle Chalk overlain by a thin layer of glacial drift. The soil contains a large quantity of free calcium carbonate, and has a pH of about 7·6 to 7·9. Although the site factors of the heath are relatively uniform, the amount of galloping in different areas varies, and Perring considers this to be the most important variable in determining the floristic composition of the heath.

Perring divides the plants of the heath into four categories, namely:

(1) Heath species, which, in East Anglia, are always associated with undisturbed chalk grassland.

(2) Pasture species, which are characteristic of swards other than chalk grassland, though they often occur in small quantity in chalk grasslands.

(3) Broad-leaved heath species, similar in character to group 1, but

which are easily recognized and could thus be used for their predictive value.

(4) Broad-leaved pasture species, similar to group 3, but which have the same recognition and predictive properties as group 2.

The heath is divided into five use categories, ranging from light use (less than 200 horses per year) to heavy (about 1350 horses per year) and very heavy use. In a graph of the percentage composition of the sward by groups of species against the use categories, Fig. 9.1(*a*), it is clearly shown that trampling by horses can substantially change the species composition of the sward.

But which species are diagnostic? Amongst the 12 sensitive heath species are rockrose, *Helianthemum chamaecistus*, thyme, *Thymus drucei*, and several species of vetches (family Leguminosae), all of which were only recorded on the lightly used areas. Two of the eight more resistant heath species, salad burnet, *Poterium sanguisorba* and red fescue, *Festuca rubra* are shown in Fig. 9.1(*b*). It can thus be seen that the presence or absence of the species in these lists is indicative of the degree of trampling. However, there are other species that become more abundant in swards that are heavily trampled. Amongst Perring's list of nine species with this characteristic are daisy, *Bellis perennis*, and bindweed, *Convolvulus arvensis*, which only appeared in the very heavily trampled areas. Amongst the four species that are listed as resistant, but becoming more abundant with trampling, is rye-grass, *Lolium perenna*, which is shown in Fig. 9.1(*b*).

Perring's study shows not only the qualitative changes in the sward, but also a measure of the quantitative change in species composition in relation to the intensity of trampling. The heavily trampled area, with 1350 horses per year, represents considerably less trampling than recreational users would give vegetation in the vicinity of car parks or other recreational attractions. What sort of human pressures can be accepted before the environmental change is too great (either in species composition or in the creation of bare ground), and how is this related to the abiotic factors of the site? There are as yet insufficient studies to enable one to quantify the amount of human trampling required to change a vegetation sward by some stated amount, or to create no more than a predetermined amount of bare ground. Recreational pressures have been more intense on the coastal areas than on inland areas, and hence an assessment of the relations between the intensity of coastal recreation use, the amount of floristic change and the environmental stability can be made.

Sand dunes

Sand dunes have often been studied since they show a succession of communities from those of the drift line to those of the dune meadows,

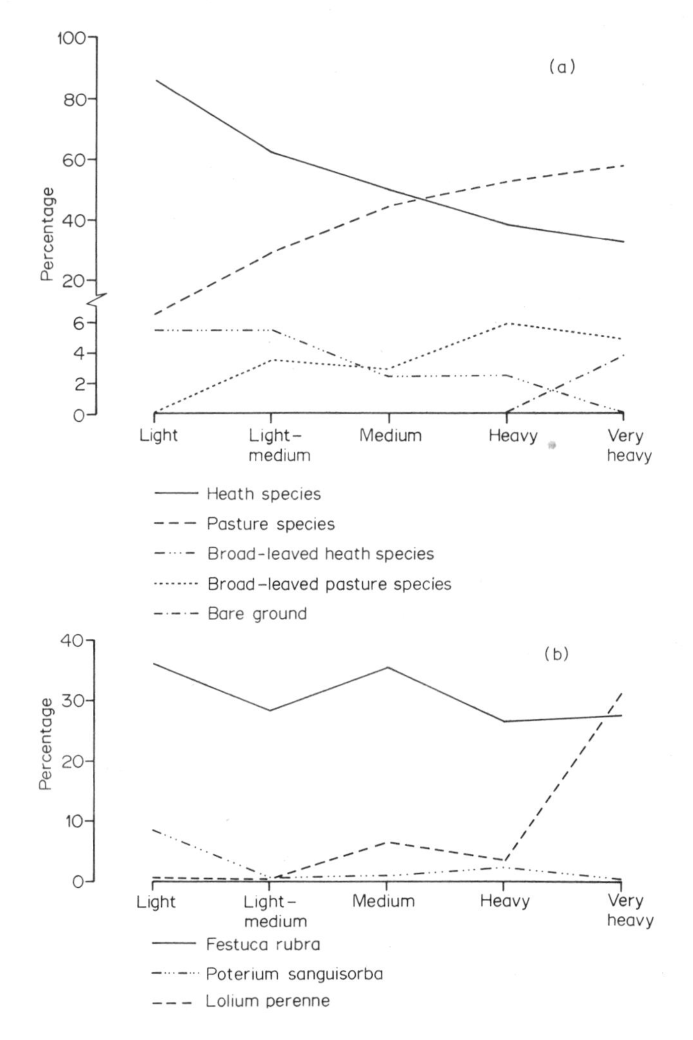

Fig. 9.1 The response of plant species to trampling by horses. (a): The four groups of plants listed in the text and the bare ground. (b): Three selected species which are discussed in the text. (Plotted from data given by Perring, 1967)

scrub and plantation (Chapter 2). A generalized section through a dune system, Fig. 9.2, demonstrates that the different stages in the succession are spatially separated at one period of time. It is only relatively recently that the effects of trampling have become important in the dune ecosystem, and research has changed from recording the succession, and its variants, to being concerned with the management of dune systems. A succession and its associated species was discussed in Chapter 2 for the dune system at Tentsmuir in Eastern Scotland, but here the more general characteristics of dune succession will be investigated in order to see what recreational potential the stages have.

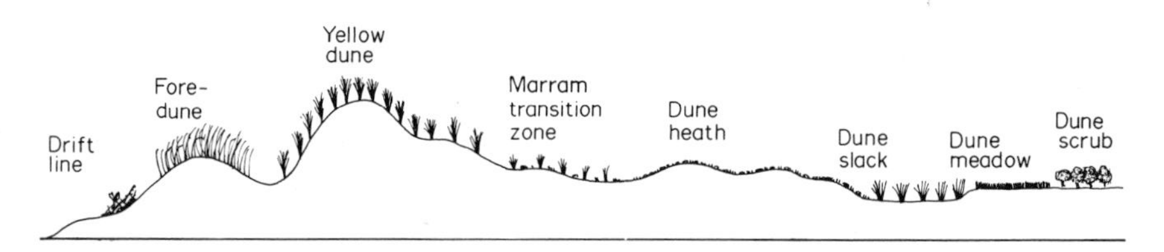

Fig. 9.2 A diagrammatic section through a dune system to characterize zones for recreational use. (Redrawn from Duffey, 1967)

Duffey (1967) gives one scheme for characterizing the zones of a dune system, Fig. 9.2, for the purposes of recording animals, according to the structure of the plants. Thus, the first zone is the drift-line, an accumulation of dead organic material, marine algae, driftwood and other debris, whose position and composition varies throughout the year. The second zone is the fore-dune, which occurs to the seaward of the main dune formations. Fore-dunes are low, usually less than 1 m in height, but varying in width from 2 to 100 m. They are floristically poor and typically dominated by sea couch, *Agropyron junceiforme*, though there will be considerable areas of exposed sand. These first two zones can thus be considered as the beach, and will be the location of the more active recreational pursuits such as beach games.

The third zone is the yellow dune, which is the main area of the marram grass, *Ammophila arenaria.* The vegetation is tussocky with wide areas of sand. The dunes commonly reach heights of 10 m, but exceptional heights of over 30 m can be recorded. This area, immediately behind the shore, is used extensively by recreational users since it provides shelter from the sea winds, and forms an ideal picnicking area, which can be relatively secluded from other people. The fourth zone is a transition zone between the marram grass and the dune heath or the grey dune. In this zone there is still a development of the marram grass, though not tussocky, with other plants and with 20 to 30 per cent of bare ground. Recreational pressure on this zone, trampling the vegetation and exposing the sand, has caused the yellow dune to spread inland.

The zones to the landward of the transition zone are of perhaps less importance recreationally, although the public has to cross them before reaching the shore. From the point of view of protecting rare species it is these zones, particularly the dune slacks, that are the most important. There is the constant threat that sand, mobilized by over-use in the forward zones, will blow in and smother these landward zones.

The experience with sand dunes in the vicinity of the town of Gullane, on the south shore of the Firth of Forth about 20 miles east of Edinburgh, will demonstrate the effects of recreational use. Gullane Bay and beach is about three quarters of a mile long and is separated from the town by a series of dunes. In 1952 about 10,000 cars visited the area, but this total had increased to between 22,000 and 25,000 in 1959–1961 (East Lothian County Council, 1962). The initial effects of this recreational pressure (East Lothian County Council, 1955) resulted in the loss of stability of the dune system. In the words of the Council's earlier report:

> 'I believe that if nothing is done the position will go from bad to worse; increased sand blow will cause it [the dunes] to become a desert and a most terrible nuisance to adjoining occupiers. Increased litter and devastation will result in fewer and fewer car owners visiting the dunes . . .'.

Two features of the recreational use are thus evident. First, the foredune and yellow dune zones, in which the recreation has been concentrated, have suffered from wind erosion. The main species of plant is the marram grass, which has an extensively creeping stem, rooting at the nodes, which binds the sand. Trampling has resulted in the breakdown of this stem system, and the north-easterly gales of the autumn and winter are for the most part responsible for the blow out of the sand. Wind erosion has meant that whilst some parts of the dune system have been eroded other parts have had sand deposition, and in both cases the plant and animal communities have been partially or completely destroyed. The area has therefore become poorer for both conservation and recreation.

Secondly, it can be seen from the County Council's report that planning is required to furnish the amenities that recreational users require, such as provision for the removal of litter. These aspects will not be discussed here since planning forms the topic for the last section of this chapter.

For quantification of the erosion caused by recreational users a sand dune area in Lincolnshire, at Gibraltar Point (Schofield, 1967) can be discussed. Here, interim figures indicate that 7500 people per season walking off a concrete path and on to a mature saltmarsh cause complete loss of the vegetation cover. A similar density walking over yellow dunes completely eliminates marram grass, sea couch, and prickly saltwort, *Salsola kali*, and results in considerable dune erosion. If protection is then afforded to the eroded areas a recovery is complete in about four years. On the grey dunes, or dune heath, between 3500 and 4000 people cause local exposure of soil and sand.

Using these data it can be seen that even the 1952 density of recreational users at Gullane would have been sufficient to cause wear on the vegetation in parts of the dune system. Since people, having parked, will tend to follow a path, a linear strip of dunes is the first to be damaged, and thereafter vulnerable to gales. Once the process of erosion has started it can be self-perpetuating for a while since sand can be eroded from beneath the undamaged stem structure of the marram grass. Thus, if both the structure of the dunes and the potential for recreational use are to be maintained, some degree of planning and control is essential.

The assessment of demand

Practices of demand assessment can be divided into two phases. First, there is the collection of data, and secondly there is the analysis of this data. The first of these phases is associated with the determination of opinions, with the elucidation of ideas that may not be clearly formed in people's minds. Such data are gained by survey methods, relying upon the completion of a questionnaire by a sample of users or by a sample of the population from which the users are drawn. The second phase, the analytic phase, is an attempt to quantify the survey data, to determine significant differences and relations, and to build the results into more general theories of recreation demand. In discussing the analysis of data later in this chapter the theoretical side of recreation demand will be stressed, and it will be assumed that the reader is familiar with the more simple statistical analyses such as the determination of confidence limits, the χ^2 analysis of contingency tables, and the fitting of linear regression equations. Text-books that discuss these subjects include those of Bailey (1959) and Sokal and Rohlf (1969). Since these theoretical analyses in general involve the economic theory of supply and demand, it is assumed that the reader is familiar with such economic concepts. A text-book that discusses the subject is that of Lipsey (1966).

Examples of both the design of the questionnaire and the analysis are drawn from a survey of visitors to Spurn Peninsula Nature Reserve, carried out during the summer of 1970 (Pitt, 1971). The reserve is situated at the extreme south east of Yorkshire, occupying the terminal $5\frac{1}{2}$ km of a curved peninsula which extends half way across the Humber estuary. A 240–250 year cycle from peninsula to peninsula via a breach, island formation and the deposition of sand has been described by de Boer (1963). The location of the reserve at this part of the eastern coast of England has led to its importance not only for military activity but also for the migration of birds and insects (Countyside Commission, 1969b). The reserve, of approximately 100 ha and with 11 km of shore, half facing the North Sea and half facing the Humber estuary, provides a popular recreational area in an interesting geographical setting. It was because of the combined conservation and recreation importance of the site that the survey was planned.

The design of questionnaires

The questionnaire employed during the survey at Spurn is depicted in Table 9.4. This design was influenced by the suggested classifications of the Countryside Commission for such surveys (Davidson, 1970), though modifications were introduced as a result of a short pilot survey.

A certain amount of information about the interviewee was obvious to the interviewer without any questions being asked, and this is recorded at the top of the questionnaire form. The position of the interview was recorded since interviewers either surveyed people as they left the reserve, or they surveyed people *in situ* on the reserve. The results of these two sub-samples do, in certain instances, show significant differences. The results show that 96 per cent of groups brought their cars on to the reserve (for which a charge of 25p was made in 1970). The group structure was recorded as a count of the numbers of adults and juveniles (subjectively assessed) in each group.

TABLE 9.4 The questionnaire for the 1970 Spurn Peninsula Survey

SPURN PENINSULA SURVEY 1970.

Serial No. of interview /
Weather
Position

Time and Date / /
State of Tide
Method of Transport
Group structure

I am carrying out a survey on behalf of the Yorkshire Naturalists' Trust to discover how Spurn Peninsula is used by visitors. Could you help by answering a few questions?

1. Are you a member of the Yorkshire Naturalists' Trust? — Yes/No
   If 'NO' – Did you know that this area belonged to the Yorkshire Naturalists' Trust? — Yes/No
2. Could you tell me which of these activities (HAND CARD) you have done while you were here today/you intend to do while you are here today?
   Picnicking
   bird-watching
   other nature-study
   fishing
   swimming
   sun-bathing
   walking
   photography
   others
   .... SPECIFY OTHERS HERE ....
3. Which of these would you say was the main object of your visit. — ....
   SPECIFY OTHERS ....
4. HAND MAP TO THOSE AT EXIT AND ASK QUESTION, OTHERWISE RECORD POSITION YOURSELF. Here is a map of the area. Could you tell me where you spent most of your time? — ....
5. How long have you been/do you intend to stay on the peninsula? — ....

Clearly some of the Trust's concern is with the conservation of natural habitats.

6. Do you think that areas of the country should be preserved in their wild state as a matter of principle? — Yes/No
7. Do you think that Spurn should be managed in such a way? — Yes/No
8. If a Nature Trail were available, would you use it?
   EXPLAIN, IF NECESSARY — Yes/No
   If 'YES': What do you feel is a fair cost for a descriptive booklet — ....
9. Here are some of the things that have attracted visitors to Spurn (HAND CARD). *a*) peace & quiet *b*) open spaces *c*) wild birds *d*) sandy beaches *e*) passing ships
   Which of these pleases you most? ....
   Does anything not on this list attract you more? — Yes/No
   SPECIFY
10. Here are some of the things that other visitors have disliked about Spurn (HAND CARD). Which of these displeases you most? *a*) litter *b*) derelict buildings *c*) wind *d*) mist *e*) driving and parking difficulties *f*) lack of toilets. — ....
    Do you find anything not on this list unattractive? — Yes/no
    SPECIFY

TABLE 9.4—*continued*

11. Here are some suggestions that have been made to improve the area (HAND CARD). *a*) litter bins *b*) more parking spaces *c*) more toilets *d*) pull down the old buildings *e*) a place to turn the car at the end. Which of these would you like most to see carried out?	....
Would you like to see any of the others carried out?	....
Would you like any improvements not on this list?	Yes/No
SPECIFY	
12. Is your visit to Spurn a day-trip or part of a holiday?	day trip/holiday
13. Where have you come from today?	....
14. Where is your home?	....
15. Have you been to Spurn before?	Yes/No
If 'YES'	
16. When was your last visit?	....
17. How many times have you been here since this time last year?	....
18. How many times did you come in the previous twelve months?	....

After a short preamble to make clear the aims of the questionnaire, the first question aimed to ascertain the visitors' knowledge of the nature reserve status and of the ownership of the area. A prominent sign at the entrance to the reserve displayed the ownership, but yet 27 per cent of the 87 per cent of visitors who were not members of the Yorkshire Naturalists' Trust were ignorant of the ownership.

The second and third questions were concerned with the activities of the visitors. Davidson (1970) makes the point that such classifications should have meaning in planning terms, but in this survey it will be seen that the recreation choices are resource-based and of short duration. Because of this restriction to the available recreation activities, the classification is a basic classification without any secondary or lower level divisions. The two questions were posed so as not only to gain an insight into the spectrum of activities, but also to determine the primary reason for visiting the nature reserve. The fourth question was related to the previous two questions since it was designed to determine if integration of two or more land uses by zoning on the reserve was possible. The fifth question was concerned with the time spent on the reserve, and a modification of Davidson's classification was adopted (Table 9.5). The fact that 75 per cent of visitors spent two hours or more on the reserve is of importance in planning for amenities.

The next three questions, 6 to 8 inclusive, were included to determine the visitors' reactions to conservation, and particularly to the

TABLE 9.5 A classification of the time spent on the Spurn Peninsula Nature Reserve. The time intervals are similar to those used by Davidson (1970)

Time spent on reserve	*Percentage of visitors*
less than $\frac{1}{4}$ hr	0
$\frac{1}{4}$hr and less than $\frac{1}{2}$hr	1·2
$\frac{1}{2}$ hr and less than 1 hr	3·8
1 hr and less than 2 hr	19·9
2 hr and less than 4 hr	38·7
4 hr and less than 6 hr	28·2
More than 6 hr	7·0
overnight	1·2

conservation interest at Spurn. Answers to question 7 (96 per cent of interviewees said 'Yes') form a biased sample, since only visitors were questioned and there was no attempt to question the non-visitors. However, it shows a general favourable public response to the conservation of natural and semi-natural habitats. 74 per cent of interviewees indicated that they would use a nature trail, a factor vital in the educational effort of conservation (see Chapter 8).

Questions 9 to 11 inclusive were included to investigate the opinions held about the reserve. Although lists (there were several lists with the headings arranged in random orders) were handed to the interviewees, the subjects on these lists had been determined in response to questions asked in the pilot survey. Such data only record the instantaneous opinions, though it was evident that these questions led directly to brief discussions between group members. The questions were purposely made 'open-ended' by asking respondents to name things not on the lists. It became evident from the answers that 'peace and quiet' was the greatest attraction, that derelict buildings were the feature most disliked, and the removal of these buildings and the provision of litter bins were the amenities most in demand.

The remainder of the questions were designed to gather data for the economic analyses. Thus, question 13 was used to determine broad zones from which visitors came, whereas questions 12 and 14 gave data on the pattern of visiting from within and without Great Britain. The questions on visitation rate, 16 to 18 inclusive, were asked in order to attempt to estimate any possible trends in recreation demand. It was found that 34 per cent of groups were on their first visit, a recruitment rate more than double those determined on other nature reserves in Yorkshire by Usher, Taylor and Darlington (1970).

Each survey must have a clearly defined objective, and the questions will be related to that objective. The Spurn survey was designed to show the opinions of countryside users towards conservation and recreation within this one clearly defined area, and the data was collected so as to form a quantitative basis for the preparation of a reserve management plan. Davidson (1970) shows how other kinds of questions may be asked, and gives methods for classifying the results. Thus, if there has been a publicity drive, it may be desired to survey the effect by asking interviewees either where they first heard of the area or where advertisements have been seen. Information may be required for a future policy of charging for amenities, such as an entrance charge, a parking charge or a charge for information. It is often useful to estimate the willingness to pay by relating the question to existing prices, for example 'are the prices too high, too low or about right?' Where there are no existing prices, the questioner can either start high and reduce the price until the interviewee agrees, or the interviewee can be asked to suggest a price. When the latter approach was adopted towards the nature trail at Spurn (question 8), 53 per cent of the potential users of the trail suggested between $7\frac{1}{2}$p and $12\frac{1}{2}$p inclusive, 23 per cent sums less than $7\frac{1}{2}$p, and 24 per cent sums greater than $12\frac{1}{2}$p. It would seem that the most acceptable charge would be either 10p or $12\frac{1}{2}$p, a cost which would be taken into consideration when a trail guide was prepared.

Thus, in any survey work the first task is to specify the objective, and the second is to devise a questionnaire such that the questions are not ambiguous and yield the information that is required. There is much to be gained from a relatively standardized approach to the classification of answers to questionnaires since this leads to comparability between surveys. An approach towards such standardization has been one of the first tasks fostered by the Countryside Commission.

Economic analysis of survey data

Recreational and conservational forms of land use cannot be assessed in direct economic terms, and hence direct economic comparisons cannot be drawn between these forms of land use and other rural uses such as agriculture and forestry. Recreation can be conceived of as a product, both a spiritual product of renewal and an aesthetic product are usually associated with countryside recreation. Conservation cannot be conceived in such terms, although it is related to the informal, resource-based recreation pursuits since both require open countryside and a degree of 'peace and quiet'. Some methods have been devised to assess the benefits of recreation, though none have been developed for an economic assessment of conservation.

One of the first of the recreation studies was that of Trice and Wood (1958). The method determines the maximum that people will pay for the recreation experience, the market price, which is multiplied by the number of visitors to give the market value. From this are subtracted the travel costs and other expenses, and the result has been called the Free Recreational Benefit (FRB).

One subjective element of this calculation immediately arises, and that is to fix the market price. This could be assessed as the maximum payment that any group made in order to come to the reserve, but in practice this would over-estimate the actual market price. Trice and Wood suggest using the average price paid by the 10 per cent of visitors who pay the most.

As an example of the analysis consider the reserve indicated in Fig. 9.3. The outer zone, zone 4, from which 10 per cent of the visitors come, delimits the market zone, and the average cost of a visitor from this zone is b_4. If we consider any other zone, say the ith where $i = 1$, 2 or 3, then the average travel cost is b_i, and for each visitor the free recreational benefit is $b_4 - b_i$, the difference between the actual cost and the market price. For

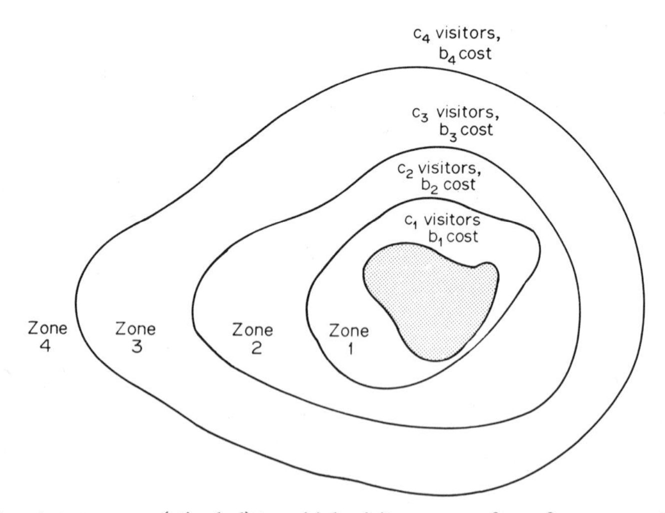

Fig. 9.3 A nature reserve (stippled) to which visitors come from four zones. The model is discussed in the text

all visitors from this zone the FRB is $c_i(b_4 - b_i)$. Summing for all visitors from all zones, the FRB is given by

$$\text{FRB} = \sum_{i=1}^{3}\{(b_4 - b_i)c_i\} \tag{9.1}$$

Question 13 of the survey determined the origin of the day visitors to Spurn, and from these replies eight zones have been determined using also the availability of census data. A map of northern England, Fig. 9.4, shows the location of each zone, and some of the data are given in Table 9.6. The cost is derived by measuring the straight line distance from

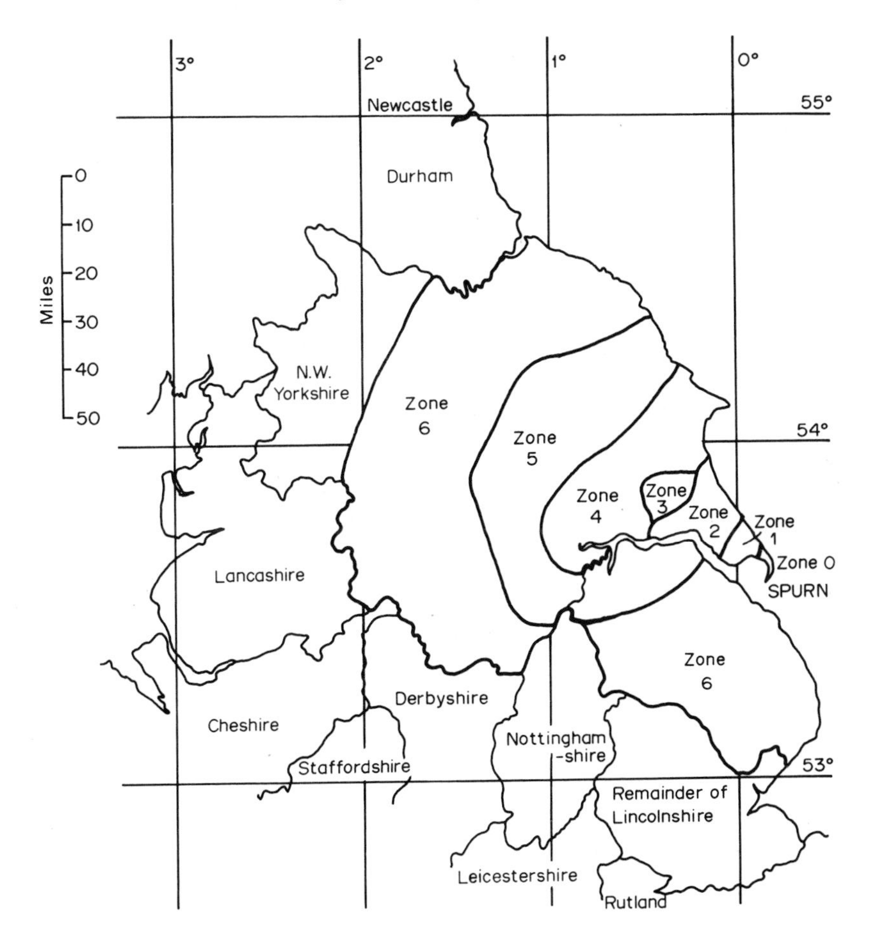

Fig. 9.4 A map of north-east England showing the zones from which visitors came to Spurn Nature Reserve

Spurn to the centre of population of the zone, and doubling this for the return journey. This distance has been multiplied by 1·15 to convert it into road miles, and further multiplied by 1·92 to convert it to the average running cost in pence of a car (Automobile Association, 1970). Zones 6 and 7 have been combined since approximately 10 per cent of the visitors come from them. This sets the market price at £4·03. Summing over zones 0 to 5 as in Equation (9.1) gives

$$\text{FRB} = £3079{\cdot}70$$

for the sample of visitors included in the survey. Division of this figure by the proportion of all the visitors who were sampled would give the FRB for the months July to September. Further division by the proportion of visitors during these three months to the number who come throughout the year would give the FRB on a per annum basis.

There are many over-simplifications in the Trice and Wood analysis. One has been the determination of the market price, but what does one include in the expenses of the recreational experience? In the analysis carried out above, Table 9.6, the cost has been assessed as the running cost of a car plus the 25p admission charge. Should the full cost of running the car, including insurance, garaging and depreciation, have been used instead of the running cost? Should some value be put on the time taken in making the journey to the reserve? Should all incidental costs, such as refreshments, also be included in the cost? There is no standard measure of the costs to include in such analyses. In addition it is assumed that all users derive equal satisfaction from their recreation experience, an assumption implying the homogeneity of the population. It is possible, for example, that visitors travelling the furthest distance come

TABLE 9.6 The data used for a Trice and Wood analysis of the visitors to Spurn Peninsula Nature Reserve

Zone	*Number of groups*	*Total cost (p)*
0	138	31·62
1	151	69·16
2	634	130·98
3	49	157·48
4	61	201·64
5	70	267·88
6 }	98 } 114	378·28 } 403·07
7 }	16 }	554·92 }

from higher income groups, and will thus pay more than the average person for their recreation. If such were the case it would artificially raise the market price. Criticisms of the analysis have been discussed by Smith (1968).

Another basis on which recreation can be assessed is due to Clawson (the methods are described by Smith (1968) and Smith and Kavanagh (1969)). The method essentially aims at deriving a demand curve, and then setting entrance fees (or other charges) so as to estimate an expected income and an expected visitation rate. The data for Spurn are given in Table 9.7. The population figures for each of the zones were derived from

TABLE 9.7 Data for Spurn Peninsula Nature Reserve used in a Clawson analysis. The population data is based on the national census of 1961, and the visitation rate is expressed as groups per 1000 of the population

Zone	*Number of groups*	*Population*	*Visitation rate* V	*Cost* (p) C
0	138	600	230·00	31·62
1	151	7,800	19·359	69·16
2	634	366,000	1·7322	130·98
3	49	39,200	1·2500	157·48
4	61	108,800	0·56066	201·64
5	70	633,400	0·11052	267·88
6	98	4,172,600	0·02349	378·28
7	16	12,760,000	0·00125	554·92

the 1961 census and are rounded to the nearest 100 people. The visitation rate is expressed as groups per 1000 of the population since the gate charge is per car and not per person.

In order to carry out the analysis, a demand curve has to be drawn and expressed by a regression equation. In Fig. 9.5 the natural logarithms of the visitation rate and the cost have been plotted, though the data for zone 7 have been omitted since they are anomalous. This is due to the fact that the origin of the day visit has been recorded in such general terms as 'Lancashire', 'Cheshire', etc., and therefore the whole of such areas have been included in the total population estimates. It is possible that only people from the east of Lancashire or Cheshire were capable of making the journey (because of the amount of time taken in travelling), and hence the population from which the visitors were drawn would be considerably smaller than those recorded in Table 9.7.

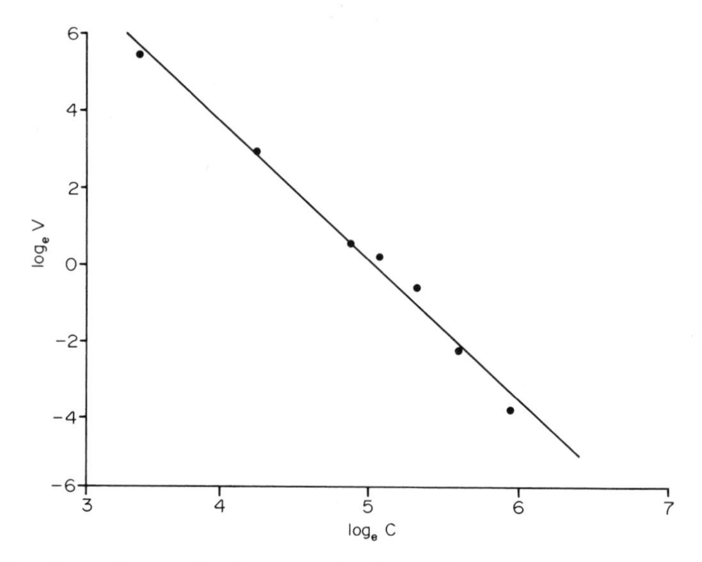

Fig. 9.5 The demand curve for the recreational experience at Spurn Nature Reserve. V represents the visitation rate and C the travel plus entry cost. Logarithmic scales are used for the axes, and the points represent the zones in Table 9.7

The regression equation is

$$\log_e V = 18{\cdot}261 - 3{\cdot}633 \log_e C \qquad (9.2)$$

or alternatively

$$V = 8{\cdot}522 \times 10^7 \times C^{-3{\cdot}633} \qquad (9.3)$$

These equations specify the demand curve for the recreation experience at Spurn, and from Table 9.6 we can postulate that a margin exists when $C = 403.07$p. When the cost of visiting Spurn is below this margin we can presume that people will travel to Spurn, and that the group visitation rate can be estimated from Equation (9.2) or (9.3). When the cost is equal to or greater than this margin we can assume that there will be no visits.

Since the sample included data for 1217 groups, each paying an entrance fee of 25p, the income is approximately £304. If, however, an entrance fee of 50p were to be charged, the application of Equation (9.3) would predict visitation rates as shown in Table 9.8. It will be seen that the cost of a visit from zone 6 has exceeded the marginal cost, and hence we can assume that there will be no visitors. Such a procedure can be

TABLE 9.8 The predicted visitation rate, number of visiting groups and income by fixing a gate charge of 50p at Spurn Peninsula Nature Reserve

Zone	*Cost* (p) C	*Visitation rate* V	*Population*	*Number of groups*	*Income* (p)
0	56·62	36·472	600	22	1,100
1	94·16	5·7468	7,800	45	2,250
2	155·98	0·91845	366,000	336	16,800
3	182·48	0·51937	39,200	20	1,000
4	226·64	0·23634	108,800	26	1,300
5	292·88	0·09311	633,400	59	2,950
6	403·28	0	4,172,600	0	0
Total	–	–	–	508	25,400

repeated for many different entrance charges, and a demand schedule drawn up which tabulates the entrance fee with the predictions on the number of visiting groups and on the income to be derived. The demand schedule for Spurn is shown graphically in Fig. 9.6, where the effects of the margin at 403.07p can be clearly seen.

Such an analysis shares some of the disadvantages of the method of Trice and Wood in that there is an assumption of homogeneity of the

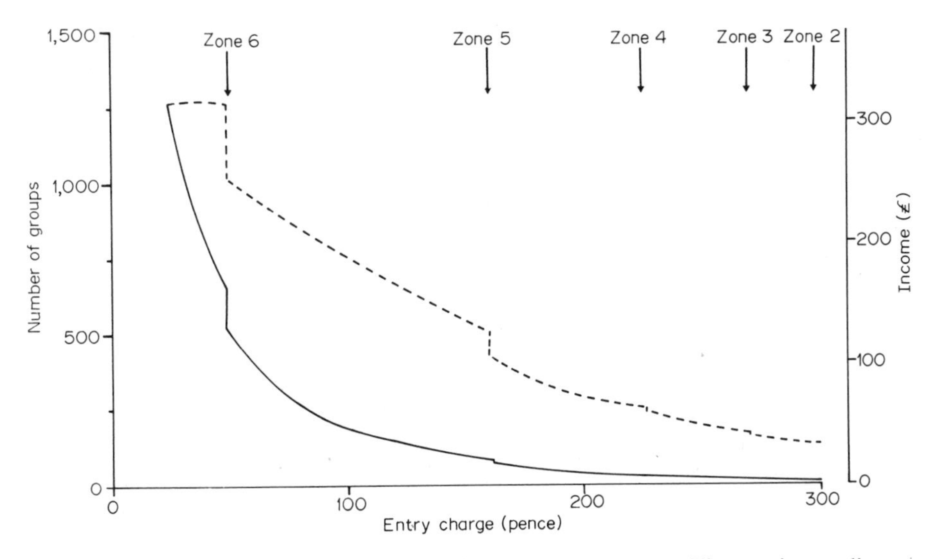

Fig. 9.6 The Clawson demand schedule for Spurn Nature Reserve. The continuous line shows the number of groups as a function of the entry charge, and the dashed line the revenue also as a function of the entry charge. The arrows indicate at what charge a zone becomes marginal and ceases to visit the reserve

population throughout the zones. The analysis also determines precisely a demand curve, Fig. 9.5, though this is liable not to be stationary but to move daily as a result of weather or the availability of other recreational areas.

However, the analysis has a direct use for the management of the nature reserve. It might, for example, be an aim of management to maximize the income to be derived from the recreational visitors. If this were the aim at Spurn, Fig. 9.6 shows that a maximum income could be expected when there was an entrance charge of 40p. However, when an entrance fee was first charged at Spurn it was not designed to bring in revenue, but it was levied in order to restrict the number of people entering the reserve. The conservation aim was to maintain the level of disturbance to the birds within acceptable limits. If a desired number of visitors can be specified, then an appropriate entrance fee can be determined from the graph in Fig. 9.6.

Surveys, since they are recording the people using the reserve, are essentially a part of recreation research and are not directly related to conservation research. It is essential, however, that some form of integration of competing, or potentially competing, forms of land use should be found. Where an aim of management is conservation, then this aim is likely to exist for maybe hundreds of years, and hence virtually all other forms of land use are potential competitors. I, personally, think that if wildlife is to be conserved in perpetuity, then it is essential to find ways of using land not just for the protection of the species but also for recreation and education. The demands of man upon the environment for living space, for his technology and for his leisure are increasing, but yet the supply of land is fixed. If the wildlife of the world is to get a slice of the land, then it must co-exist with man, and man must co-exist with it. The Spurn survey showed that the greatest recreation benefit of this nature reserve was the 'peace and quiet', a factor that is essential if the wildlife, particularly the diversity of bird species, is to be preserved. Recreation and conservation thus aim at maintaining the same site conditions. Survey and analysis has given an indication of how this can be achieved, hence avoiding the competition of these two forms for land use in the future.

Recreational planning

In the first sections of this chapter the ecological implications of recreation have been considered, together with the means of assessing and analysing 'demand' for countryside recreation. These form the two sets of data that are required for recreation planning, since the one determines what, if anything, is needed, and the other, the effects if implementation occurs. A third aspect, however, of the planning process is provision. It is unlikely that provision will completely satisfy the demand in all its forms, and hence provision must attempt to achieve a balance between the demand and the ecological effects of complete satisfaction of that demand.

In order to demonstrate recreational planning within a conservation framework, a coastal nature reserve in Southern England has been selected. Southern England has a large population, and, having a milder climate than other parts of Britain, the region attracts considerable numbers of holiday makers. The coast provides a resource-based recreational amenity, and wherever there is access it can be assumed that a demand will exist. The coastal survey of England and Wales (Countryside Commission, 1969b) has highlighted the many potential users of the coast, and the conflicts that can arise over multiple use. The Commission's chart of human impacts upon the coastal wildlife lists 19 categories subdivided into 61 sub-categories, only one of which concerns conservation operations. Amongst the recommendations in the Commission's report are several concerning planning (paragraphs 111 and 112), stressing the view that coastal conservation within only a narrow strip is unrealistic, and that planning should be based on wider zones.

Studland Heath National Nature Reserve

This reserve of 174 ha is situated seven miles west of Bournemouth at the southern entrance to Poole harbour on South Haven Peninsula (Fig. 9.7). The ferry from Sandbanks to South Haven Point gives the population of Poole and Bournemouth easy access to Ferry Road, and the Peninsula. There is easy access from Swanage as well. Access on to the reserve is mainly either from the north, people walking round the shore from Shell Bay, or from the south, since a large car park is situated a quarter of a mile to the south of the reserve's seaward boundary.

The reserve can conveniently be separated into two areas, the Little Sea and land to the west of it, and the land to the east of the Little Sea. The former area is used for recreation on only a very limited scale, as is indicated by Teagle's (1966) survey of visitors. The area to the east of Little Sea, composed mainly of dune ridges and dune slacks, is where the

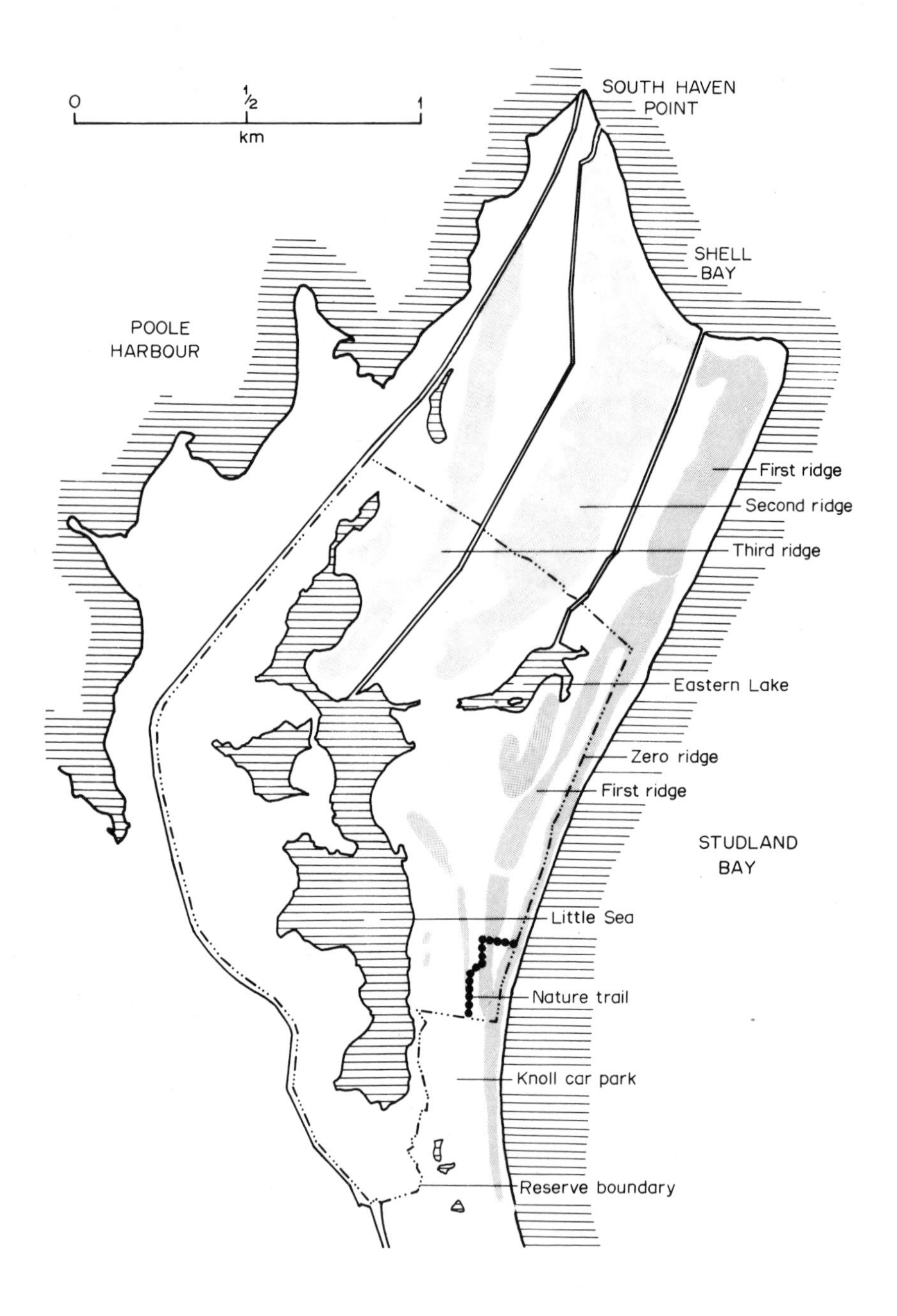

0
1/2
1
km
SOUTH HAVEN
POINT
SHELL
BAY
POOLE
HARBOUR
First ridge
Second ridge
Third ridge
Eastern Lake
Zero ridge
First ridge
STUDLAND
BAY
Little Sea
Nature trail
Knoll car park
Reserve boundary

majority of the recreational users are to be found. Four dune ridges appear on the Peninsula, as indicated in Fig. 9.7. On the fore dunes the sea couch, *Agropyron junceiforme* and marram, *Ammophila arenaria* are dominant. Marram is the dominant plant on the first dune ridge, and some of it persists on the Second Ridge, but it is entirely absent on the Third Ridge except in locally disturbed areas. On the Second and Third Ridges the heather, *Calluna vulgaris*, is the dominant species. The dune slacks support an acid marsh community, with thickets of sallow, birch and sweet gale, *Myrica gale*, together with tussocky purple moor grass, *Molinea caerulea.* As the succession progresses the birch becomes more abundant, eliminating the ground flora.

During the war the reserve was used for the experimental testing of weapons, and part of the beach was taken over as an experimental site for testing anti-invasion devices. The area was cleared in 1946–47, when extensive areas of heath and scrub were burned. Since 1945 there has been an annual increase in the number of visitors to the area, until by the early 1960s there were up to 10,000 people using the shore on one day. The Nature Conservancy negotiated a lease, and the area became a National Nature Reserve in 1962. The Nature Conservancy were faced with integrating the situation of indiscriminate recreational use with their overall policy of conservation and protection of areas of scientific interest.

A management plan (Hemsley and Copland, 1963) was prepared, which stated amongst the five principal objectives of management:

> 'To adopt a positive and progressive policy towards the visiting public, including provision of special facilities for organized use of the area for natural history teaching and general biological training'.

The public use is concentrated on the beach which adjoins the reserve's eastern boundary, with, on a fine summer day, there being more than 10,000 people per km of coastline. A survey of visitors during 1965 (Teagle, 1966) shows the zonation of use on the peninsula to be closely related to the parking facilities (Fig. 9.8). But, what are the recreational demands on the reserve? The prime demand is the natural resource of sea, sandy shore and a dune system for shelter and privacy. The amenities which are associated with such a resource-based recreation are few, but

Fig. 9.7 The Studland Heath National Nature Reserve. The map shows the main features of the reserve and the dune and slack systems. The reserve boundary is demarcated thus —···—···—···

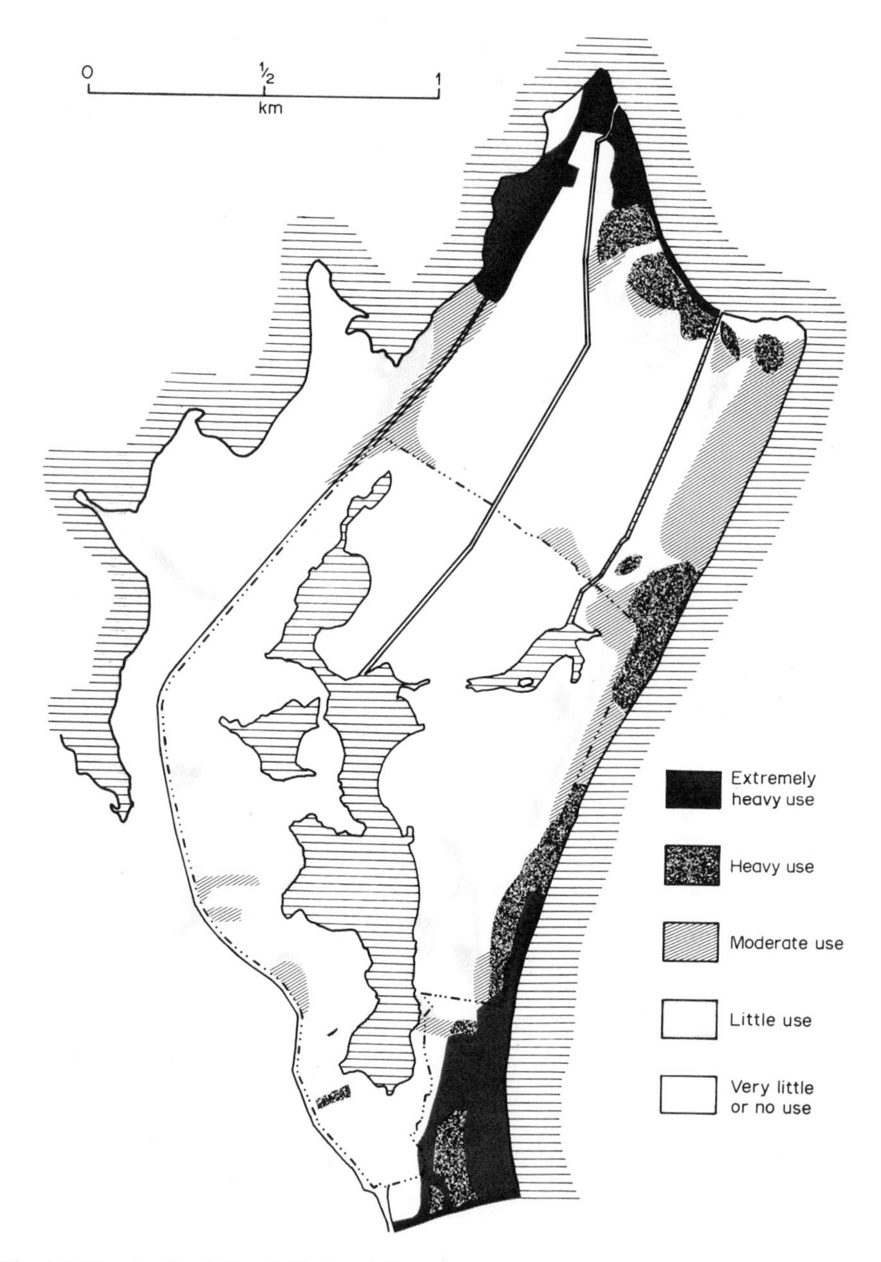

Fig. 9.8 The Studland Heath National Nature Reserve, showing the intensity of recreational use. (Redrawn from Teagle, 1966)

include parking facilities, litter collecting facilities, fire prevention and information. In general terms it can be seen that the car park acts as a 'honey pot', and within its vicinity a demand exists for the associated amenities. In this way some areas of the reserve are subject to heavy use, whilst areas more remote from car parks remain isolated. Since the Knoll car park was already in existence outwith the reserve boundary, the Nature Conservancy's planning of recreational amenities and controls must be influenced by this factor. It should be noted that the Nature Conservancy does not have the provision of recreational amenities as a primary objective, but it does provide such facilities within an overall planning framework with a nature conservation objective. The planning for fire, litter collection and information, three of the amenities previously listed, will demonstrate the means of planning and its relation with the ecology of the area.

The reserve management plan makes provision both for the purchase of fire-fighting equipment and for a fire-watching post on the highest ground in the reserve. The need for protection arises both from the recreational use of the area (three children were killed in a gorse fire in 1960) and for protection of the ecosystems. Teagle shows that there is a fire hazard throughout most of the year, and his map indicates that much of the peninsula has been burnt since about 1932. Fires have usually been started either by smokers or picnickers in the vicinity of Ferry Road, or by barbecues in the dunes. But what effects does fire have on the ecosystem? Though these have hardly been studied, it is evident that many of the animal species are destroyed if burning is uncontrolled. When the plant communities have been burnt, experience at Studland indicates that about 15 years elapse before the heath returns to its original condition, though in *Molinia* areas regrowth was quite rapid. The length of time will, however, also depend on the maturity of the heathland succession prior to burning, e.g. pine/birch woodland, mature gorse, young *Calluna*, etc. However, the associated fauna of the heathlands have differing habitat requirements and the planned use of fire as a management tool for obtaining varying age stands of heath may be desirable. Overall, the uncontrolled burning of the reserve is undesirable, and hence the lighting of fires, cooking stoves and barbecues is prohibited. A patrol of wardens can be mounted, and a look-out is kept particularly at times of the greatest fire hazard. Fire-fighting equipment is on hand for whenever it is needed.

Secondly, the reserve management plan states that the provision of litter bins is a priority, and Teagle (1966) considers that litter is the greatest of the human impacts on the reserve. Until the Nature Conservancy leased the reserve there was no collection of litter, and one of their first tasks was to clear the reserve of an accumulated backlog. The removal of litter was considered essential not only to preserve the amenity value of the reserve but also because of its ecological effects. Scavenging species, particularly the gulls and the fox, have been attracted to areas where there is litter accumulation, as in the car parks. Evidence at Studland Heath suggests that paper breaks down in about four months, whilst other articles of litter are not decomposable. Of these glass is perhaps the most harmful, not only because of fire danger but because small mammals can enter bottles without being able to escape. In the 15 bottles collected in 1965 that contained mammals there were 49 specimens of four species (a mean of 3·26 mammals per bottle), though the variance was large since one bottle contained 14 mice and shrews. What provision has to be made for this litter? In the reserve, mainly in the south eastern corner where the recreational use is more intense, the Nature Conservancy has installed 12 litter bins, each containing a bag that can easily be removed and replaced with an empty bag. At times of greatest public use these bags can be changed each day, and the Nature Conservancy has arranged for the collection of the bags for disposal. Similar litter bins are also provided in the car park outside the reserve area. However, the aim of the Nature Conservancy is to explore ways of reducing the litter problem, and by 1971 the number of litter sacks needed had been reduced to six.

Thirdly, there is the recreational demand for information. On the edge of a nature reserve it is inevitable that visitors will want to see some of the scientific interest of the area, and it is also useful to educate such people with ideas on the conservation and protection of the wildlife of the habitat, with whom they are sharing a recreational experience. Thus, a nature trail has been planned, starting and ending in the Knoll car park. The trail passes over three dune ridges and through two dune slacks, and hence the processes that have been at work during the last century or more can be explained. The trail also takes the opportunity of showing that wildlife is not necessarily in direct conflict with man. Thus, although Teagle reports that there is a general hostility to all snakes (all three British species are found on the reserve), he is hopeful that the educational programme will result in people not attempting to kill any snake that they see. The design and aims of the Studland Heath nature trail will not be discussed here since the plan of such trails was discussed in Chapter 7. A second nature trail, a woodland trail, has been laid out in 1971 and also starts in the vicinity of the Knoll car park. A small observation hut over looks Little Sea. This hut is open by appointment only and is much used by bird watchers expecially for observing wildfowl during the winter.

What then are the planning lessons to be learnt from Studland Heath National Nature Reserve? First, there are the basic facts of the reserve. There is a considerable scientific interest, as well as the presence of rare species, and hence the primary objectives of management are to retain an adequate representation of the major habitat types. The other basic fact is that there is a considerable and growing recreational demand for the resources of the sea-shore and the adjacent dune system. These two basic facts provide the framework within which planning can proceed.

In general people visit the reserve with a car, and few people will move far from their car, particularly over rough country such as *Calluna* heath. Thus, the provision of parking places ensures that some areas are readily accessible, whilst other areas are kept relatively isolated. On South Haven Peninsula the parking at South Haven Point has little influence on the reserve since there is a walk of about $1\frac{1}{2}$ km around the shore to reach the reserve's northern boundary. The Knoll car park, not owned or managed by the Nature Conservancy, is thus the main factor determining the recreational use of the shore. Planning has therefore focussed attention upon this area, with the nature trails starting and ending here and the routes of the trails lying within the south eastern corner of the reserve (see Fig. 9.7). The density of litter bins is greater in this area, and it is here that one needs to watch for fire, and in the event of an outbreak put it out before it spreads inland towards the more isolated areas. In addition to these main parking areas cars may be parked freely along roadside verges adjacent to the reserve.

The recreational pressures are thus directed towards some particular zones, and the integration of intensive recreational use with the maintenance of undisturbed habitats is achieved by informal zoning. Watch can be kept within the zones of highest recreational intensity for the effects of wear on the environment, and indeed Teagle does suggest that wear on the dunes in the vicinity of Knoll car park leading to vegetation erosion is so great that the sea is encroaching at a rate of 1·7 m an^{-1}. The informality of the zoning is essential, since it not only allows for the occasional recreational user to penetrate the reserve, but it does mean that the recreational users, in sharing a resource with the wildlife, can become aware of the conservation operations.

PART THREE
Planning

10. The Management Plan

Introduction

In Part One, 'Ecological theory', of this book we examined the dynamic nature of the plant and animal species which make up the ecosystem. Although at first glance many of the processes being discussed seemed to be precise, yet one becomes increasingly aware of the random influences acting to 'blur' this precision. In Chapter 2, for example, we discussed ecological succession and defined clear-cut successful stages, nine for the shingle succession at Dungeness. Such a process is in reality an over-simplification, since succession is far more complex and less precise. On any piece of land, arbitrarily chosen, it is more than likely that there will be some abnormality in the succession, either by one stage being missed out, or by a reversal to a previous stage, or by the chance introduction of another species causing a temporary variation of the stage. The clear-cut stages that are put forward are in reality averages or means in the development of a community, forming part of a continuum of variation both lengthwise from the initial to the final form of the community and breadthwise in that each stage has its own variation. In ecological studies we usually attempt to define the mean, whereas it might be that the variance attached to the mean is of as much importance.

In Chapter 5 when we discussed the results of exploitation of populations the results were not as clear-cut as the means indicated. In the cultures of *Folsomia* a mean population size of, say, 500 animals per culture may have been quoted, but in the replicates of the experiments the actual population sizes may have varied between 350 and 750 animals. Once again an ecological study has defined a mean, making the result appear precise, but the variance is of considerable importance. So as to demonstrate statistical significance in ecological experiments one requires a large number of replicates in order to 'iron out' the effects of these large experimental variances.

In Part Two, 'Application', of this book we have seen some of the aims of conservation management. If we define conservation in terms of the 'wise use of biological resources', then as we have seen in Chapter 6 we have to define 'wise' in terms of the aim of management. An element of imprecision is introduced here since the management of ecosystems based on one aim may be detrimental to the management based on another aim. We might, as frequently happens, wish to manage a grassland so as to gain a maximum yield of hay whilst ensuring the preservation of various species that occur in that grassland. As we saw in Chapter 5 one must define the aim and then determine the exploitation rate that will best satisfy that particular aim. If the species that we wish to preserve in our grassland are those that are associated with grazed areas then it may be possible to integrate maximum yield with preservation. If, however, the species that we wish to preserve are those that are only found in ungrazed areas, then these two legitimate aims of conservation management are directly opposed to each other and integration of the two on the same parcel of land is impossible.

Management is thus an imprecise science, the lack of precision being contributed both by the inherent variability of organisms and ecosystems, and by the very nature of the limitations imposed by man. Management is itself concerned with intervention within the ecosystem, either to alter the *status quo* in a wanted direction, or to prevent the *status quo* altering in an unwanted direction. In order to undertake such actions one is almost always faced with a lack of specific data about the biological mechanisms involved, and hence one will have to estimate what is likely to happen, another source of imprecision. Resources, especially financial, required by the manager are likely to be limited and hence decisions will have to be taken about the priority of action.

All these sources of imprecision make it essential that records should be kept and routine observations made so that the effect of management activity, both successes and failures, can be assessed. In this way a data bank can be built up for each resource, and future management can be more precise. It is therefore necessary to plan for conservation management.

The classical approach to management has been by foresters, who for over a century have been writing and implementing management plans. Descriptions of such plans and the methods employed in preparing them can be found in text-books on forest management, for example Brasnett (1953), Knuchel (1953) and Osmaston (1968). More recently, with the focus of conservation being aimed at nature reserves and wildlife sanctuaries, plans for their management have been compiled. Eggeling (1964) has described such a plan for the Island of Rhum, and Usher (1967) has prepared a plan for a Local Nature Reserve (L.N.Rs are reserves that are controlled by Country or Borough Councils). It has become accepted practice within the Nature Conservancy for management plans to be prepared for National Nature Reserves, and during the last few years many organizations such as the County Trusts are preparing and implementing management plans. Anyone who becomes involved with the conservation movement is increasingly likely to have to study conservation planning, and the relations between conservation and other competing forms of land-use.

A *pro forma* for a management plan

The *pro forma* that is used in this chapter is based upon the structure of management plans that have been implemented by the Nature Conservancy in Britain. The plan is set out in four sections, namely descriptive (Chapters 1–3), a statement of the objectives (Chapter 4), prescriptive (Chapters 5–11), and lists of species and record forms (Appendices). In the *pro forma* below there is a typical set of headings. These may not be applicable for all nature reserves and wildlife sanctuaries, and other headings may be required for some reserves (for example the *pro forma* contains no marine headings).

INTRODUCTION

This normally sets the reserve into its geographical area, and compares the reserve with series of reserves of similar nature elsewhere in the Country/Vicinity.

CHAPTER 1 NAME AND GENERAL INFORMATION

(a) Name
(b) Location
(c) Brief description
(d) Area and boundaries
(e) Access
(f) History of establishment
(g) Bye-laws
(h) Permits
(i) Grid references
(j) Maps
(k) Collections of museum material
(l) Collections of photographs

CHAPTER 2 REASONS FOR ESTABLISHMENT

(a) General. Reasons associated with the overall policy of the body managing the reserve.

(b) Specific. Reasons associated with the establishment of the particular reserve.

CHAPTER 3 SURVEYS AND SCIENTIFIC INFORMATION

(a) Topography
(b) Drainage and hydrological regime
(c) Geology
(d) Climate
(e) Soils
(f) Vegetation
(g) Fauna
(h) Land-use history
(i) Archaeology and ancient monuments
(j) Research projects
(k) Public and recreational interest
(l) Sporting rights
(m) Pest control

CHAPTER 4 AIMS OF MANAGEMENT

The major aim will often be wildlife conservation, with subsidiary aims such as education, research, recreation and amenity. It may be necessary to specify short term aims of public relations where there is hostility to the presence of a reserve, or of legal changes in boundaries, leases, etc.

CHAPTER 5 MANAGEMENT PROGRAMME

(a) Scientific management. An interpretation of the aims of management into practice.

(b) Estate management. Details of the maintenance of the estate, for example, buildings, fences, paths, bridges, walls, signs, hides, shelters, nest-boxes, etc.

CHAPTER 6 PUBLIC ACCESS

(a) Bye-laws
(b) Public right of way
(c) Permits
(d) Any other rights or privileges

CHAPTER 7 WARDENING

Details of the present position and the likely demand for wardening facilities.

CHAPTER 8 TIME SCHEDULE AND FINANCE

The items on the management programme will need to be costed, and in the light of any limitation on financial resources a time schedule will be drawn up and the priorities for attention will be indicated.

CHAPTER 9 DIVISION OF RESPONSIBILITIES

This chapter lays down the responsibility for implementing the whole plan and/or specific parts of it.

CHAPTER 10 ADVICE AND RECORDS

Statements of any committee responsible for management advice, and of organizations or individuals from whom specific information can be obtained. The records will include periodic or progress reports, as well as records of research activity designed to monitor the effects of management practice. It is also important to record detailed statements of management decisions that have been taken, paying particular attention to recording the reasons why that decision was taken.

CHAPTER 11 RENEWAL, AUTHORSHIP OF THE PLAN AND REFERENCES

A date will be specified for the renewal of the plan. The references will include all literature quoted in the plan. Authorship of each section (where the plan has been written by several people) should be given. Acknowledgments may also be included.

APPENDICES

(A) *With all copies of the plan*

Lists of plants
Lists of animals
Map(s) showing items detailed in the scientific management programme
Map(s) showing items detailed in the estate management programme

(B) *With the master copy of the plan only*

Copies of legal documents relating to the reserve
Records of committee decisions
Records of expenditure
Records of authorization of research and survey projects

Looking first at the descriptive section of the plan (Chapters 1–3) one can see that this defines the locality factors such as the geographical position, geology, climate and hydrological status. This is the basic framework within which the ecosystems have developed, and it allows for broad comparisons to be made between areas in order to base predictions for the reserve on experience gained elsewhere. The classification of the ecosystems represented on the reserve will usually depend upon the plant communities and on the soil types, both of which are static and relatively easily identified. Animal species tend to be of less use in classification since they are so numerous that few people could identify them and many of the species are mobile. After the classification the dynamics of the situation should be assessed. What are the forces holding the ecosystem in its present form, or in what direction is it changing? What is the distribution of species in the ecosystem, or what is the effect of time going to be upon the wildlife that we wish to conserve today? These are biological attributes, and in the first two chapters of this book we discussed analyses that attempted to answer such questions. But equally important is the effect of man upon the ecosystem. What are the research, educational and recreational potentialities of the reserve, and what will be the effect of exploiting such potential on the ecosystems? The first section of the management plan is thus descriptive, containing a summary of the environmental and biological attributes of the reserve and of the human interest. It is also predictive in that the experience gained from research and study of similar ecosystems can suggest what changes may occur and how to plan the exploitation of the wildlife in the broadest sense.

The second section of the plan deals with the aims of management. Since a nature reserve forms an integral part of the pattern of land-use in a region or country, its management cannot be divorced from the wider issue of using the environment to benefit mankind. During 1971 approximately 0·7 per cent of the land surface in Britain is covered with nature reserves, but in overall land-use planning decisions, what is the role of wildlife conservation? This is very difficult to demonstrate since there is no way to assess the value of conservation in economic terms. Until such economic theories have been produced and tested, judgements must of necessity be subjective, often slanted towards the policy and resources of the administering organization. National organizations will inevitably limit their attention to the most important sites, again based on some value judgement of the importance of these sites. Since the finance for their operations is nationally based there is the possibility of more flexibility in the management decisions. Local organizations, however, will want rather more in the form of immediate returns from wildlife conservation. Unless they are very far sighted this local benefit will become an important aspect of the aims of management, and is one that is likely to conflict with the very long-term aim of sustained conservation management. Thus, this section of the management plan will always attempt to find methods of integrating wildlife conservation with the other forms of land-use, helping in forming a balance in the overall land-use planning decisions of the area in which the reserve is situated.

The third section of the plan is prescriptive, in which the aims of management are interpreted in a practical manner bearing in mind the locality, environmental and human factors. The formulation of the management strategy requires experience to be drawn from many disciplines, since there are not only the biological aspects of the reserve to consider but also the sociological and economic factors. Examples of some of the techniques that can be used in management have been discussed in Part Two, 'Application', of this book. However, one other aspect of this section of the plan is of importance. Wildlife conservation is a long-term form of land-use. Although the management plan is written for only a limited period of time (often five years, less often 10 years), the implications of management practice will not always be felt within the period of time covered by the plan. For example, an operation such as scrub clearance may result in a changed ecosystem for half a century. Thus, during the time taken for the effects of a management decision to follow through the various stages of an ecological succession the men responsible for making those decisions may have nothing further to do with the management of the reserve. It is therefore essential that there is good documentation. The minimum that is required is a statement of the reasons why a management decision was taken. Why, for example, were sessile oaks to be planted instead of any other species? It is, however, better to document both sides of a decision, stating why the opposite view was rejected. After taking a decision it is invaluable to study its implementation by periodic recording. For example, two or three permanent quadrats could be unobtrusively marked by corner posts, and a system of annual recording of the vegetation clearly specified in the management plan. The amount of work required for this is minimal, but the effect of the management prescription can be assessed provided that some form of control recording is also specified.

The Aberlady Bay Local Nature Reserve management plan is reproduced in order to demonstrate the preparation and structure of these plans. This has been made possible by the kind permission of the East Lothian County Council, in whose county the reserve is situated.

The Aberlady Bay local nature reserve management plan

The complete plan, with the exception of the appendices, is reproduced here. As a preface to the first printing there is a statement that the various proposals for the future management of the reserve have been evolved after consultation with interested people and organizations, but that they are purely the recommendations of the author of the plan. The County Council considered the management programme outlined in Chapters V to IX of the plan in the light of recommendations of the Aberlady Bay Nature Reserve Management Committee and the General Purposes Committee of the County Council. Their decisions, which are recorded in the second printing of the plan, are as follows:

They agreed to the following recommendations:

(1) The appointment, for the summer months only, of a full-time warden whose duty will include the maintenance of paths, etc., control of buckthorn, hogweed and hawthorn, and general supervision of the public subject to getting a grant under the Countryside (Scotland) Act to cover this expense, Chapter VII, para. 3.

(2) All recommendations for new notices or resiting will be considered in detail by the Biological Sub-Committee, 2(b) (iv) and 3(b) (vi).

(3) Once a warden is appointed, they were prepared to provide an informative leaflet or booklet or both, 2(b) (i).

(4) The North Berwick District Council, who are responsible for the maintenance of the footbridge, will make periodic inspection of the state of the bridge and carry out repairs as necessary, 3(b) (iii).

(5) The question of a special marked ride for horses was continued

meantime because it was thought that it might lead to an increase in horse riding within the reserve, 3(b) (v).

(6) Provision of more litter bins and their clearance were made the responsibility of the County Sanitary Inspector, 3(b) (viii).

(7) The recommendation for the eastern boundary of the reserve was noted and will be kept in mind for the future but no immediate action will be taken, 5(a), (b), (c).

(8) The County Council also noted the need to appoint one of their officials to have overall responsibility for the reserve and would consider this in the future, Chapter IX.

They were not prepared at that stage to take any action on the following, although it is understood that certain of these matters have been considered further since the initial discussions of the management programme:

(1) Extend the car park at Peffer Burn footbridge, 3(b) (i).
(2) Arrange for additional supervision of breeding birds, 1(a) (i).
(3) Dredge again the open water area at the Marl Loch, 1(c).
(4) Inspect for or carry out any anti-erosion work on the dunes, 1(d) (iii).
(5) Divert the lower path from the Marl Loch to the sand bar, 3(b) (iv).
(6) Form a shallow artificial pond to increase feeding for waders, 1(a) (ii).

As mentioned in the reproduced plan, the area within the boundaries of the Aberlady Bay Nature Reserve is not owned by the County Council.

In the following reproduced version of the management plan the plates, illustrations and tables have been re-numbered so as to conform to the layout of this book.

ABERLADY BAY LOCAL NATURE RESERVE MANAGEMENT PLAN

INTRODUCTION

Aberlady Bay Local Nature Reserve is important for three reasons. Firstly, it is situated very close to a centre of population, and it therefore has great potential educationally, both for school teaching and for teaching at a more advanced level. It is suitable for teaching such subjects as physical geography, geology, ecology and general natural history.

Secondly, the Reserve provides a winter refuge for wildfowl. Together with Caerlaverock National Nature Reserve, wildfowl shooting is controlled by permit. Aberlady Bay was the first place in Scotland where any control of wildfowling had been attempted. The following is an extract from a Nature Conservancy report (*Wildfowling at Caerlaverock National Nature Reserve*, The Nature Conservancy, Edinburgh, second edition, November, 1964): 'Shooting, except by permit, has been prohibited since 1953 and in March, 1955, the validity of Bye-Laws to control shooting on the foreshore was confirmed by the Lord Justice-Clerk. Although permits to shoot were issued before 1957–58, it was not until that season that a formal system of issuing was introduced, requiring a return on birds shot and the number of visits to the Reserve.... Permits are generally issued to shoot wild duck only, although in 1959–60 and 1962–63 geese, mainly pink-footed, were also permitted to be shot. There are no short period permits, all permits being issued for the whole season which, because the shooting area is entirely below H.W.M.O.S.T., extends from 1st September to 20th February.'

Thirdly, the sand dunes near Jovie's Neuk, in the north of the Reserve, should be considered as an example of one of a series of dunes occurring around the coasts of Scotland. The following extracts are taken from a Nature Conservancy report (*General Introduction to Dune Reserves in Scotland*, Tentsmuir Point National Nature Reserve Management Plan, Nature Conservancy):

'The main scientific interests of the dune systems are physiographical and botanical, particularly the closely linked processes of sand accretion and plant colonisation. These are controlled to a greater or lesser extent by the following factors which determine the variations of topography and vegetation in each area:

(1) geographical position (including structure of adjacent coastline, aspect, exposure to direction of maximum wave fetch, past changes in sea level);
(2) supply and composition of sand, especially shell content;
(3) climate (especially wind direction and force, and rainfall);
(4) land use (modification of natural habitat by cultivation, afforestation, grazing or military use).

'A broad division can be drawn between the east and west coasts. Much of the east is composed of sedimentary rocks and is low lying where traversed by the firths or estuaries of the main rivers which transport large quantities of sand and other deposits. The most extensive dune systems of Scotland occur here, near the Forth, Tay, Moray and Dornoch Firths and in parts of Aberdeenshire. The west coast, formed of more resistant igneous and metamorphic rocks, is generally more rugged and lacks large estuaries with offshore sandbanks or bars.... Differences in the topography of east and west coast dunes are due mainly to wind factors. With full exposure to the westerlies there is continued movement of sand inland, a complex system of unstable dunes and blow-outs and often high dune ridges. On parts of the east coast where the prevailing west winds oppose the easterly dune-building winds, parallel dune ridges develop, with lower relief, greater stability and more rapid lateral accretion.

'Differences in the vegetation of east and west coast dunes are due mainly to differences in the composition of the sand. The shell-sands of the Western Isles contain more than 50% calcium carbonate, compared with less that 2% in the mineral sands of the east coast. This has little effect on the unstable dunes which in all cases are colonised mainly by Marram (*Ammophila arenaria*), but inland of these an acid heath rapidly develops on the east coast dunes, in contrast to the rich dune pasture of the Hebrides. Lichens are characteristic of the acid area, giving rise to the "grey" dunes of the east coast where bryophytes are important colonisers....

'Twelve dune systems around the Scottish coast have been chosen to illustrate the range of dune forms and vegetation. The location of these is....

1. **Aberlady Bay (East Lothian).** The Aberlady dunes, rising to 200 feet, are part of a long stretch of sand-dunes formed along the south side of the Firth of Forth. The dunes are higher than those of Tentsmuir in Fife due to exposure to west winds, which are causing some erosion at the seaward edge. There is a sizeable area of salt marsh and hardly any development of parallel dunes and slacks. The landward dune areas have suffered from public pressure and transformation to golf courses.'

It can thus be seen that Aberlady Bay Local Nature Reserve has two characters – education and conservation. The conservation aspect makes the Reserve particularly important in that it is a link between areas managed for wildfowl refuges and for preservation and research on sand dune systems.

CHAPTER I
GENERAL INFORMATION

The Aberlady Bay Nature Reserve (National Grid Reference NT 455814) is situated on the south shore of the Firth of Forth, in the west of the County of East Lothian. It is approximately fourteen miles east-north-east of Edinburgh, and six miles north-north-west of Haddington.

To the north-west the Reserve is bounded by the Firth of Forth. The southern boundary is a track, near the high water mark of ordinary spring tides, to the north of the Kilspindie Golf Links and Aberlady. To the east, the Reserve is bounded by the Luffness and Gullane Golf Links. The Reserve comprises two sections. The largest, about three quarters of the area, is the Gullane Sands or Aberlady Bay, an area of sand and mud, exposed at low tide, forming the estuary of the Peffer Burn. To the east of this bay is an area of dunes, grassland, fresh-water marsh and saltings, which is from one eighth to one third of a mile broad, and just under two miles long. To the south of the bay the foreshore contains many interesting geological exposures.

The Reserve was declared by the East Lothian County Council on 14 July, 1952. The minutes record '. . . the Council resolved to make the following declaration to be known as the Aberlady Bay Nature Reserve Declaration:

> that the land extending to 1439 acres or thereby at Aberlady Bay in the parishes of Aberlady and Dirleton and County of East Lothian, as described in the appendix to the Principal Minutes, is the subject of agreements entered into in terms of Sections 16 and 21 of the National Parks and Access to the Countryside Act, 1949, and is managed as a Nature Reserve in terms of Sections 19 and 21 of the said National Parks and Access to the Countryside Act, 1949.'

The agreements were made with the owners of the land and their tenants. The northern section of the land east of Aberlady Bay is owned by Gullane Golf Club (Secretary: R. L. Balfour-Melville, Esq., Gullane Golf Club, Gullane); and the southern section of this land by Colonel A. J. G. Hope, Luffness Estate Office, Aberlady (Factors: Messrs. Blair, Cadell & Macmillan, W. S., Edinburgh), and the tenants are the Luffness (New) Golf Club. The strip of land to the south of Aberlady Bay is owned by Wemyss Landed Estates Co., Longniddry, and the Kilspindie Golf Club are tenants.

Bye-Laws were confirmed by the Secretary of State on 7 October, 1952. Following a legal dispute arising from the wildfowling interests, two further Bye-Laws were confirmed by the Secretary of State on 12 February, 1954, following a public enquiry at Haddington.

There is no restriction on public access to the Reserve. There is a footbridge over the Peffer Burn (Plate 9) from the A198 about half a mile east of Aberlady (NT 472806). A car park adjacent to this bridge has a maximum capacity of twenty-three cars, and cars are frequently parked on the verge of the A198 when this car park is full. Access to the landward section of the Reserve can also be obtained from the Gullane dunes near the Hummel Rocks (NT 465831); from the Gala Law quarry, with very limited car parking facilities, on the A198 approximately one and a half miles east of Aberlady (NT 476815); and from the Gullane and Luffness Links. Access to the southern side of Aberlady Bay is by a footpath running from Aberlady Green (NT 465802) to Gosford Bay (NT 449789).

A limited number of permits to shoot duck on a section of the Reserve are issued during the shooting season in accordance with Bye-Law 14. The permits were first discussed at the meeting of the Local Management Committee on 10 January, 1953. It was then decided to recommend that permits to wildfowlers resident in East Lothian should carry the following conditions: (1) only mallard and wigeon to be shot; (2) shooting to be forbidden south of the Peffer Burn and east of a line fifty yards west of the concrete blocks south of the Marl Loch; and (3) shooting to be confined to the hours between sunset and sunrise. At times since 1953, 1959–60 and 1962–63, the shooting of geese has been allowed, but this is now forbidden. In recent years, 26 permits have been issued to residents in East Lothian for the duration of the shooting season. The present conditions under which a permit is issued are: (1) the permit is not transferable; (2) the permit must be produced on demand; (3) shooting is only allowed between sunset and sunrise; (4) shooting is not allowed between the southern boundary of the Reserve and the Peffer Burn and within that part of the Reserve lying east of a line fifty yards to the west of the remaining pillars south of the Marl Loch; (5) shooting by any weapon other than by ordinary smooth bore shot-gun is not allowed; (6) the permit is valid from 1 September to 20 February of the following year; (7) the Management Committee reserves the right to withdraw the permit at any time; and (8) the holder of the permit will inform the Management Committee not later than 15 March of the number of days on which the permit has been used and the number of species of duck shot.

Ordnance Survey Maps

1 inch (Scotland), 7th Series; Sheet 62.

1 inch (Scotland), Popular Edition; Sheets 68 and 74.

2½ inch; Sheets NT 47, NT 48.

6 inch; Sheets NT 47 NE & NW, NT 48 SE.

25 inch; Sheets NT 4479, NT4480, NT 4580, NT4680, NT 4681, NT 4682, NT 4683, NT 4780.

Geological Survey Maps

¼ inch, Geological Survey of Scotland; Sheet 15.

Air Photographs

Vertical:	540/801	4303–04	4.7.52
	OS/62/33	072–77, 082–85, 137–39	1.5.62
	OS/63/233	022–23	21.9.63
Oblique:	Dr. St. Joseph	RF 57–65	
		RF 68–74	

Maps Accompanying Plan

1. Map of Reserve showing boundaries. Fig. 10.1
2. Map of Reserve showing proposed revised boundary. Fig. 10.2
3. Map showing main vegetation types. Fig. 10.3
4. Map showing paths and possible pony riding routes. Fig. 10.5

Ground Photographs

A collection of black and white photographs is held by the County Council.

CHAPTER II
REASONS FOR ESTABLISHMENT

Aberlady Bay and a coastal strip of land were first recommended as a Nature Reserve in the final report of the Scottish Wild Life Conservation Committee in 1949 (Command 7814). This Command states:

'NNR 16 Coastal Strip, Aberlady–Gullane, East Lothian

This coastal strip with a small area of salt marsh, dune and raised beach on the southern shore of the Firth of Forth, has a characteristic maritime flora. It is one of the best areas of its kind in South Scotland, and should be protected against any development which would spoil it. The reserve includes the estuary of the Peffer Burn, and Aberlady Bay, which is a fine wintering place for waders and wild fowl.'

The Nature Conservancy decided that the area was more suitable as a Local Nature Reserve, and the Deputy County Clerk of East Lothian was informed of this decision on 30 October, 1951. However, in November of that year, the Nature Conservancy agreed to give scientific advice about the area whilst the County Council were to be responsible for management. The East Lothian County Council announced their intention of establishing a Local Nature Reserve on 24 January, 1952.

Thus the primary reason for establishing the Reserve was conservation of the habitats associated with a stretch of Southern Scottish coastline. It was considered that the habitats at Aberlady, and their associated flora and fauna, were amongst the most important in Southern Scotland. It was also considered to be an essential feature of the County Development Plan then under preparation.

CHAPTER III
FACTORS OF THE LOCALITY

1. Topography above H.W.M.O.S.T.

(a) *Landward Boundaries:* From Aberlady the Reserve boundary follows the northern verge of the A198 to a small track about 550 yards east of the footbridge across the Peffer Burn. The boundary follows this track and the

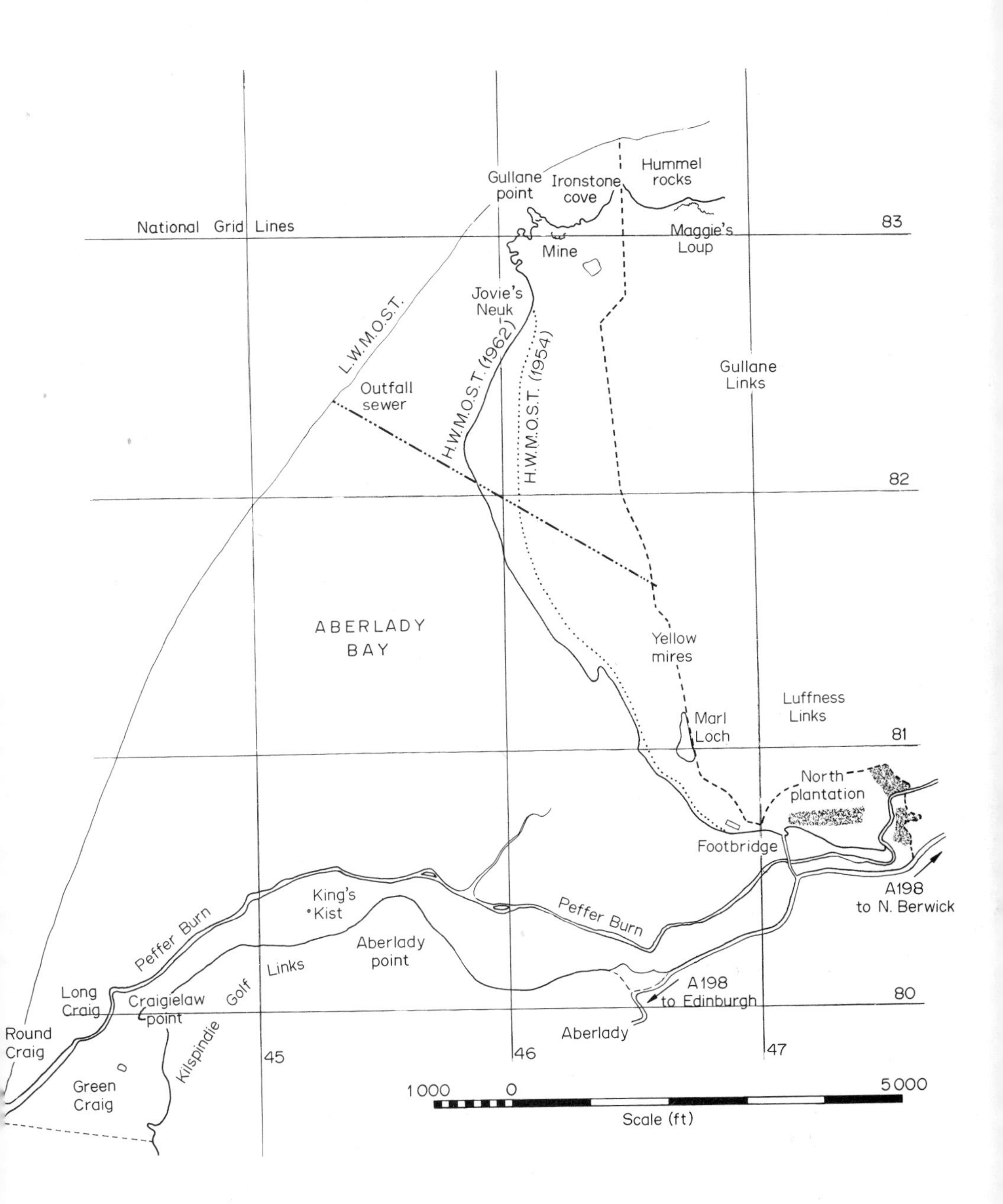

Fig. 10.1 Plan of the reserve showing boundaries and physiographic features

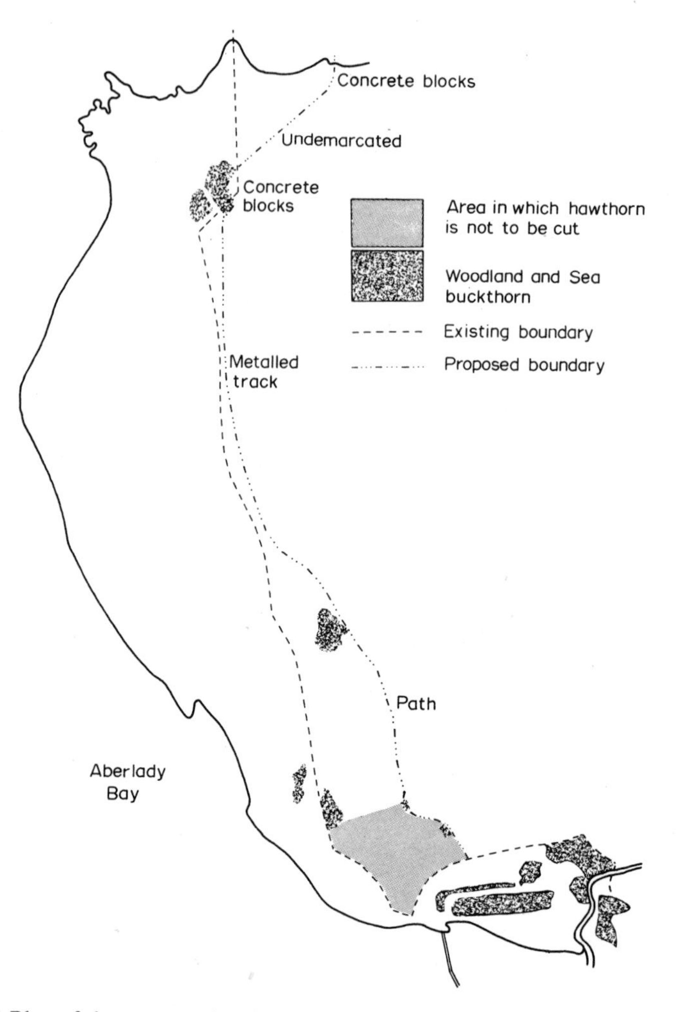

Fig. 10.2 Plan of the reserve showing the proposed revision of the eastern boundary

eastern and southern margins of the adjacent mixed woodlands, past a small refuse tip, to a drain, northwards along the drain, across the Peffer Burn, and along the eastern margin of North Plantation, a wood of mixed deciduous tree species. From here the boundary follows westwards along the line of a sunken stone wall to a track about 40 yards from H.W.M.O.S.T. near the northern end of the footbridge.

From here the boundary follows a path to the south of the Marl Loch, along the eastern edge of the Loch, along a path running just east of the fresh-water marshes, to the first of the remaining concrete defence blocks. The boundary follows the line of defence blocks for about three quarters of a mile till they veer westwards. The boundary, undemarcated, crosses from west to east of the metalled track to a short line of defence blocks, and from these it is a straight line, undemarcated on the ground, running out to the centre of the Hummel Rocks.

(b) *Relief:* The landward area of the Reserve can be divided into five sections. (i) The estuary of the Peffer Burn is flat, with the development of saltings, areas of grassland, just above H.W.M.O.S.T., intersected with pools and gullies mostly nine to twelve inches deep. Near the footbridge there are no clearly defined banks to the Peffer Burn. However, an old bank, about three to four feet high, can be discerned immediately south of the plantation of Scots pine. About 300 yards above the footbridge, and further upstream, the Peffer Burn flows between unstable muddy banks, and in places reed beds have developed. About 800 yards above the footbridge the Peffer Burn is crossed by an iron bridge, but this has fallen into disrepair, and it is now very dangerous. (ii) The remainder of the Reserve east of the footbridge is probably on raised beach. It is rather flat, with a height of 10 to 12 feet above Ordnance Datum (Newlyn Datum). Some sedge swamps and wet meadows occur between the new Scots pine plantation and North Plantation (Plate 10). (iii) The majority of the Reserve lying between the footbridge in the south (Plate 9) and the dunes in the north (Plate 11) is flat, and is either on raised beach or blown sand. Ordnance Survey spot heights on some of the paths show the area to lie between 12 and 16 feet O.D. A slight depression, running approximately north and south, is about 700 yards long and 50 yards broad. To the south of this is the Marl Loch (Plate 12), and the remainder forms the Yellow Mires, a collection of fresh-water marshes. (iv) The dune system is quite well developed, and runs approximately north and south for a distance of about half a mile. There is evidence of the development of parallel dunes, with a small ridge about 20 to 30 feet high, about 250 yards inland from the shore. Some small depressions, which are ponds in winter and mostly dry in summer, represents the partial development of a system of dune slacks. One of these slacks has a luxuriant growth of water plants, and is moist in nearly all summers.

The forward dunes can be divided into three zones. Firstly, just behind the present beach there are small hummocks, one to two feet high and eight to ten feet across, which represent the embryo dunes. Behind these are larger dunes, and the Marram Grass becomes plentiful, together with some Sea Lyme Grass. These rise to a maximum height of 40 or 50 feet. Inland of these other plants become plentiful, bare sand surfaces are infrequent, and the dunes are more or less stable. These dunes have a height of 25 to 30 feet. (v) At the north of the dunes a small valley separates dunes from teschenite rocks and soils derived from these rocks. The rocks give rise to a gently undulating topography, with a few rock faces, particularly near the shore. The height of the outcrops is from about 30 feet near the dunes to 75 feet at the highest point of the Reserve. This point is just to the south of a large clump of Sea Buckthorn, and from here the structure of the sand dune system can be seen clearly. A slightly higher ridge of teschenite runs north-eastwards from the Reserve to the rocks by Maggie's Loup. Near the Hummel rocks a flat area on the teschenite has been cleared, and is used as a football pitch.

(c) *Drainage:* The Peffer Burn drains a large area of East Lothian. This area is bounded by Gullane and Dirleton in the north, North Berwick in the north-east, East Fortune in the east, and Athelstaneford, Drem and Luffness in the south. This water is sometimes polluted, and is never clear.

Drainage of water from the Gullane Links would appear to be by seepage through the ground. Two open drains flow into the Yellow Mires about 200 yards north of the Marl Loch, but these only carry water in the winter. Presumably, the ground water from the Links feeds the Marl Loch and Yellow Mires. Hence, any attempt to drain the grassland to the east of the Marl Loch would greatly influence the habitats of the fresh-water marshes. During the winter, water flows out of the Yellow Mires by a drain from the south of the Marl Loch.

The dune slacks are fed by ground water seeping from the dunes. Only one of the slacks is moist all the year round. During the winter, most of the ground between the Marl Loch and the dune slacks is waterlogged, and very little seepage occurs into Aberlady Bay.

2. The Reserve below H.W.M.O.S.T.

The largest area of the Reserve lies below H.W.M.O.S.T., forming the estuary of the Peffer Burn. The Peffer Burn flows westwards for about two miles from the footbridge before discharging into the Firth of Forth. The area south of the Peffer Burn is mainly mud flats, except to the north of the Kilspindie Golf Links, where a number of rocks are exposed. Amongst these rocks, which are submerged at high tide, are The King's Kist and Green Craig, both of which are slightly above H.W.M.O.S.T. Only Green Craig carries any vegetation.

The mud flats are found near the Peffer Burn, particularly in the area from Aberlady to Aberlady Point. Mussel-beds are associated with stretches of the mud flats.

The area north of the Peffer Burn is mostly covered with sand. This is almost flat, but it rises very slightly towards the north, thus creating a sand bar at the entrance to Aberlady Bay. As the tide comes in water follows the course of the Peffer Burn, and then flows northwards and eastwards over the sands, last covering a long spit of sand projecting south-west from the dunes. The sand bar has been enlarging and gradually turning more towards the north during recent years. Part of the sand bar is now a few feet above H.W.M.O.S.T., and this is used by the Terns for nesting. As this process has taken place the salt-marsh south of the bar has filled in with blown sand, and the H.W.M.O.S.T. marked on Ordnance maps has advanced westwards by about 50 yards, and by nearly 120 yards just south of the bar, between 1954 and 1962. Sand is blown off the bar at low tide by south-westerly and westerly winds, and is deposited at the western end of the dunes. The dunes have advanced towards the north-west by nearly 200 yards between 1954 and 1962, and the structure of the beach is thus changing. When the Reserve was declared in 1952, the shore, from Jovie's Neuk to south of where the sand bar now is, was nearly straight. Now, since the west end of the dunes are being built up faster than the east, the coastline is becoming a bay, running from Jovie's Neuk to the sand bar, with a north-westerly aspect.

Drainage in Aberlady Bay is by slight depressions exposed in the sand at low tide. These are continually altering their position. A spring issues from the sand about 700 yards west-south-west of the Marl Loch. This drains south-westwards into the Peffer Burn.

3. Geology.

The geological interest of the Reserve is considerable, but unevenly distributed. The Reserve includes the whole of the Aberlady Bay, and the western half of the Gullane Point–Hummel Rocks, Geological S.S.S.Is., but outside these there is little of significance. The geology of the western section of the Reserve has been described by Dr. P. McL. D. Duff (*Edinburgh Geology: An Excursion Guide*, Edinburgh Geological Society, 1960)

The oldest rocks seen in the reserve are sediments, belonging to the Calciferous Sandstone Series, which are exposed at the north of the Reserve around Ironstone Cove. The sediments are sandstones, shales and fireclays and are bounded by the intrusions which form the eastern fringe of the Hummel Rocks and Gullane Point. Their inland extension is obscured by blown sand, and their relationship with the younger strata to the south totally obscured.

An extensive exposure of sedimentary rocks belonging to the Carboniferous Limestone Series occupies the foreshore from Kilspindie to Long Craig. This section, the most complete in the district, is shown in Table 10.1.

The rocks of this sequence are disposed in two folds with a north-west/south-east trend. An anticline occurs in the west, and is followed by a syncline to the east. On a small scale the structures are complex, there being a number of minor reversals of dip and the beds are cut by three small faults parallel to the trend of the folds. The dip of the strata is nowhere high, and marine erosion has etched the rocks into a scarp and dip slope topography. The hard beds, predominantly of four limestones, form prominent features and tend to build pavements bounded on the seaward margin by cliffs. The softer intervening beds are not so well exposed and have frequently been worn into notches. The false bedded sandstone which occurs between the Skateraw Lower and Long Craig Upper Limestones builds a small stack, known as The King's Kist; the shales which form the base of this stack are noticeably undercut relative to the sandstone.

In places, the limestones are traversed by 'veins' of sandstone along their joints. These have been formed by the injection of still unconsolidated sandy

TABLE 10.1 The Carboniferous Limestone Series seen on the foreshore to the south of the Reserve

Description of Rocks	*Depth in feet (unless otherwise stated)*
Dark shales with sandy beds and a thin limestone	–
Skateraw Middle Limestone. A hard, grey, thick-bedded limestone with crinoid remains and *Productus giganteus*	12
Grey nodular shales	20
Skateraw Lower Limestone. A hard, grey limestone with *Productus giganteus*	2–4
Coal	3 in
Sandy shales	4–8
Yellow sandstone with contorted bedding	4
Shaly beds, sandy above and limy below	8
Long Craig Upper Limestone. Nodular and dolomitic, consisting chiefly of crinoid fragments	12–18
Coal	10 in
Papery shales and seatearth	6
Long Craig Middle Limestone. Consisting largely of colonial corals and containing nests rich in brachiopods	8
Yellow sandstones and grey papery shales	20

material into fissures opened in the more consolidated limestones by earthquake activity.

A teschenite sill is intruded into the sediments of the Calciferous Sandstone Series at Gullane Point, and another forms the north-eastern part of the Hummel Rocks, just beyond the eastern boundary of the Reserve. The latter sill is the more interesting in that it contains sandstone 'veins' which broaden downwards from its upper surface. It has been suggested that these veins occupy contraction cracks formed during the cooling of the sill. A third group of exposures of teschenite, probably part of another sill, occurs on the foreshore to the north and north-east of Kilspindie Castle. A fourth teschenite sill forms the rock outcrops around Green Craig in the extreme south-west of the Reserve, and has been intruded some feet above the Middle Skateraw Limestone. Although its lower contact can not be seen, baked sediments can be found within a foot or two of its expected position. The contact is not planar, for exposures of teschenite, often in the highly altered form known as white trap, protrude through the sand to the east of the main outcrop of the sill, from which they are separated by exposures of altered sediments. The precise form of the lower contact of the sill can not be determined, but it seems probable that irregular tongues of much-altered teschenite protrude below the lower contact of the sill proper.

The remainder of the Reserve is occupied by superficial deposits of which the oldest is the 25 ft Raised Beach. This is seen from place to place on the landward side of the Reserve from Green Craig to Luffness, and northwards from that point to just beyond the Marl Loch. The remainder of the area is occupied either by blown sand, which obscures both the solid rocks and the Raised Beach, or by the present day beach.

4. Soils.

The soils of the County of East Lothian have been surveyed by the Macauley Institute of Soil Sciences but the results are, as yet, unpublished. No intensive soil survey has yet been carried out on the Reserve itself. Most of the soils are 'young', since with sand being blown on to the area characteristic profiles have not developed. The information already collected about the soils is included in Section 6 (Vegetation) of this Chapter.

5. Climate

(a) *General:* The nearest climatological station is approximately six miles away from the Reserve to the north-east, near North Berwick. Although North Berwick differs somewhat in aspect from Aberlady Bay, the climatological observations at North Berwick can be considered as being reasonably representative of the conditions at the Reserve.

Meteorological Office Stations are situated at Drem and East Fortune, approximately three and six miles east-south-east of the Reserve respectively. As both of these are more inland sites they are probably less representative of conditions at Aberlady Bay than is North Berwick. A general summary of the climate of East Lothian and Berwickshire has been prepared by the Meteorological Office about 1966 (*The Climate of East Lothian and North Berwickshire*, Meteorological Office Climatological Services (Met O 3) Climatological Memorandum No. 49).

TABLE 10.2 The average maximum, minimum and mean temperatures, and the temperature extremes, recorded at North Berwick (six miles north-east of the Reserve)

	Average temperatures			*Extreme temperatures*			
	Max.	*Min.*	*Mean*	*Maximum*		*Minimum*	
Months	°*C*	°*C*	°*C*	°*F*	*Year*	°*F*	*Year*
January	6·0	0·5	3·3	58	1957	9	1941
February	6·6	1·0	3·8	60	1926	6	1929
March	8·7	2·1	5·4	69	1945	14	1947
April	11·5	3·8	7·6	69	1946	23	1944
May	13·9	5·9	9·9	75	'39, '48	26	1942
June	16·9	8·6	12·8	85	1933	34	1932
July	18·8	10·8	14·8	88	1943	39	1954
August	18·4	10·6	14·5	84	1955	38	1932
September	16·5	9·0	12·8	80	1949	28	1954
October	12·8	6·4	9·6	75	1959	25	1926
November	9·4	3·6	6·5	63	'27, '46	21	'47, '52
December	7·3	2·1	4·7	60	1948	16	1927

(b) *Temperature:* The temperature recorded at North Berwick (118 ft.) is summarised in Table 10.2. The average annual mean temperature at North Berwick during the period 1931–60 was 8·8°C (48°F). The warmest month is July, but it is only fractionally warmer than August. January is the coldest month, although the extreme minimum temperature was recorded in February.

In an average year an extreme minimum of about −8°C (18°F) and an extreme maximum of about 25°C (77°F) can be expected. On average, only one or two days per year have a maximum temperature of 0°C or less, and about 45 days have a minimum temperature of 0°C or less. The average period without air frosts extends from about the end of April until about the beginning of November. During the period 1923–64 the average number of days with air frost in March was 5·0, in April 1·8, in October 0·3 and in November 3·7. More than ten days, on average, were recorded with air frosts for the months December, January and February, and no air frosts have been recorded during the period May to September inclusive.

(c) *Precipitation:* The mean monthly rainfall at North Berwick during the period 1916–50 is shown in Table 10.3. The average annual is thus 25·7 inches at North Berwick, though it is possible that it is slightly less than this at Aberlady Bay. Rainfall was relatively evenly distributed throughout the year, but there is a tendency towards dryness in spring when easterly winds prevail. The approximate average annual duration of rainfall during the period 1931–40 was slightly less than 600 hours. The average number of days per month with 0·1 inches or more of rain during the period 1923–64 is given in Table 10.3.

TABLE 10.3 The mean monthly rainfall in inches and the mean number of days with 0·1 inch or more of rain, recorded at North Berwick

Months	*Jan.*	*Feb.*	*Mar.*	*Apr.*	*May*	*June*	*July*	*Aug.*	*Sept.*	*Oct.*	*Nov.*	*Dec.*
Rainfall	2·12	1·48	1·61	1·45	2·11	1·94	2·68	3·06	2·43	2·65	2·34	1·82
Days	16·1	13·5	13·2	12·7	13·5	12·8	15·4	16·1	15·0	15·6	16·2	16·4

During the period 1923–64 snow fell on an average of 13·8 days in the year, but it only lay for ten days. Most snow fell during January and February (3·9 and 3·7 days respectively). Hail falls on an average of only 3·2 days per year, and thunder is heard on an average of 5·0 days per year, mainly during the months May to September inclusive.

(d) *Water Balance:* Using average values of potential evaporation, which are probably underestimates, it would appear that potential evaporation exceeds rainfall from April to July inclusive. It also appears that a total potential water deficit, by calendar months, of at least 2·77 inches occurs during that period. Reckoning by shorter wet and dry periods would give a P.W.D. considerably higher than this.

(e) *Wind:* The average annual wind speed at 33 feet above ground is about 14 m.p.h., and the number of days with gales during the year averages about seven. The marked preponderance of winds from between west and south-west and, particularly in the spring, from between east and north-east, shows the funnelling effect of the Forth–Clyde gap. The westerlies are stronger on average in the Firth of Forth than elsewhere along the East Coast of Scotland. During severe westerly gales gust speeds of 70 to 75 m.p.h. are not unduly rare in East Lothian. Aberlady Bay is fairly well exposed to the west.

(f) *Miscellaneous:* Fog, i.e., visibility of 1100 yards or less, occurs on an average of 7.1 days per year. It commonly takes the form of cold, wet sea fog, or haar. Haars begin in March or April, are worst from May to July, and are sometimes frequent in September.

The average annual duration of bright sunshine for the period 1923–50 at North Berwick was 1345 hours, with a mean daily duration of 3·68 hours. This represents slightly more than 30% of the possible.

6. Vegetation

(a) *General:* There are six main plant communities occurring within the Reserve. The distribution of these is shown in Fig. 10.3, and a complete list of all the plants within the Reserve is given in Appendix III. The communities are:

(1) the mud flats,
(2) the salt marshes,
(3) the fresh-water marshes and the Marl Loch,
(4) the dunes,
(5) the grasslands, and
(6) the woodlands.

Two further sets of plant communities are more localised and are not included in Fig. 10.3. These are the driftline, and the roadside – the area of often disturbed soil just to the north of the A198.

(b) *Mud flats:* Most of the areas of mud are devoid of communities of higher plants. In one bay south of the Peffer Burn, the eel-grasses (*Zostera marina* and *Z.angustifolia*) are abundant. As the mud gives way to sand, species of glasswort (*Salicornia*) become plentiful. Four species of this genus have so far been found within the Reserve, and it appears that all are abundant. The glasswort never forms a closed canopy over the sand or mud, and is only very rarely found above H.W.M.O.S.T.

(c) *Salt Marshes:* The Reserve demonstrates salt marsh accretion, with a gradual rise in ground surface and change in vegetation towards the land. Near the sand and mud flats the salt marsh contains a few plants of glasswort, but the sea meadow grass (*Puccinella maritima*) and sea milkwort (*Glaux maritima*) are the main colonizers of the sand, though sea sandwort (*Honkenya peploides*) is locally dominant. Since sand and mud accumulate wherever plants are established the ground level rises, and other species enter the community. The creeping fescue (*Festuca rubra*) and thrift (*Armeria maritima*) are dominant species, and are associated with an abundance of herbs, such as sea plantain (*Plantago maritima*), sea aster (*Aster tripolium*), sea spurrey (*Spergularia media*), sea arrowgrass (*Triglochin maritima*), and two species of sedge (*Carex extensa* and *C.distans*). Three plants of sea lavender (*Limonium vulgare*) have

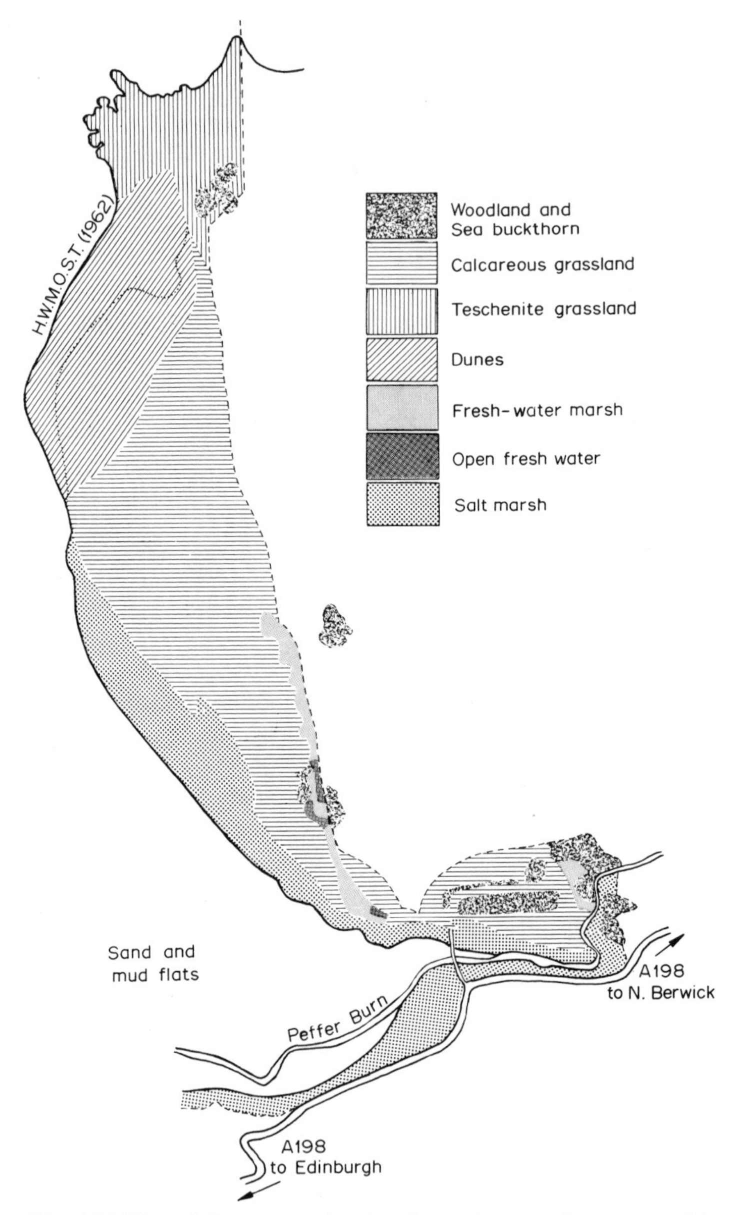

Fig. 10.3 Plan of the reserve showing the main vegetation communities

recently (1965) colonized this zone of the salt marsh near the footbridge, and it has been suggested that the seeds were either washed up by the tide or brought by birds. The nearest recorded native plants are at Culross, 30 miles up the Firth of Forth, and the Farne Islands, about 60 miles away.

Further inland of this herb-rich zone the conditions become less saline, and there is a rapid transition to grassland communities. However, in some localities, the ground remains brackish, the thrift becomes less abundant, and the mud rush (*Juncus gerardii*) and narrow blysmus (*Blysmus rufus*) become locally dominant. These two plants are associated with sections of the salt marsh–grassland transition zone that are moist at all times of the year.

(d) *Fresh-water Marshes:* Fresh-water vegetation communities have developed in four sections of the Reserve.

(i) THE DUNE SLACKS: Only one of these slacks is moist throughout the year, and hence this is the only one with a typical marsh vegetation. The other slacks tend to be wet in winter, dry in summer, and to contain only remnants of fresh-water plant communities. The most abundant plants in the main slack are amphibious bistort (*Polygonum amphibium*), water mint (*Mentha aquatica*) and the water and marsh horsetails (*Equisetum fluviatile* and *E.palustre*). The buckbean (*Menyanthes trifoliata*), which is so abundant near the Marl Loch, is absent from the dune slacks. There are only six species of sedge (*Carex* spp.) in the slacks, and these are all more abundant in the other marsh areas.

(ii) THE MARL LOCH AND THE YELLOW MIRES: The area of the Marl Loch (Plate 12) and its associated marshes, the Yellow Mires, is botanically one of the most important plant communities in the whole Reserve. The Marl Loch had become covered with a continuous mat of vegetation, mostly sedges, and after the war there was no open water. In late 1963 the concrete defence blocks were removed (see Chapter III, section 8) and, through the co-operation of the contractors, Messrs. Carmichael & Son, vegetation and dead organic matter was removed from the Marl Loch by a drag-line excavator. Since 1963, the natural biological production of the water plants has caused the depth of clear water to decrease. There has been a very rapid growth of stoneworts (*Chara* spp.) and bladderwort (*Utricularia vulgaris*), and evidence suggests that open water will only be maintained for a period of eight to ten years. Some sedges (*Carex* spp.), yellow iris (*Iris pseudacorus*) and bur-reed (*Sparganium erectum*) are already colonizing the open water.

The Marl Loch is the most northerly habitat in Britain of the marsh stitchwort (*Stellaria palustris*). This plant is confined to base rich marshes and fens, and its distribution in Great Britain is decreasing due to drainage of its habitats.

The Marl Loch and Yellow Mires are particularly rich in species of sedges (14 species of the genus *Carex*, and 6 species in other genera), and of horsetails (3 species of *Equisetum*). The buckbean (*Menyanthes trifoliata*) is particularly plentiful, and three types of marsh orchids (*Dactylorchis incarnata* ssp. *incarnata*, ssp. *coccinea*, and *D. purpurella*) can be recognized by their flesh, red and purple colours. Grasses are almost absent from these marsh communities.

(iii) THE CURLING POND AND THE PONDS CREATED BY EXTRACTION OF DEFENCE BLOCKS: This set of ponds and marshes are grouped since there is a tendency for the water to be slightly brackish, as shown by the occurrence of the water crowfoot, *Ranunculus baudotii*, in this group of marshes and not in the marshes previously described. Floristically, they contain fewer species than the marshes associated with the Marl Loch (11 species of *Carex* and 4 other sedges), but they are nevertheless interesting. The series of ponds left by the removal of the defence blocks has each developed differently, and now some are almost devoid of vegetation, others are covered with floating leaves of bog pondweed (*Potamogeton polygonifolius*) or amphibious bistort (*Polygonum amphibium*), whilst others are crowded with a number of upright spikes of mare's-tail (*Hippuris vulgaris*). One area of these marshes contains a large stand of cotton-grass (*Eriophorum angustifolium*).

The Curling Pond is particularly interesting. This contains many plants of glaucous bulrush (*Schoenoplectus tabernaemontani*), as well as an abundance of the lesser water-plantain (*Baldellia ranunculoides*) growing in one of its most northern British habitats. The occurrence of a particularly lush form of shoreweed (*Littorella uniflora*) is also outstanding. The bladderwort (*Utricularia vulgaris*), so common in the Marl Loch, is very rare in the Curling Pond, and the stoneworts (*Chara* spp.) are absent. The brookweed (*Samolus valerandi*), a plant associated with marshes near the sea, is abundant in this group of marshes.

(iv) THE PEFFER BURN MARSHES: Upstream from the footbridge, a series of water meadows and marshes are associated with the north bank of the Peffer Burn. The water meadows are particularly rich in species, and the meadow sweet (*Filipendula ulmaria*), ragged robin (*Lychnis flos-cuculi*) and marsh marigold (*Caltha palustris*) are abundant.

Beside the Peffer Burn is a narrow belt of reed-beds, with the sea clubrush (*Scirpus maritimus*) just beside the burn, and the reed (*Phragmites communis*) being the dominant species. This quickly gives way to a community found nowhere else in the Reserve, an almost pure stand of the great pond sedge (*Carex riparia*). This sedge grows to a height of about four feet, and hence the few plants that are associated with it tend to be tall. Examples of these are hemlock water-dropwort (*Oenanthe crocata*) and valerian (*Valeriana officinalis*). This marsh community stretches from the reed-beds to the margin of the woodlands.

(e) *Dunes:* Varying amounts of seaweed and other drifted material along the tideline cause accumulation of sand into embryo dunes, and these are colonized by sand couch-grass (*Agropyron junceiforme*). The sand composing these dunes was found to have a pH of 8·85 and to contain 10·8% of shell material (July 1967). At about 25 yards from the shore two other grasses occur, the chief being the marram grass (*Ammophila arenaria*) and the sea lyme grass (*Elymus arenarius*) being less frequent. These three species of grass, together with scattered plants of coltsfoot (*Tussilago farfara*), creeping thistle (*Cirsium arvense*) and orache (family *Chenopodiaceae*), form the community for the first 80 to 120 yards of the dune system. This area of dunes is termed the 'fore-dune'.

As one passes further inland *Agropyron junceiforme* becomes rare, and one passes through an area of about 100 to 150 yards which can be termed the 'intermediate-dune' or 'grass/moss-dune'. This is characterized by an abundance of marram grass, and the occurrence of a large number of grasses and herbaceous species, such as cock's-foot grass (*Dactylis glomerata*), false oat-grass (*Arrhenatherum elatius*), sand sedge (*Carex arenaria*), daisy (*Bellis perennis*), dandelion (*Taraxacum officinalis*) and silverweed (*Potentilla anserina*). Mosses such as *Brachythecium albicans*, *B. rutabulum* and *Hypnum cupressiforme* are frequent, and the lichen *Peltigera canina* is abundant.

As the dunes become older, and, due to sand accretion, further away from the sea, rain water washes away the nutrients, which were formerly contained in the shell fragments. The dunes behind the intermediate-dunes, known as the 'grey-dunes' or 'lichen-dunes', contained only 2·4% shell material in the surface inch of sand, and 5·7% shell material at a depth of six inches. The pH of the layers was 6·85 and 7·80 respectively (July, 1967). There is thus a tendency for the surface of the grey-dune to become slightly acidic, giving rise to plant communities containing a lesser number of plant species. The grey-dunes are characterized by the growth of lichens (grey in colour), and the main species in the Reserve are *Cladonia impexa*, *C. foliacea* and *Cetraria aculeata*. Marram grass is less frequent, and the plants less robust than in the fore- and intermediate-dunes. Bird's-foot trefoil (*Lotus corniculatus*) and wild thyme (*Thymus drucei*) are both abundant.

(f) *Grasslands:* Two distinct types of grassland occur in the Reserve – those associated with soils derived from teschenite rocks, and those associated with calcareous sandy soils. The sand in the dry grassland east of the Marl Loch contains 4·8% shell material in the surface inch and 13·2% shell material at a depth of six inches. The pH was 6·85 and 7·85 respectively (July, 1967). The Department of Botany, University of Edinburgh, have used these grasslands for teaching purposes, and have collected data on a transect running from teschenite to calcareous grassland. Table 10.4 summarizes some of their data collected in 1966 and 1967.

The classification of a single 'calcareous grassland' is an over-simplification. However, it is locally modified (Plate 13), and contains a diversity of herbaceous species, many of them uncommon as far north as Scotland. Such species as the viper's bugloss (*Echium vulgare*), hound's tongue (*Cynoglossum officinale*), musk thistle (*Carduus nutans*), and centaury (*Centaurium erythraea*) are all found in the grasslands. With the lack of rabbit grazing pressure the hawthorn (*Crataegus monogyna*) is becoming established (Plate 14), and, with scattered bushes three to five feet high, the calcareous grassland could eventually become a thicket.

Locally, the grass has been disturbed for turfs for the golf courses. Such areas develop very interesting and specialized floras. Where the sand is dry colonization by perennial plants is slow, and communities of spring-flowering annual plants are to be found. These species are very small, often less than one inch in height, and include the rue-leaved saxifrage (*Saxifraga tridactylites*), early forget-me-not (*Myosotis ramosissima*), lamb's lettuce (*Valerianella locusta*),

TABLE 10.4 Edaphic factors and vegetation cover along a transect running from teschenite to calcareous grassland

	Teschenite		*Intermediate*		*Calcareous*	
Soil:						
Depth	5 cm	20 cm	5 cm	20 cm	5 cm	20 cm
Moisture content	15%	20%	21%	14%	11%	7%
Solid $CaCO_3$	0·2%	0%	0·8%	0%	2·8%	4·0%
pH	6·7	7·7	7·6	7·6	7·7	8·3
Vegetation:						
Grasses						
Ammophila arenaria	absent		rare		abundant	
Arrhenatherum elatius	absent		rare		frequent	
Festuca rubra	absent		scarce		abundant	
Festuca ovina	abundant		abundant		absent	
Anthoxanthum odoratum	frequent		scarce		absent	
Herbs						
Galium verum	absent		absent		scarce	
Astragalus danicus	rare		scarce		frequent	
Thymus drucei	abundant		frequent		absent	
Leontodon hispidus	abundant		frequent		absent	
Bellis perennis	frequent		absent		absent	
Mosses	rare		scarce		abundant	
Lichens	abundant		scarce		absent	

whitlow grass (*Erophila vernia*) and the moss *Tortello flavovirens*. The first of the perennial plants to colonize these are bird's-foot trefoil (*Lotus corniculatus*), purple milk vetch (*Astragalus danicus*) and biting stonecrop (*Sedum acre*). In the moister areas a continuous plant canopy develops, but it contains many unusual species. These include the lesser clubmoss (*Selaginalla selaginoides*), variegated horsetail (*Equisetum variegatum*) and a dwarf variety of the grass of parnassus (*Parnassia palustris* var. *condensata*).

(g) *Woodlands:* There are three blocks of woodland within the Reserve. A block of 3·1 acres, a young Scots pine plantation, has been planted near the north end of the footbridge. The pine stand is so dense that there is virtually no development of a ground flora.

There are two blocks of deciduous woodland, of 2·2 and 1·7 acres, situated to the north and south of the Peffer Burn in the extreme south-east of the Reserve. These woodlands are composed mainly of oak, wych elm, birches and sycamore, though there are one or two trees of each of the following species: yew, Scots pine, lime, Norway maple, horse chestnut, laburnum, hawthorn, common elm, alder, beech and white poplar. There is a moderately well developed ground flora containing the herbs and ferns usually associated with deciduous woodlands on rich soils. The white bryony (*Bryonia dioica*), a plant which climbs over bushes and low trees, is abundant, although the species has been introduced into Southern Scotland.

(h) *Comparison with Other Areas:* Tentsmuir Point National Nature Reserve is an area similar to Aberlady Bay in that it contains both sand dunes and an area for wildfowl roosting and feeding. The list of higher plants for each Reserve contains 349 species, though Aberlady has in addition 2 sub-species. Aberlady Bay does, however, have a particularly fine collection of sedges (*Carex* spp.); 18 species have been found on the Reserve, whilst only 10 are recorded from Tentsmuir.

One plant that is absent from the dunes in the Reserve is the burnet rose (*Rosa spinosissima*). This species is frequently associated with dunes, and occurs at Dalmeny about 20 miles up the Firth of Forth.

7. Fauna

The reserve has a diversity of animal habitats, and 54 habitats are specified on the chart (Appendix 6). Most of these habitats are influenced by the maritime nature of the Reserve, except possibly the woodlands and water meadows upstream from the footbridge beside the Peffer Burn. Aberlady Bay compares favourably with the 39 animal habitats specified on the Tentsmuir Point National Nature Reserve.

(a) *Vertebrates:* The bird life of the Reserve has been studied for a long while, and the preservation of the area where migratory birds feed was initially one of the main features of the Reserve.

(i) BIRDS: The list of birds recorded from the Reserve contains 199 species, five further species of doubtful status, and five additional sub-species (Appendix 7 contains this list). A total of 51 species have nested within the Reserve.

The shore-nesting birds form one of the main conservation interests of the area. A colony of terns nests in the vicinity of the sand bar, and during the 1950s about 30 to 40 pairs nested each year. However, during the 1960s the sand bar has enlarged and become slightly higher, and the part-time Warden has prevented too much disturbance. The tern colony has increased, and, in 1966, 110 pairs nested, of which there were six pairs of little terns. There has always been about 10% arctic terns and the remainder common terns in this colony. From about mid-July, terns begin to roost in the Bay, and gradually they build up to a very large roost in August and September. A count on the 2nd September, 1958 revealed that about 14,000 terns came into the Bay from the west, 4,000 from the east, and about 8,000 were already in the Bay. Normally, however, the flocks number between 3,000 and 7,000 birds, and contain five or six species (common, arctic, little, roseate, sandwich, and a few black tern).

The saltings were formerly used by a large number of waders for breeding. In the 1950s the numbers of each species were higher than in the mid-1960s. The decline is possibly caused by increased disturbance by the public, but mainly by a change in the habitat. Since myxomatosis destroyed the majority of the rabbits, the vegetation has grown up, and the waders are known to prefer nesting in situations where they can see over the top of plants surrounding the nest. The following figures illustrate the decline over the period from the mid-1950s to the mid-1960s:

Ringed plover – 12 or 16 pairs to 3 or 4 pairs.
Lapwing – 20 or 30 pairs to 10 or 12 pairs.
Dunlin – 7 pairs to 2 or 3 pairs.
Redshank – 7 or 9 pairs to 6 or 7 pairs.

The snipe also decreased, but this is thought to be due to the hard winter of 1962–63, and evidence suggests that the number of breeding snipe is now starting to increase. The location of the nesting birds is also changing. The ringed plover nested in the saltings about 200 yards south of the sand bar, but now they nest on the bar with the terns.

The eider and shelduck also nest within the Reserve. The eider nest anywhere in the grasslands, but the number of young reared each year is very few, probably due to dogs and predation by crows, gulls and long-eared owls. The shelduck nest in burrows, and it is not yet ascertained whether they are excavating their own burrows now that the rabbits have become so scarce. They hatch 40 to 60 young per year, but predation by gulls means that only about 20 of them are fledged.

The grasslands also provide breeding grounds for many of the smaller birds. Skylarks and meadow pipits are very numerous, and reed bunting, sedge warbler and whitethroat are frequent. The woodlands provide habitats for many other species of passerines to nest.

The establishment of open water on the Marl Loch has had little effect on the bird life. A pair of mute swans have nested and reared young on the Loch. A coot was seen in the 1967 nesting season, but it apparently did not breed. Garganey have tried to nest on several occasions, but they have never succeeded.

The Bay provides roosting and feeding grounds for ducks, geese and waders. Many rare waders are seen during the migratory period (September to March), and the list includes such birds as the cream-coloured courser, avocet, lesser yellowlegs, Temminck's stint and pectoral sandpiper. 19 species of duck and 8 species of geese have been recorded from the Reserve. The habits of the duck and geese differ, for the ducks feed by night on fields and water meadows, and lie out in the Bay during the day. The geese feed by day on fields often far inland, and fly out to the sand bar at night.

The wildfowl provide sport for a group of 26 licensed people. Geese are not allowed to be shot, and a return on the number of times that a permit has been used and the number and species of duck shot has to be lodged with the County Council.

Table 10.5 is taken from an appendix to '*Wildfowling at Caerlaverock National Nature Reserve: The First Seven Years*', the Nature Conservancy, Edinburgh, and also abstracted from the minutes of the Management Committee. The ducks that are shot are mallard (47%), wigeon (42%), teal (10%) and others including scaup and goldeneye (1%).

(ii) MAMMALS: The mammals of the Reserve have not been studied in detail,

TABLE 10·5 A summary of shooting by permit during the ten years 1957–67

Year	Numbers of permits: Issued	Not returned	Used	Analysed	Av. No. of visits per Permit	Numbers of birds shot: Total Duck	Total Geese	Av. No. shot per used Permit
1957–58	24	0	18	21	16	154	0	8·6
1958–59	24	2	18	20	10	131	0	7·3
1959–60	26	0	24	23	9	154	34	7·8
1960–61	24	3	21	19	15	171	0	8·2
1961–62	25	2	22	22	18	292	0	13·1
1962–63	26	0	25	19	15	257	34	11·6
1963–64	26	0	26	15	15	223	0	8·6
1964–65	26	1	25	19	16	289	0	11·6
1965–66	26	1	23	19	14	226	0	9·9
1966–67	26	0	25	21	10	160	0	6·4

but yet 16 terrestrial species have been recorded. The grey and common seals are to be seen off Gullane Point and the Hummel Rocks. Many members of the order *Cetacea* (whales, porpoises, and dolphins) have occurred in the waters of the Firth of Forth near the Reserve, and the occasional corpse has been washed ashore.

The rabbit had the greatest effect on the Reserve. The hill to the east of the Reserve had been a rabbit warren for hundreds of years but, since the myxomatosis outbreak, the rabbit has been scarce. The Reserve falls within a rabbit clearance area. The brown hare is said to have become more abundant since the decline of the rabbit but it has not exerted the same grazing pressure on the vegetation as the rabbits. The result has been the growth of vegetation, the spoiling of the breeding grounds of some species of birds, possibly the causing of some herbs to become scarcer due to increased competition with grasses, and the spreading of sea buckthorn and hawthorn.

The mole is common in the Reserve, and has caused damage to the golf greens within and without the Reserve. Trapping is used to eliminate this pest, and no evidence of strychnine has been seen. Foxes have used the sea buckthorn clumps, but no steps appear to have been taken by farmers to eliminate this animal.

(iii) AMPHIBIANS AND REPTILES: Five species of amphibian are found within the Reserve. The three species of newt (common, palmated, and great crested) are found, but they are apparently becoming scarcer. It has been suggested that this is caused by too many children collecting them in the few ponds in which they occur. The common frog is abundant and breeds most frequently in the dune slacks. These areas contain water at the time of spawning, and hence development of the frogs relies upon a wet spring keeping the dune slacks moist. The toad is scarcer and its main area for breeding is the small ponds, created by the extraction of the defence blocks, between the Marl Loch and the Curling Pond. Very few tadpoles are ever observed in the Marl Loch.

No reptiles have been recorded from the Reserve.

(iv) FISH: No survey has been carried out on the fish inhabiting the Peffer Burn, the sand and mud flats, or the rocks either north of Kilspindie or off the Gullane Point and Hummel Rocks.

(b) *Invertebrates:*

(i) MARINE: A survey of the marine invertebrate fauna of Aberlady Bay is included in a thesis from the Geology Department, University of Edinburgh (*Some Aspects of the Sediments and Organisms of Aberlady Bay, East Lothian*, Norah J. Allman and Frank Simpson, B.Sc. Thesis).

Two species of mollusc, *Hydrobia ulvae* and *Littorina littorea*, are particularly abundant in the Bay. To the south of the Peffer Burn and on the old midget submarines, where a firm substratum is available, the barnacle, *Balanus balanoides*, becomes established. Near the Peffer Burn, there are colonies of the mussel (*Mytilus edulis*), and amongst these fronds of the seaweed (*Fucus* spp.), barnacles, *L. littorea*, and valves of *Cardium edule* and *Scrobicularia plana*.

The coiled castings of the lug-worm (*Arenicola marina*) can be seen at low tide in areas of the Bay that are sandy and the occurrence of this species indicates sediments that are rich in organic matter. Two other polychaetes are found in the Bay – these are *Nereis diversicolor* and *Pectenaria koreni*. The estuarine amphipod *Corophium volutator*, which makes small U-shaped burrows about 5 cm. deep, is also abundant. A number of burrowing lamellibranchs are also found, the most frequent being *Macoma baltica* in the sands and *Scrobicularia plana* near the Peffer Burn.

Two distinct animal communities are recognized, the epifauna and the infauna. The epifauna are animals attached to rocks, seaweed or other animals, and occur in exposed places. These animals tend to be transported after death and this group contains such species as *Mytilus edulis* and *Littorina littorea*, that have probably moved into the Bay from adjacent rocky coasts. The infauna are animals living in sheltered areas or on the soft sea-floor, and, since these tend to be deposition sites, the animals remain *in situ* after death. *Arenicola marine*, *Pectenaria koreni* and *Cardium edule* belong to the infaunal community. A few species are intermediate in character between the two communities, and an example of this is *Hydrobia ulvae*.

(ii) TERRESTRIAL: No full-scale survey has been carried out on any group of terrestrial invertebrates. A certain amount of collecting of insects and spiders has been undertaken by Mr. E. Pelham-Clinton (Royal Scottish Museum) and Dr. Crowson (University of Glasgow). Mr. R. Waterston (Royal Scottish Museum) has collected fresh-water molluscs.

No list of insects has been prepared but it would undoubtedly be very long. An estimate of the number of species of beetles would be about 1500, and there is also a large number of flies and moths. The Reserve is particularly noted for the number of Southern species which are occurring towards the northern limit of their distribution, although some of them also occur on warm coastal areas to the north such as Tentsmuir, St. Cyrus and Findhorn. Particularly prominent is the cinnabar moth (*Callimorpha jacobaeae*) which feeds on the ragwort (*Senecio jacobaea*). The red and black moths and the yellow and black striped caterpillars are obvious features of the Reserve. In some years the larvae are so abundant that all the ragwort is eaten and skeletal plants can be found on the Reserve.

The flies of the Reserve are also of particular interest. There is a number of rare species associated with the dunes. One species of fly, *Spilogona compuncta* (family *Muscidae*), is only known to occur in Great Britain at Aberlady and possibly on Arran (although its locality there is unknown). As previously mentioned, the beetles are very numerous but one weevil, *Cleonus piger*, is very rare in Great Britain, but occurs frequently in the dunes. Its larva burrows into thistle stems.

Despite the general scarcity of butterflies in recent years, Aberlady still has a very large population. The most frequent are the meadow brown (*Maniola jurtina*), small heath (*Coenonympha pamphilus*), common blue (*Polyommatus icarus*), small tortoiseshell (*Aglais urticae*) and green-veined white (*Pieris napi*). At least two other species always breed in the Reserve, the dark green fritillary (*Argynnis aglaia*) and small copper (*Lycaena phlaeas*); and others, mostly migratory, visit the Reserve and possibly breed during suitable seasons, the red admiral (*Vanessa atalanta*), painted lady (*Vanessa cardui*), small white (*Pieris rapae*) and large white (*Pieris brassicae*). The Microlepidoptera (rather small moths) are peculiar on the East Lothian coast in that they occur as smaller, darker forms. The most interesting species which occur within the Reserve are of the genus *Crambus*.

Thus, the insect life of the Reserve is extremely rich. It has been suggested that if lists of insects in selected habitats on the East Coast of Scotland were compiled then the list from Aberlady would be the longest.

8. History.

(a) *General:* The history of the area has been studied by Mr. N. Tranter, who has written much of it into a series of articles entitled '*Footbridge to Enchantment*'. However, the aim of this section is not to survey the characters concerned but rather to review the land use of the area prior to its declaration as a Nature Reserve.

The hill to the east of the Reserve always seems to have been associated with rabbits and the area just north of the Peffer Burn with sports. The north end of the Reserve contained ironstone and, in this connection, the name of Jehova Gray has been handed down in the name Jovey's Neuk, Jovie's Neuk or Jova's Neuk. Jehova Gray lived in a small two-roomed cottage by Gullane Point, used water from a small fresh-water spring that is still running to-day and delivered the ironstone that he mined to ships which were beached, presumably in Jovie's Neuk. The ironstone went up the Forth to the Carron Iron Works and was used for making carronades.

The area just north of the footbridge was laid out as a golf course in the 1860s after the laird to the west of the Bay, the Earl of Wemyss, had used the Aberlady course at Kilspindie for a shooting-range. The course on the Reserve was used for 20 or 30 years till the inhabitants of Aberlady once again established a course to the west of the Bay. Then, in 1926, a guide book to the Aberlady district advertised two grass tennis courts, a curling pond, a pleasure ground and a golf club-house (stone-built). None of these, with the exception of the curling pond, can be seen to-day.

The wartime saw considerable change to the Bay. In 1940, about 800 concrete defence blocks were erected. These are 4′ × 4′ square, and are about 4′ above ground and 1′ below and are either corrugated or smooth according as iron or wood shuttering was used. Irish labourers were employed and, during construction, they went on strike for more pay; it was estimated by Aberlady inhabitants at the time that each block cost £5 to build. A track was built to bring trucks on to the area with aggregate and stone chips and sand which was taken from Gullane. The blocks were covered with turfs which were dug from around the bases of the blocks. Most of the blocks, which were known locally as 'Hitler's Stepping Stones', were removed in 1963, by the County Council, for the sea-wall round a power station being built at Cockenzie.

Poles, sections of tree trunks, were set in concrete and erected all over the Bay to prevent enemy aircraft landing. These poles are about three to four feet long and are still visible in many parts of the Bay. Out on the sand bar, one defence block was built and two Japanese midget submarines were beached. These were used for practice bombing attacks from the air.

Grazing of the Reserve grasslands has been continuous until recent times. It is probable that the Heritors of Gullane have common rights to graze cattle over the area now owned by the Gullane Golf Links. Not long before the war one lady in Gullane grazed a cow on the golf course. During the war and until just before the Reserve was declared in 1952, the butcher in Gullane grazed sheep over the whole area. Since declaration, no grazing by domestic animals has taken place.

(b) *Bye-Laws and Court Cases:* The initial factors in the declaration of a Nature Reserve on 14 July, 1952, have been detailed in Chapter II of this Plan.

The Bye-Laws were confirmed by the Secretary of State on 8 August, 1953, without a Public Enquiry, and the first meeting of a Local Management Committee took place on 10 January, 1953. They met to consider, *inter alia*, the issue of licences to shoot wildfowl, and on 6 March of that year they recommended to the County Council that the shooting of all birds at all times within the Reserve should be prohibited. This was rejected by the County Council on 12 May, 1953. On 12 and 13 August, two wildfowlers were charged with shooting without a permit but these charges were subsequently dropped.

A change in the Bye-Laws was proposed and a Public Enquiry was held at

Haddington from 9 to 11 November, 1953. The new Bye-Laws were confirmed by the Secretary of State on 12 February, 1954. Late in the summer of 1954, Mr. Barclay was again charged with shooting and the dispute centred around whether foreshore shooting rights are *inter regalia majora*, and thereby held inalienably by the Crown on behalf of the Public. Sheriff Middleton, in court at Haddington, held '. . . that the Public had in law a general right to use the foreshore in any part of the kingdom for the purpose of recreation and that this right was under the protection of the Crown and was superior to the right of any private proprietor in the foreshore.'.

However, the County Council appealed and the Lord Justice Clerk, Lord Thomson, together with Lord Patrick and Lord Birnam, said 'The Crown has power to consent to the public right of resorting to the foreshore for recreation or wildfowling being prohibited or restricted in appropriate places or circumstances, so that nature reserves which include foreshore may effect their statutory purpose of preserving the wild life of the country.'.

Since 1955, there have been no further Bye-Laws, no further legal disputes, and wildfowling has been controlled by issuing permits to about 26 people.

9. Public Use

(a) *General:* There are two aspects of public use that have to be established. Firstly, the people who are using the Reserve at the moment have reasons for coming into the area and these should be determined. Any trends in the pattern of usage over the last few years help to make management prescriptions simpler to formulate. Secondly, the future development of the whole of the coastal zone of East Lothian should be considered and it should be an aim of this Plan to demonstrate how the Reserve can become an integral part of the attractions of this coastline.

(b) *Present Use:* Surveys of people using the Reserve, entering and leaving by the footbridge, were carried out over the weekend 29 and 30 July, 1967. The weather on the Saturday was changeable, mostly cloudy with some sunny periods and two short showers. On Sunday, the weather was sunny and this was reflected in the greater number of people using the Reserve.

The times when people and cars entered the Reserve, and the times of people leaving the Reserve, are illustrated in the histograms in Fig. 10.4. These show the pattern on Saturday but the Sunday pattern was very similar except that more people came during the lunch period. Due to the warmer weather on Sunday activity on the Reserve continued later into the evening than on Saturday.

The people using the Reserve can be divided into four groups. Firstly, there are those who come because of the natural history aspects of the Reserve. These include both the really interested people – counting the birds, Natural History Society members, etc. – and people who are casually interested, often describing themselves as taking a 'nature walk'. Secondly, there are people who use the

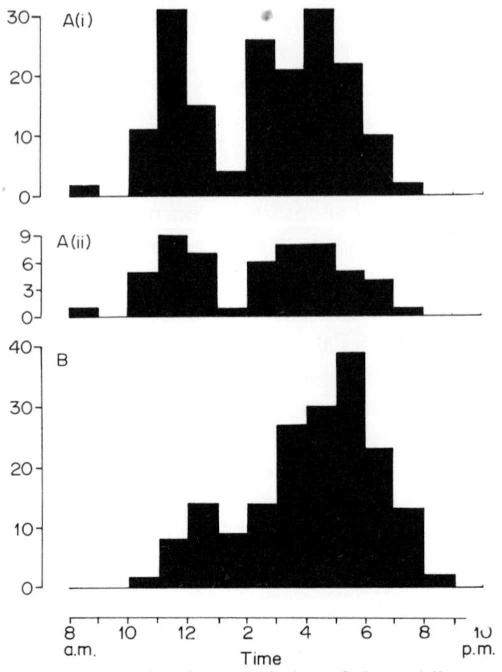

Fig. 10.4 Histograms showing the characteristics of the public usage of the reserve on Saturday, 29 July 1967. (a): Histograms showing the arrival of people (i) and cars (ii). (b): Histogram to show the departure of people

whole stretch of the Reserve, walking out to the dunes or Gullane Point. These people tend to be either making for the beach or taking a dog for a walk. Thirdly, there are those people who just want to stroll, they cross the footbridge, and either sit down or walk no further than the Marl Loch. Lastly, there is a group of people who just use the car park. Many of these people picnic in or by their cars and they tend to be families with more than two children. After about 1800

TABLE 10.6 A summary of public use of the Reserve on the week-end 29 and 30 July 1967

	Number of people						
Period	*Natural history interest*	*Walking far*	*Staying by bridge*	*Staying in car park*	*Total*	*Number of cars*	*Number of dogs*
Saturday	32	33	39	71	175	55	18
Sunday	76	99	51	65	291	90	30
Week-end	108	132	90	136	466	145	48
Percentage of people	23	28	19	30	100	–	–

hours a number of young couples use the car park but never leave their cars. The results of an analysis of people using the Reserve on the two days of the survey are given in Table 10.6.

Pooling the data collected on the two days, the following facts emerged. About 60% of visitors to the Reserve come from Edinburgh and about 20% from the County of East Lothian. Of the remaining 20% about 16% are from other parts of Scotland and 4% from England and Wales. About 15% of the people interviewed were on holiday in the Lothians. The frequency with which people visit the Reserve was also determined and categories and percentages of people are:

Frequency	Daily	Weekly	Monthly	Yearly	First Visit
Percentage	1	20	26	30	23

There is, therefore, a substantial number of people who regularly visit the Reserve and their opinions are more important than those of the casual visitors. Nearly all of the regular visitors walk further into the Reserve than the Marl Loch and many of them (*c.* 70%) welcomed the idea of having a small informative booklet or leaflet on sale. No one disagreed with the idea. Also, most of the regular users of the car park said that they would not object to paying a small charge for the use of the park. Many comments on the area such as 'quietness', 'quiet scenery', and 'solitariness' were expressed and many people came back because of this quiet nature of the Bay. A similar sentiment was expressed by Robert Louis Stevenson when, in *Catriona*, he referred to the area as: 'Few parts of the coast are lonelier . . . such a shining of sun and sea, such a stir of wind in the bent-grass and such a bustle of down-popping rabbits and up-flying gulls, that the desert seemed to me like a place alive' (quotation taken from Nigel Tranter's '*Footbridge to Enchantment*').

(c) *The East Lothian Coast:* The Nature Reserve should be considered in relation to the Development Plan for the East Lothian. This section of the Management Plan is based on the 'County of East Lothian Development Plan: Tourist Development Proposals', which was approved by the County Council on 12th April, 1965, and by the Secretary of State on 6th April, 1966.

Paragraphs 78 and 79 of the '*Coastal Survey 1961*', which is included as an appendix to the above mentioned Plan, sets the scene for action affecting Aberlady Bay Nature Reserve. The paragraphs state:

'78. First, that the coastline has great variety and wealth of features. It is the current planning policy to preserve and maintain this variety, keeping some areas "remote", protected and undeveloped, while providing positive facilities in others and this policy should be re-affirmed.

'79. Second, that the dune and beaches along the A198 Longniddry, Gullane and North Berwick are suffering from severe erosion and have reached the limit of their capacity. Any vast increase can but result in the complete breakdown of the coastal dunes as is occurring at Gullane and so access rights to alternative beaches must be obtained.'

Paragraphs 27 and 28 of the *Coastal Survey 1961* are, in some ways, a first approximation to a Management Plan. The only difference between the outline in the Survey and the prescriptions of the Plan is that this Plan does not envisage the provision of a second car park for the Reserve. This Plan underlines the threat to the Reserve mentioned in the last sentence of Paragraph 28: 'There is also the need for some measures to control the incursion into the Nature Reserve of a large number of the holiday crowds that arrive at Gullane Bents.'

In accordance with Paragraph 78, quoted above, Aberlady Bay must be one of the areas to be kept 'remote', whilst the adjacent beaches and dunes of Longniddry and Gullane are developed as recreational areas.

(d) *The Future:* The previous sections of this chapter have highlighted two features of the public usage of Aberlady Bay Nature Reserve. Firstly, there is the natural history of the area and survey shows that about a quarter of the people visiting the Reserve are interested in natural history. Public usage does and will exploit the natural history aspects of the Reserve and hence conservation projects will be required to ensure that none of the habitats are destroyed by over use. On the other hand many of the visitors to the Reserve come here because of its loneliness and quietness. This character of the Reserve must also be conserved and its very nature implies that it will be destroyed if too many people endeavour to use it.

CHAPTER IV
OBJECTS OF MANAGEMENT

The first attempt to define the objects of Management are contained in the minutes of the meeting of the Biological Sub-Committee on 9 January, 1964 (see appendix to minutes of the Management Committee meeting on 14 April, 1964). These record:

'The Management Plan would state the objectives of the Reserve. The Sub-Committee consider that the Reserve serves primarily as a winter refuge for wildfowl but is also an area of great biological and geological interest with increasing educational and recreational uses. The Plan would, therefore, make prescriptions for integrating wildlife conservation and public use.'

The main object of management stems both from this initial suggestion and from the results of a survey of people visiting and using the Reserve. It can be stated as:

1: To conserve the fauna, flora and habitats of the Reserve in order to provide an area of high educational value.

One has, therefore, to steer a course between encouraging people to use the

Reserve and to understand what is in it, while yet not making it too popular. This is necessary so that both the natural history (Chapter III, sections 6 and 7) and the sense of 'space, quietness, and loneliness' (Chapter III, section 9) are conserved. The prescriptions of Chapter V aim at establishing a form of 'biological open-air museum' and maintaining the sense of quietness and sky which is so highly valued by its visitors. The main object of Management can be broken down into a series:

Conservation

1a: To conserve the bird life, particularly by supervision of the shore-nesting species and by conservation of the salt-marsh as a feeding ground for migratory birds in passage.

1b: To conserve the various plant communities, particularly the fresh-water communities associated with the Marl Loch and the Curling Pond.

1c: To maintain an area of open water on the Marl Loch.

1d: To guard against erosion of the sand dune system in the north of the Reserve.

Education

1e: To encourage public usage of the Reserve for appreciating natural history, and to inform and interest the public in all aspects of the Reserve.

1f: To encourage schools, particularly in Edinburgh and the Lothians, to take an interest in the natural history, physiography and geology of the Reserve.

The other objects of management can be grouped into three categories:

Recreation

2a: To maintain a system of controlled wildfowl shooting.

2b: To control areas subjected to heavy pressure by the public, e.g. re-routing of paths.

Research

3a: To encourage biological and physiographical surveys of the Reserve.

3b: To ensure that scientific research relating to the Reserve, by individuals or institutions, can be pursued.

Miscellaneous

4a: To maintain good relations with the owners of the ground.

4b: To consider revision of the eastern landward boundary of the Reserve.

CHAPTER V
MANAGEMENT PROGRAMME

1. Management for Conservation

(a) *Ornithological:*

The bird life of the Reserve falls into three groups – (i) the shore-nesting birds, (ii) the birds feeding in the salt marsh and the Bay, and (iii) the smaller breeding birds of the grasslands and woodlands. Management prescriptions are required for the first two groups.

(i) SHORE-NESTING BIRDS: The factors influencing the growth of the tern colony and the change in breeding ground of the waders has been discussed in Chapter III, section 7(a) (i). At the advice of the R.S.P.B. the part-time Warden erected a line of poles and suitable advisory notice boards so as to canalize people away from the centre of the breeding area. Terns are attractive birds, and so long as a colony continues to attempt to nest within the Reserve a part-time Warden should be employed to patrol their breeding ground.

The possibility exists that the terns might move their site from year to year, or that another species of bird may require special supervision during the breeding season. Recommendations about additional supervision of breeding birds should be made by the Biological Sub-Committee.

(ii) WADERS FEEDING IN THE BAY AND MARSHES: At present the salt marsh provides a certain amount of food, animal life associated with brackish water, for migratory birds. Experiments were tried in 1959 to create shallow 'flashes' in the salt marsh south of the sand bar, but these were on a limited scale and have met with little success. In order to increase the feeding and to facilitate observation a further shallow artificial pond, three to nine inches deep, should be considered. Before such a pool is established, a scientist, with experience of such areas, should be consulted to ascertain whether the area could support a large enough population of animal life for food for the birds. A survey of the animal life inhabiting the brackish water of the Reserve might be useful.

It remains possible that such a pool would have an enhanced educational value if a permanent hide were constructed. This work would not be required, however, during the period covered by this plan, since the pool will require some time before it becomes a major feeding ground for waders.

(b) *Botanical:*

(i) FRESH-WATER MARSHES: The Marl Loch and the Yellow Mires have a very rich flora (Chapter III, section 6(d) (ii)). Threats to these areas come from drainage and toxic chemicals. Management should aim at avoiding drainage schemes in the vicinity of the Marl Loch and an area running about half a mile to the north of the Loch. The effects of toxic chemicals on the plant and animal life of the fresh-water marshes could be serious, and their use anywhere within the Reserve should be discouraged.

(ii) DISTURBED SOIL: The Golf Course has, in the past, cut areas of turf within the Reserve. This practice should not be stopped, since the bare ground thus created gives rise to two types of specialized plant communities (Chapter III, section 6(f)). If no new turf cutting operations have been started by 1970, the Biological Sub-Committee should recommend two or three areas, not more than a quarter of an acre each, from which the turf should be cleared.

(iii) GRAZING: The problem of allowing grazing on the Reserve was raised by the factor of Luffness Estate and the Tenant Farmer in 1964 and 1965. The advice of the Nature Conservancy, and a prescription of this Plan, is that there should be no grazing in the vicinity of the Marl Loch, the Curling Pond, or the marshes and moist ground associated with either of these areas. In order to graze this ground, herbicides would have to be used to destroy the horsetail (*Equisetum* spp.), which occurs in one of the most interesting plant communities in the Reserve, and which is poisonous to cattle. The damage to the saltings and to the banks of the Marl Loch by the cattle's hooves would also be considerable. The area considerably to the east of the Marl Loch, a small part of which was enclosed and cut for silage in 1967 (Plate 13), may be suitable for grazing. However, part of this area would probably require a ground drainage system, which would influence the seepage of water into the Marl Loch and its associated marshes. Hence, it is recommended that no ground drainage schemes are carried out within the Reserve. The erection of fences, especially to the west of the Marl Loch, should be opposed, since this would destroy the 'sense of space'. The advice of the Nature Conservancy and the Biological Sub-Committee should be sought in all aspects of grazing.

(iv) HORSES AND PONIES: A consideration of the damage to plant communities by the riding of horses and ponies, and prescriptions for routing these animals, will be found in section 3(b) (v) of this Chapter.

(v) CONTROL OF SEA BUCKTHORN: There are three main areas of sea buckthorn, *Hippophaë rhamnoides* – beside the Marl Loch, at the north of the Yellow Mires, and near the Hummel Rocks. With the lack of rabbit grazing pressure, it is estimated that the forward edge of these clumps is advancing at two to three feet per year (Plate 15), and, particularly on the more open teschenite soils, self-seeded bushes are becoming established. If it is possible, the three clumps of sea buckthorn should be maintained at their 1967 size by frequently cutting all outspreading shoots. It is also prescribed that all other plants of sea buckthorn should be dug out before they become too large to control.

(vi) CONTROL OF OTHER SPECIES: The giant hogweed, *Heracleum mantegazzianum* (Plate 16), an alien that is spreading in the Lothians, is undesirable because it forms dense stands in which few other plants are able to grow. It is, therefore, prescribed that this plant be exterminated from the Reserve.

Hawthorn is not desirable in the Reserve since it causes the open character of the Reserve to be destroyed (Plate 14). Later in this Chapter, section 5, the boundaries of the Reserve are discussed, and an area which the hawthorn would be allowed to colonize is demarcated. Over all other areas of the calcareous grassland, the hawthorn should be cut.

(vii) BURNING: It has been suggested that burning of the ground vegetation might create a short turf habitat more suitable for shore-nesting birds. Whether burning would have the desired result is not known, but there are many possible undesirable results (e.g. the fire getting out of hand). Therefore, burning as a management tool is not recommended.

(c) *The Marl Loch:* Some of the changes in the Marl Loch since it was dug out in 1963 have been described in Chapter III, sections 6(d) (ii) and 7(a) (i). In order that this fresh-water fauna and flora should be conserved, and in order to provide a focus for a possible nature trail, it is recommended that the open water area should be maintained. However, just as the open water became overgrown with marsh vegetation in the past, so it will become re-overgrown, and any chance to use mechanical apparatus in clearing away more of the marl and vegetation should be taken. It is likely that during the period of this Plan that work on the Aberlady and Gullane sewage schemes will commence, and this will involve the use of some heavy machinery on the Reserve. Since it is considered that clearing operations will have to take place every eight to twelve years, it might be useful to take this opportunity to dredge the area of open water again. A botanist should always be consulted before dredging commences to ascertain that no damage will occur to the rarer plants.

(d) *The Sand Dunes:*

(i) GENERAL: At the present time, natural processes are causing a build up of a part of the sand dunes as was described in Chapter III, section 2.

(ii) FACTORS AIDING EROSION: Three factors can help to cause erosion. (1) Wind: if there are a large number of westerly or north-westerly gales during the winter, large quantities of sand, particularly near Jovie's Neuk, may be blown away (Plate 17). This occurred during the winter of 1966–67, and in order to re-establish the marram grass on the bare sand slopes some thatching with sea buckthorn branches would be useful. (2) Public pressure: any increase in the public usage of the dune system would encourage erosion (Plate 18). Public pressure on this coastline is increasing, but as the dunes are at least one and a half miles from the nearest car park, any damage is unlikely to be as severe as at Gullane. The effects of hooves of horses and ponies activate erosion (Plate 19), and all animals with iron-shod hooves should be forbidden to enter the dune system. (3) Fire: Fire damages the grasses and in the burnt state they cannot hold the sand against wind erosion. When the public pressure on the dunes increases, it might be useful to have one or two racks of fire-brooms situated in prominent positions near the dunes. Notices drawing the public's attention to the fire risks and mentioning the provisions of Bye-Law 9 should be exhibited.

(iii) INSPECTION: At the present time, there is little immediate danger of large-scale erosion of the dune system. However, in view of the erosion at Gullane Dunes, it is prescribed that the dunes on the Reserve should be inspected annually, preferably in early spring. Any anti-erosion work that has to be undertaken should follow the principles used on the Gullane Dunes.

2. Management for Education

(a) *General:* Three categories of people visit the Reserve. Firstly, there are organized parties from schools, universities, or societies, and these would normally be supervised. They would usually be interested in some aspect of the natural history, physiography or geology, and would probably require no special information about the Reserve.

Secondly, there is the general public who comes on to the Reserve with the purpose of following some aspect of field studies. To the regular visitor (e.g. the bird-watcher, wildfowler) no provision is required for education. However, a large number of people in this second category visit the Reserve without any clear idea of what they will see, but nevertheless they come here to look for birds and flowers since it is a Nature Reserve. Such people have expressed the wish for some guidance, and this would have to be provided at or near the Reserve.

Thirdly, there are the people who visit the area for recreation – walking the dog, swimming near the Point, picnicking in or by the car park, etc. Many of these people are not interested in the Nature Reserve, some in the 'sense of space' and quietness, and some in finding a free car park.

It should be stressed that the Bye-Laws apply to any person entering the Reserve, and that parties of persons using the Reserve do require a permit to collect any plants or animals (Bye-Laws 10, 11 and 14).

(b) *Aspects of Education:*

(i) BOOKLET OR LEAFLET: Most people who visit the Reserve for a natural history excursion consider that this is essential. A small publication that describes the land form of the Reserve, discusses the plant communities and some of the plants, possibly in relation to the physiography, and describes the animals associated with the Reserve, particularly some of the species of birds, and in general concentrating more on ecology than listing species, should be prepared. This could probably be put on sale in a shop in Aberlady, and it might be possible at some time to sell copies of this to visitors entering the Reserve at the footbridge over the Peffer Burn. Schoolmasters feel that it has some potential use in schools, and the local schools should be encouraged to buy copies for field teaching. The presentation should not be too specialized, and it should probably be pitched at fourth or fifth year school level.

(ii) NATURE TRAIL: There is already a nature trail at Yellowcraig, about three miles to the east of the Reserve. School teachers have expressed the view that a nature trail is only of limited use to them. At the present time, therefore, there appears to be little need to lay out a nature trail on the Reserve.

However, this should be kept under constant review. When a booklet about the Reserve is prepared, there might be a demand from the public for some of the aspects of the natural history and physiography of the Reserve to be demonstrated. This could be achieved either by a nature trail or by informative notices (see Chapter V, section 2(b) (iv)). A nature trail can effectively keep a high proportion of people away from parts of the Reserve that require special attention, e.g. nesting areas.

If a trail is eventually laid out on the Reserve, an explanatory leaflet, sold in conjunction with the booklet, would be required.

(iii) FILM STRIPS: School teachers have expressed their opinion that film strips, on selected subjects, are the most useful aid to teaching prior to a field visit. The preparation of film strips or loop films is expensive, and falls outwith the scope of a prescription of the Management Plan. However, if the Education Authorities of Edinburgh or the Lothian County Councils consider that the Reserve has a sufficiently great teaching potential, they could be encouraged to finance the preparation of suitable educational films. All filming work of this nature on the Reserve should be encouraged and a record of any films made should be kept with the management Plan.

(iv) INFORMATIVE NOTICES: Notices informing the public about the dune reclamation work at Gullane have proved successful, and it is possible that informative notices could be used to advantage at Aberlady. For example, a notice on the highest point of the Reserve, just south of the sea buckthorn clump near the Hummel Rocks, could demonstrate the physiography of the dune system. Relying on natural curiosity it might serve to keep people off the dune paths and on to the metalled track, thus reducing erosion danger. A similar informative notice by the Marl Loch, demonstrating some of the plants growing there, would provide interest.

These notices should be situated beside paths, and should be marked on a map, possibly with the Bye-Laws, and also in the booklet.

(v) ADVERTISING: Talking to people who use the Reserve, it has been suggested that the Reserve should advertise itself and that periodicals and newspapers should sometimes carry short articles about the Reserve. There is no doubt that Mr. N. Tranter has stimulated an interest in the Reserve by his short articles discussing both the biological and historical aspects of the area.

3. Management for Recreation

(a) *Wildfowling:* No changes in the present system of wildfowl shooting are prescribed. Under this system, 25 permits have been issued to residents of East Lothian for the duration of the shooting season and an additional permit has been given to Inspector David White, of the Dunbar Police Station. It is the responsibility of the Management Committee to issue these permits.

The permits, issued in terms of Bye-Law 14, have conditions attached to them that are listed in the minutes of the Management Committee meeting on 5 July, 1966. These are:

1. The permit is not transferable.
2. The permit must be produced on demand.
3. Shooting is only allowed between sunset and sunrise.
4. Shooting is not allowed between the southern boundary of the Reserve and the Peffer Burn and within that part of the Reserve lying east of a line fifty yards to the west of the remaining pillars south of the Marl Loch.
5. Shooting by any weapon other than by ordinary smooth bore shot-gun is not allowed.

6. The permit is valid from 1 September, 1966, to 20 February, 1967.
7. The Management Committee reserves the right to withdraw the permit at any time.
8. The holder of the permit will inform the Management Committee not later than 15 March, 1967, of the number of days on which the permit has been used and the number and species of duck shot.

In future conditions attached to the permits, conditions 6 and 8 would be modified to include the dates appropriate to the current year. It was also agreed by the Management Committee that no geese were to be shot.

(b) *Control of Public:*

(i) CAR PARKING: The only car park for the Reserve is immediately west of the footbridge over the Peffer Burn. This has a maximum capacity of 23 cars and, in order to keep the Reserve 'remote', it is the opinion of the Biological Sub-Committee that this park should not be enlarged. The County Police should be encouraged to prevent car parking along the roadside of the A198 near the footbridge. It might be useful if this road were a 'Clearway' as the 'Police: No Parking' signs are often thrown over the wall south of the A198. If roadside parking could effectively be stopped, neither the number of people on the Reserve nor the rate of turnover by the Reserve would be increased if the car park were extended to hold a maximum of 36 cars. This might be useful if a full-time Warden, as discussed in Chapter VII, section 3, is to be appointed.

An increase of parking facilities at Gala Law Quarry, about half a mile east of the Reserve and holding four cars at present, should be opposed. Any enlargement of parking facilities at Gullane, by increasing the car park towards the Nature Reserve, should also be opposed, since this would allow an increased number of people into the dune area of the Reserve. No scheme to allow vehicular traffic into the Reserve should be considered except under Bye-Law 14 when it is required for special purposes (e.g. a film unit).

(ii) FOOTBRIDGE: The footbridge over the Peffer Burn (Plate 9) should be kept in a good state of repair. In the winter of 1966–67, it was damaged by gales and, in the summer of 1967, one of the planks forming the carriageway was missing. Nothing in the Bye-Laws states that the County Council cannot be held responsible for loss, damage or injury whilst on the Reserve. If the footbridge is in a state of disrepair, injury to people, particularly children, is possible. It is recommended that the footbridge is inspected regularly, possibly monthly or fortnightly, for any sign of structural weakness.

(iii) MAINTENANCE OF PATHS: As the number of people using the Reserve increases, so also does the need to canalize their movements in order that all parts of the Reserve are not disturbed (Plate 19). Canalization is achieved by a good system of footpaths but footpaths on sand crossing marshy places are notorious for getting in a bad state. This must be overcome by laying a few drainage pipes at the level of a ditch and building up the path to the height of the surrounding ground. Any such work should not damage the vegetation forming the sides of the ditches on either side of the path.

(iv) DIVERSION OF FOOTPATHS: In places where footpaths cross the dunes, excessive use by the public may lead to erosion. Sections of footpath that are liable to erosion should be planted with marram grass and an alternative path cleared of vegetation. If there is an existing path, this should be used but where this facility is lacking a path should be cleared by cutting the grass along its line during the spring. Planting of marram grass over paths to be closed would best be done in autumn or early spring when public pressure on the Reserve is slight.

Paths may also have to be diverted to protect special areas of the Reserve. The lower path from the Marl Loch to the sand bar causes concern during the tern nesting season and it should be diverted. Most people using this path are making for the sand dunes and beach and hence a diversion to the right, about 50 yards before the shore is reached, would channel most people to the dunes and away from the sand bar. There is already a small path forking off to the right, which is marked on the map and this should be enlarged before marram grass is planted over the old path. The paths are indicated in Fig. 10.5.

(v) PONY RIDING: Pony riding in the Reserve should be restricted to areas where erosion danger is slight and to paths where the ponies will not cause danger or annoyance to other persons using the Reserve (cf. Bye-Law 3). In Chapter V, section 1(d) (ii) it was recommended that animals with iron-shod hooves should be forbidden to enter the dune system. In order to prevent disturbance to shore-nesting birds, they should also be forbidden to ride closer than 50 yards to that part of the sand bar which is exposed at high water. Riders should be encouraged to keep to the remainder of the foreshore wherever this is possible.

It has been suggested that one or two paths should be demarcated for use by ponies. Two suggested routes, if the eastern boundary of the Reserve is re-declared, are illustrated in Fig. 10.5. The first of these paths allows riders access to the foreshore just south of Jovie's Neuk. It runs from the football pitch near the Hummel Rocks and follows a small valley between the teschenite rocks and the dunes. The second path would run from the football pitch to the ford beside the footbridge and would follow the re-declared eastern boundary if this is achieved. It should also be noted that the rocks immediately east of Maggie's Loup are impassable to ponies at high tide and hence the construction of a small slope here, just outside what might be the revised Reserve boundary, would be a contribution to controlling ponies within the Reserve.

When these two Pony Tracks are established, riders will be obliged to keep to them whilst on the area of the Reserve above H.W.M.O.S.T. The two paths should be demarcated with small posts, perhaps about two feet high. The risk of erosion of the dunes starting on the path from the football pitch to Jovie's Neuk should always be remembered.

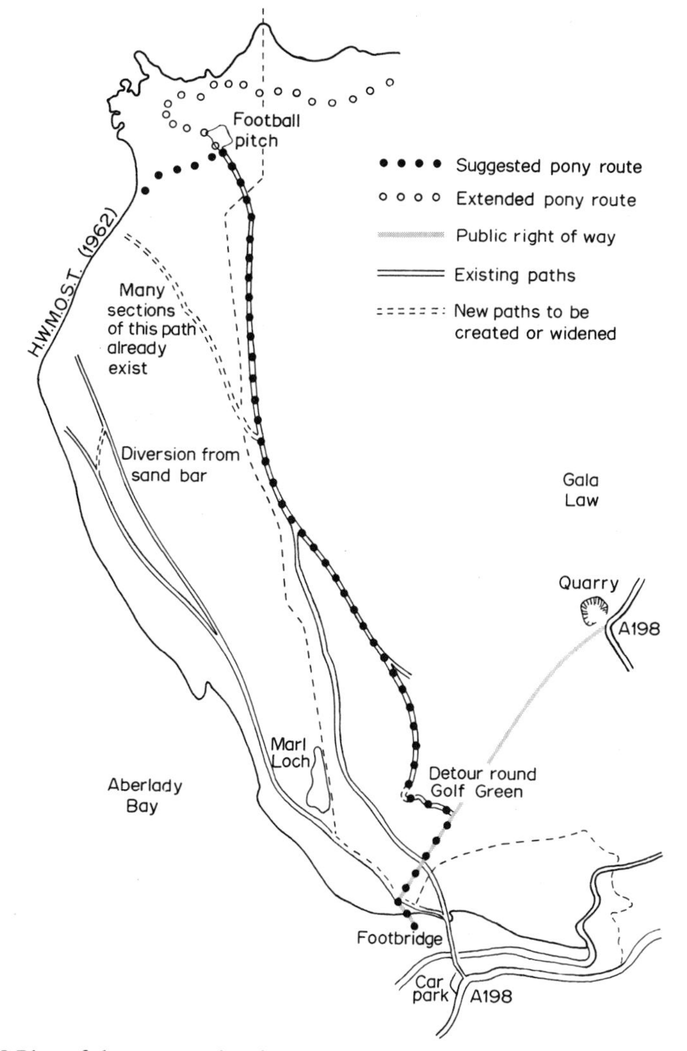

Fig. 10.5 Plan of the reserve showing management proposals for paths and pony riding routes

(vi) WARNING NOTICES: The notices showing the Bye-Laws need to be re-sited. The notice by the footbridge should be moved from the roadside to near to the bridge so that it can be seen by the public walking from the car park to the Reserve. The survey showed that few people have ever walked from the car park to the roadside to read this notice. A second notice is situated on the concrete blocks at Maggie's Loup. This is considerably to the east of the Reserve boundary but it would be correctly sited if the boundaries are re-declared. Two notices are required on the western section of the Reserve, situated by the

H.W.M.O.S.T. near the village of Aberlady, and at the west end of the Kilspindie Golf Course.

The survey showed that whereas about 95% of people entering the Reserve at the footbridge knew that they were on a Nature Reserve, less than 15% said that they had ever read the Bye-Laws. It would, therefore, be useful to have two notices, one by the footbridge to the north of the Peffer Burn and another near Maggie's Loup, that give a summary of the Bye-Laws. Such a notice could be worded in the following manner:

Aberlady Bay Nature Reserve

Please do not pick the wild flowers or disturb the birds. You are asked to keep to the paths, particularly in the nesting season.

Additional notices might be required by a part-time Warden asking the public to remain outside a demarcated area.

All notices should be inspected, at least annually, and kept clean and painted. Many of the notices on the Reserve in 1967 are shabby and the paint and wording is peeling off. This is a situation that should be avoided. Similarly, notices should be placed where they must be seen, i.e., beside paths, and not ten yards from a path in long grass.

(vii) LITTER: Litter presents a problem in that it cannot be collected from the Reserve. There is an increasing number of bottles being left among the marram grass in the dunes and this increases fire danger in the area. If any labour becomes available, it would be useful to gather and dispose of as much litter as possible. A litter bin should also be provided at the car park and arrangements made for this to be regularly emptied. Drums containing oil and other pollutants, washed up by the tide, should also be disposed of if this is at all possible.

(viii) SEWAGE: No prescriptions can be made in regard to sewage but the County Council should be encouraged to take all possible steps to decrease pollution by sewage in the Reserve. Proposals have been drawn up by Messrs. Carter & Wilson, Edinburgh, for the County Council and their report is entitled *Proposals to Remedy Sewage Pollution at Aberlady Bay.* The following are extracts:

'There are three areas where pollution is evident – in the Bay close to Aberlady village, along a stretch of 600 yards over the existing West Gullane sewer through the links, and at the sea outfall of this sewer. There is a fourth area – the West Peffer Burn – where pollution occurs intermittently through the operation of a vegetable washing plant at Luffness Mains.

'We recommend that the first three areas be dealt with by piping the sewage from Aberlady to the West Gullane sewer, constructing a screening and settlement plant for the combined sewage, constructing a new sewer from the plant to high water mark, and extending the present West Gullane sea outfall. The fourth area should be dealt with by strict control of the operations at Luffness Mains . . .'

'Aberlady. Our proposals here are for a gravity sewer laid underground along the foreshore eastwards for a distance of 800 yards to the footbridge over the Peffer Burn, an underground balancing tank and pumping station beside the car park at the footbridge, and a pumping main 950 yards long to join the West Gullane sewer . . .'

Treatment works are proposed to be situated in or near to the clump of Sea Buckthorn, east of the Reserve boundary, near the Yellow Mire. It is envisaged that work will commence on the new sewage scheme during the period covered by this Plan and the Biological Sub-Committee should ascertain that a minimum amount of damage results to the habitats of the Reserve as a result of this scheme.

4. Research Management

(a) *General:* It is not envisaged that the County Council will, itself, take an active part in research or survey work. However, it can encourage other bodies or persons to undertake scientific work in relation to the Reserve and the presence of a Nature Reserve in the vicinity of Edinburgh implies the potential use of the ground for scientific research. At the present time, no prescriptions can be made as to the form that research work should take. However, the Management Committee and the Nature Conservancy, who undertook to advise the County Council in scientific matters relating to the Reserve, can control research by the issuing of normal permits to collect samples or specimens.

(b) *Surveys:* Survey work should concentrate on groups of animals or plants that are not well known and on physiographic changes in the Reserve. The results of completed surveys are included as appendices to this Plan. An intensive survey of the soils of the Reserve is required.

A record of any person applying for a permit to collect for survey work during the period of this Plan should be added to Appendix 10 of this Plan and details of results of the survey should be included in later editions of the Management Plan.

(c) *Scientific Research:* The details of all research projects that are started during the period of this Plan, and of permits that are issued in order to undertake a research project, should be added to Appendix 10 of this Plan. Records of publication of research results or the location of unpublished reports or theses should also be added to Appendix 10.

The following research projects are known to be taking place, or to have taken place on or near the Reserve:

Miss N. J. Allman and Mr. F. Simpson (Geology Dept., University of Edinburgh) – Studies on the sedimentation and marine organisms of the Bay (completed – B.Sc. thesis topic).

Dr. J. Dodds and Dr. P. Myerscough (Botany Department, University of Edinburgh) – Studies on the physiographic changes in the dune system; on the populations of spring flowering annual plants; and on the vegetation associated with the transition from teschenite to calcareous sandy soils (part of the work was undertaken by Dr. J. K. Marshall).
Dr. Roy Watling (Royal Botanical Gardens, Edinburgh) – Studies on the distribution and rate of growth of fairy rings.
Miss P. J. Watson (Scottish Plant Breeding Station, Roslin, Midlothian) – Breeding of the grass *Poa pratensis* for pasture improvement.

(d) *Permits:* No prescriptions are made in relation to the present system of issuing permits by the County Clerk, under Bye-Law 14. The Nature Conservancy and the Scottish Wildlife Trust are asked to submit their comments on each application and it is the responsibility of the Management Committee to agree to the issue of a permit. It is a condition of a wildfowling permit, see Chapter V, section 3(a), that a return has to be made. Permit holders for research or collecting should be asked to submit brief details of their work on the Reserve within some defined period of time after expiry of the permit.

5. Revision of the Eastern Boundary

(a) *General:* When the Reserve was declared in 1952, the boundaries were not based on any scientific knowledge of the biological or educational nature of the terrain. During the last fifteen years, much more has been learnt of the ground to the east of Aberlady Bay and from this it appears that a revision of the eastern (landward) boundary of the Reserve would be beneficial to conservation and education. A number of points can be listed as:

1. The existing boundary is not demarcated in various places, existing as a line on a map or along the line of a feature that no longer exists. A clearly defined boundary would allow for more precise management.
2. Many people have considered areas outwith the Reserve to be included. As an example, the Bye-Laws are posted on the concrete blocks at Maggie's Loup, some 320 yards to the east of the actual boundary.
3. A large proportion (60–70 per cent) of visitors to the Reserve spend most of their time on paths to the east of the reserve boundary.
4. The sites most noted for their wild flowers, those with the grass of parnassus, with the frog orchid or with the spring flowering annuals, are to the east of the boundary.
5. The sites of special research interest, both for higher plants and fungi, are outwith the Reserve.

(b) *Procedure:* It, therefore, seems that an attempt should be made to re-declare the eastern boundary. An approach should be made by the County Council to the owners of the ground for their views on this slight enlargement of the Reserve.

If the new boundary can be agreed, the Bye-Laws will require amendment. Bye-Law 1(b) would have to be reworded so as to include this extra strip of land.

(c) *Proposed Boundary:* The suggested eastern boundary is marked in Fig. 10.2. From north to south, the boundary follows the following features:

1. The line of concrete defence blocks at Maggie's Loup and their projection seaward (approx. 105 yards on land).
2. A straight line, undefined on the ground, from the southernmost block at Maggie's Loup to the northernmost block in the sea buckthorn clump to the south of the Hummel Rocks (approx. 450 yards).
3. The eastern and southern boundary of this clump of sea buckthorn until its junction with a partially metalled track (approx. 145 yards).
4. The metalled track from the sea buckthorn clump until it curves sharply eastwards at the junction of a small path (approx. 1490 yards).
5. This small path from its junction with the track to a golf green just beside a small clump of sea buckthorn (approx. 480 yards).
6. The west boundary of this sea buckthorn clump and along a small path running south-east to a larger golf green beside another clump of sea buckthorn (approx. 240 yards).
7. The east boundary of this sea buckthorn clump and along a straight line to the nearest point on a sunken wall (approx. 115 yards). This wall forms part of the existing boundary of the Reserve.

(d) *Area:* This revision of the eastern boundary of the Reserve would increase the area of the Reserve by 59 acres. This can be divided as follows:

1. The triangle between the Hummel Rocks, Maggie's Loup and the sea buckthorn clump south of the Hummel Rocks

   Area added – 11·8 acres
   Area excluded – 1·0 acres

   The area that is lost is part of the golf course.
2. The narrow strip between the metalled track and the line of concrete defence blocks from the above-mentioned sea buckthorn clump to just north of the sea buckthorn clump where the track and the line of blocks diverge

   Area added – 4·1 acres
3. The area of grassland from the above defined point to the wall at the south of the revised boundary (this includes the large clump of sea buckthorn near the Yellow Mires, and the sea buckthorn clump on the east of the Marl Loch)

   Area added – 44·2 acres

(e) *Preservation of Hawthorn:* In Chapter V, section 1(b)(vi), it was prescribed that the hawthorn should be cleared from the grasslands of the Reserve. In order that the development of a hawthorn scrub (Plate 14), and possibly the natural establishment of tree species, can be studied, an area of approximately 15 acres at the southern end of the area added to the Reserve by re-declaration should be reserved for hawthorn. In this Reserve no hawthorn is to be cut. In this situation, the outline of a hawthorn thicket will merge with the nearby Scots pine plantation and will not destroy the open character of the Reserve. The area is demarcated in Fig. 10.2.

CHAPTER VI
PUBLIC ACCESS

Locations where the public can enter the Reserve have been described in Chapter I. There is one public right of way across the Golf Links to the east of the Reserve and across the southern section of the Reserve to the east of the Bay. This is the track running from Gala Law quarry to the ford across the Peffer Burn near the footbridge. It is also possible that a right of way exists from either Gullane or Aberlady to Gullane Point, where the ironstone mine was situated, but this has not been ascertained.

Within the Reserve, there are no restrictions on pedestrians or on boats. Vehicles, as defined by Bye-Law 1(d), are not allowed except by special permission. Riders of horses, ponies and pedal cycles are controlled by Bye-Law 3, and further control of animals with iron-shod hooves is proposed in Chapter V, sections 1(d) (ii) and 3(b) (v) of this Plan.

Bye-Laws: The Reserve became subject to Bye-Laws on 7 October, 1952; and to two further Bye-Laws on 12 February, 1954. The Bye-Laws are contained in Appendix 2.

Permits: Permits can be obtained from the County Clerk who acts in conjunction with the Management Committee. The conditions attached to permits for wildfowling are stated in Chapter V, section 3(a) of this Plan. Permits to collect plants and animals have been outlined in Chapter V, section 4(d).

Hazards: Although two midget submarines, beached on the sand in the centre of Aberlady Bay, were used for bombing practice by aircraft during the war, there is no risk of unexploded bombs anywhere within the Reserve.

One fatal accident occurred near Jovie's Neuk in 1953 when a child fell into loose sand over a rabbit hole. There is no dangerously soft sand anywhere within the Reserve below H.W.M.O.S.T.

CHAPTER VII
WARDENING

1. Part-time Warden

The minutes of the Biological Sub-Committee meeting of 21 December, 1959, record that grave disquiet was expressed by members at the decline in the number of nesting birds and they recommended that a Warden should be appointed. The Management Committee minutes on 29 March, 1960, record the initial concept of a part-time Warden, to be appointed for the period 1 May to 15 July, and that an honorarium should be paid.

During the eight breeding seasons 1960–67, Mr. William Watt, of Gullane, has acted as part-time Warden. He has worked from 1 May until the young terns have flown, usually towards the end of July. His appointment is one factor helping the tern colony to increase and a part-time Warden should continue to be employed.

2. Voluntary Wardens

Members of the Biological Sub-Committee are voluntary Wardens and some have been issued with official arm-bands by the County Council.

3. Full-time Warden

No full-time Warden has been employed on the Reserve. The voluntary Wardens can only visit the Reserve in their spare time and the part-time Warden is on the Reserve for less than three months per year. The bird life requires special attention during the nesting season. The plants require protection over the whole of their flowering season (April to September or October). Of 181 people walking off the Reserve by the footbridge on the 29 July, 1967, 15 of them were carrying picked flowers and one had dug up plants. Two boys coming off the Reserve on the same day had over 80 tadpoles and two newts. The Sea Lavender, so recently established in the Reserve (see Chapter III, section 6(c)), has not been able to fully open its flowers in 1967 since all the flowering spikes have been picked as soon as they showed purple. The indications are that a section of the public feel that an unsupervised Nature Reserve is an area where a lot of flowers are growing and where these can be picked by anyone. It is, therefore, proposed that the Bye-Laws should be more rigidly enforced.

In this respect, a full-time Warden, aided by a part-time Warden during the nesting season, would be able to more effectively control the public. In order to reduce the expense of a full-time Warden, it is proposed that there should be a reasonable charge for using the car park. The charges are 2/– at Longniddry and 2/6d at Gullane. About 85 per cent of drivers, entering the car park by the Peffer Burn during the week-end at the end of July, 1967, said that they would not mind paying a charge of 2/6d. A yearly average of at least 50 cars using the park per week-end could be expected and about 150 cars use the park during summer week-ends.

The proposition is that a Warden/Odd-job-man should be employed. He could collect charges in the car park on days that this is moderately to heavily used; he could enforce the Bye-Laws more effectively by stopping people bringing plants or animals off the Reserve; he could sell the booklet or leaflet about the Reserve and the possible Nature Trail; and he could carry out work on the Reserve prescribed in this Plan, such as control of sea buckthorn, maintenance of paths, collection of dangerous litter and anti-erosion work in the dune system.

A Warden would require a small shelter in which to keep tools and a supply of publications on the Reserve. This could probably be sited with the balancing tank and pumping station of the Aberlady Sewage scheme just beside the car park.

CHAPTER VIII
PRIORITIES, TIME SCHEDULE AND FINANCE

The following are the priorities for the next five years:

1. The appointment of a full-time Warden. Estimated capital cost £80. (Chapter VII, section 3)
2. Re-declaration of the eastern landward boundary of the Reserve. Estimated cost £100. (Chapter V, section 5)
3. The preparation of a booklet or leaflet describing the Reserve. (Chapter V, section 2(b) (i))
4. Maintenance of a system of wildfowl shooting by the issuing of permits. (Chapter V, section 3(a))

All expenditure on the Reserve should be recorded in Appendix 9 of this Plan.

CHAPTER IX
DIVISION OF RESPONSIBILITIES

In 1952, it was agreed that the East Lothian County Council was to be responsible for the management of the Reserve and that the Nature Conservancy would give scientific advice about the Reserve. Two Committees have been established.

The Management Committee is a committee of the County Council and consists of seven elected members, the Convener and the County and District Councillors for Dirleton West, Aberlady and Gladsmuir North Electoral Divisions; seven co-opted members who have particular knowledge relating to management problems of the Reserve; and an Assessor from the Nature Conservancy.

The Biological Sub-Committee was set up to advise the Management Committee on the biological aspects of the Reserve. This Sub-Committee has consisted of about ten members and is under the Chairmanship of Mr. George Waterston. During 1967, it has been agreed that this Sub-Committee should become a Sub-Committee of the Lothians and East Stirlingshire Branch of the Scottish Wildlife Trust.

This Plan does not prescribe the responsibility for carrying out the prescriptions contained within it. It is hoped that the County Council will determine the division of responsibilities.

CHAPTER X
PROGRESS REPORTS AND ROUTINE OBSERVATIONS

Members of the Management Committee and the Biological Sub-Committee have never made reports on the Reserve. An annual report on the work done might be useful.

The part-time Warden submits a report to the Management Committee during his period of Wardenship and this should be continued.

If a full-time Warden is appointed monthly reports should be submitted by him to the Management Committee.

The preparation of the first quinquennial revision of this Plan is due in 1972.

CHAPTER XI

AUTHORSHIP OF PLAN

The section on the geology of the Reserve (Chapter III, section 3) is based on that prepared by Dr. G. P. Black. The section on climate is a modification of one prepared by Mr. A. Millar in 1964. Appendix 3 (Higher plants) was prepared by Miss E. P. Beattie, Appendix 4 (Bryophytes) by Mr. P. Chamberlain, Appendix 5 (Fungi) by Dr. Roy Watling, Appendix 7 is adapted from a paper published by Messrs. F. D. Hamilton and K. S. Macgregor and Appendix 8 is modified from a list prepared by Mr. F. H. W. Green. The Appendices 11 and 12 were compiled by Dr. and Mrs. R. A. Crowson. All other sections have been written by Dr. M. B. Usher (Department of Biology, University of York; and formerly Department of Forestry and Natural Resources, University of Edinburgh).

APPENDICES

Appendix I. Matrix Operations

This appendix contains only a very brief summary of matrix operations, and there is no attempt to discuss matrix theory. The text-book by Searle (1966) discusses the biological uses of matrix algebra, and contains an introduction to the theory of matrices.

Definition and names

A matrix is a rectangular array of numbers. Thus, the matrix

$$\begin{bmatrix} 0 & 1 & -5 \\ 4 & 2 & 8 \end{bmatrix}$$

is a matrix of two rows and three columns, or a 2×3 matrix, or a (2,3) matrix. Each of the six numbers in the matrix is called an *element.* If the matrix is called $\mathbf{Q}$, then the element q_{ij} is taken to be the element in the ith row and jth column of $\mathbf{Q}$. The convention that is used is that a capital bold-face letter denotes a matrix and the corresponding lower-case letter with double subscripts is used for the element.

If a matrix has the same number of rows as columns it is called a *square matrix.* Three particular kinds of square matrices are

$$\mathbf{I} = \begin{bmatrix} 1 & 0 \\ 0 & 1 \end{bmatrix} \qquad \mathbf{O} = \begin{bmatrix} 0 & 0 \\ 0 & 0 \end{bmatrix} \qquad \mathbf{A} = \begin{bmatrix} 8 & -3 \\ -3 & 6 \end{bmatrix}$$

where $\mathbf{I}$ is called the *unit matrix*, $\mathbf{O}$ the *null matrix* and $\mathbf{A}$ is a *symmetrical matrix.* The elements in the first row and first column, second row and second column, etc. of a square matrix are called the *principal diagonal* of the matrix. In the unit matrix the principal diagonal consists only of 1s and all of the other elements are 0s. In a symmetrical matrix the principal diagonal elements can take any numerical value, but the element in the ith row and jth column is identical to the element in the jth row and ith column.

If a matrix has only one column it is referred to as a *column vector*. The column vector can be written in either of the forms

$$\begin{bmatrix} 1 \\ 2 \\ 3 \end{bmatrix} \quad \text{or} \quad \{1, \quad 2, \quad 3\}'$$

A vector with only one row is referred to as a *row vector*, and it can be written as

$$\begin{bmatrix} 1, & 2, & 3 \end{bmatrix}$$

It is usual to denote vectors by lower-case bold-face letters.

Addition and subtraction

If two matrices both have the same number of rows and columns, then they can be added or subtracted. The operation is carried out by adding or subtracting the corresponding elements in each matrix, thus

$$\begin{bmatrix} 2 & 0 \\ 4 & 8 \\ -6 & 1 \end{bmatrix} + \begin{bmatrix} -3 & 0 \\ 5 & 0 \\ -2 & 4 \end{bmatrix} = \begin{bmatrix} -1 & 0 \\ 9 & 8 \\ -8 & 5 \end{bmatrix}$$

Multiplication

Two matrices can be multiplied together if the number of columns in the first matrix is equal to the number of rows in the second matrix. Defining

$$\mathbf{A} = \begin{bmatrix} a_{11} & a_{12} \\ a_{21} & a_{22} \end{bmatrix} \quad \mathbf{B} = \begin{bmatrix} b_{11} \\ b_{21} \end{bmatrix} \quad \mathbf{C} = \begin{bmatrix} c_{11} & c_{12} \end{bmatrix}$$

then

$$\mathbf{AB} = \begin{bmatrix} a_{11}b_{11} + a_{12}b_{21} \\ a_{21}b_{11} + a_{22}b_{21} \end{bmatrix}$$

BA in undefined

$$\mathbf{BC} = \begin{bmatrix} b_{11}c_{11} & b_{11}c_{12} \\ b_{21}c_{11} & b_{21}c_{12} \end{bmatrix}$$

$$\mathbf{CB} = b_{11}c_{11} + b_{21}c_{12}$$

Thus, it can be seen that the order of multiplication is important. In the first example we can say that **B** is pre-multiplied by **A**, or, alternatively, that **A** is post-multiplied by **B**. In the product matrix the element in the ith row and jth column is equal to the sum of the products of the elements in the ith row of the pre-multiplying matrix and the jth column of the post-multiplying matrix. Multiplication by scalars is carried out by multiplying each element, thus

$$k \begin{bmatrix} a_{11} & a_{12} \end{bmatrix} = \begin{bmatrix} ka_{11} & ka_{12} \end{bmatrix}$$

where k is any scalar.

Inversion

This is equivalent to division, and the operation can only be carried out on square matrices. If **P** and **Q** are both square matrices, and

$$\mathbf{PQ} = \mathbf{QP} = \mathbf{I}$$

then **Q** is the inverse of **P**, or conversely, **P** is the inverse of **Q**. In matrix notation these relationships can be written as

$$\mathbf{P} = \mathbf{Q}^{-1} \quad \text{and} \quad \mathbf{Q} = \mathbf{P}^{-1}$$

If **P** is a 2×2 matrix, then

$$\mathbf{P} = \begin{bmatrix} p_{11}p_{12} \\ p_{21}p_{22} \end{bmatrix}, \mathbf{P}^{-1} = \frac{1}{(p_{11}p_{22} - p_{12}p_{21})} \begin{bmatrix} p_{22} & -p_{12} \\ -p_{21} & p_{11} \end{bmatrix}$$

For larger matrices the standard text-books on matrix algebra, numerical analysis or computer algorithms, should be consulted. The expression $(p_{11}p_{22} - p_{12}p_{21})$ is referred to as the *determinant* of **P**, and if it is zero the inverse of **P** cannot be found.

Latent roots and vectors

These are sometimes referred to as *eigenvalues* and *eigenvectors*. They satisfy the equation

$$\mathbf{Av} = \lambda \mathbf{v}$$

where **A** is an $n \times n$ square matrix, **v** is a column vector and λ is a scalar. Alternatively, if **w** is a row vector, there is also an equation

$$\mathbf{wA} = \lambda \mathbf{w}$$

In general, if **A** is an $n \times n$ matrix, then n values of λ can be calculated, though some of these values may be repeated, negative or imaginary. Take as an example the matrix

$$\begin{bmatrix} 8 & 1 \\ 5 & 4 \end{bmatrix}$$

The values of λ can be calculated algebraically by solving the determinant equation (known as the *characteristic equation*)

$$|\mathbf{A} - \lambda \mathbf{I}| = 0$$

or

$$\begin{vmatrix} 8-\lambda & 1 \\ 5 & 4-\lambda \end{vmatrix} = 0$$

whence

$$\lambda^2 - 12\lambda + 27 = 0$$

and the solutions are $\lambda = 9$ and $\lambda = 3$. Alternatively, the values of λ can be solved numerically. Taking as an initial approximation the column vector $\{2, 1\}'$, repeated iteration by matrix multiplication yields the results in Table AI.1. It can be seen from the Table that if the initial approximation

TABLE AI.1 The numerical solution of the dominant latent root of a matrix

Approximation	*Estimated latent root*	*Estimated better approximation*
{2, 1}′	8·5	{2, 1·6471}′
{2, 1·6471}′	8·8236	{2, 1·8800}′
{2, 1·8800}′	8·9400	{2, 1·9597}′
{2, 1·9597}′	8·9799	{2, 1·9865}′
{2, 1·9865}′	8·9932	{2, 1·9955}′
{2, 1·9955}′	8·9978	{2, 1·9985}′
{2, 1·9985}′	8·9992	{2, 1·9995}′
{2, 1·9995}′	8.9998	{2, 1·9998}′

is premultiplied by the matrix the answer is the column vector $\{17, 14\}'$ or $8{\cdot}5\{2, 1{\cdot}6471\}'$, where the first element has been adjusted so that it retains the value of two. The scalar (8·5 in this case) is taken as an approximation to the latent root, and the vector is used as the next approximation to the latent vector. The process of iteration continues until the values of the latent root at two successive iterations differ by less than some small arbitrary amount. Thus, if in Table AI.1 it was desired to calculate λ to two places of decimals, the process would terminate when two iterations differed by less than, say, 0·005. On the fifth iteration the value of λ is estimated as 8·9932 and on the sixth as 8·9978. Since these differ by only 0·0046, the final result would be used to give λ correct to two places of decimal.

With a latent root of nine the column vector is $\{1, 1\}'$ and the row vector is $[5, 1]$. The other latent root is three, and this is associated with the column vector $\{1, -5\}'$ and the row vector $[1, -1]$.

In population mathematics the column vector is usually used to represent the structure of a population, and it is the dominant latent root that has biological interest and meaning. Also, in population mathematics the square matrix is *non-negative*, since all of the elements are either positive or zero. If the theorems of Brauer (1957a,b, 1961, 1962) are applied to these models, then there are four results of importance. First, the dominant latent root of the matrix is positive, being neither negative nor imaginary. Secondly, associated with this dominant latent root is a vector whose elements are all positive (since the vector is determined except for an arbitrary scalar, it might be that all of the elements were negative, but multiplication by the scalar -1 would change all of the elements to being positive). Thirdly, there is only one latent root with this property of having a latent vector of all positive elements. Fourthly, the dominant latent root is larger than or equal to the smallest row sum and is smaller than or equal to the largest row sum. The same inequalities apply to column sums.

Applying these four theorems to the example, the dominant latent root was nine, a positive number. The latent vector was $\{1, 1\}'$ which contains only positive elements, and the second latent root (there are only two latent roots for a 2×2 matrix) contains elements of different signs. The smallest column sum is five and the largest is 13, but since the two row sums are equal to nine the dominant latent root could have been written down immediately.

These theorems do have one very useful result. If a Leslie-type matrix is used in modelling populations, then there is always one, and only one, solution to the model that is biologically meaningful. If solutions to the model were likely to have negative or imaginary numbers then the result would have no biological meaning and the model could not be relied upon.

Appendix II. Some Commonly Used Statistics

These short tables include values of Student's t (t), chi-squared (χ^2) and the correlation coefficient (r) at two probability levels, $p = 0{\cdot}05$ and $p = 0{\cdot}01$. The values are tabulated for two-tailed tests. More extensive tables are given by Fisher and Yates (1963) and Rohlf and Sokal (1969).

Degrees of freedom	*Probability* = 0·05			*Probability* = 0·01		
	t	χ^2	r	t	χ^2	r
1	12·706	3·841	0·997	63·657	6·635	–
2	4·303	5·991	0·950	9·925	9·210	0·990
3	3·182	7·815	0·878	5·841	11·345	0·959
4	2·776	9·488	0·811	4·604	13·277	0·917
5	2·571	11·070	0·754	4·032	15·086	0·874
6	2·447	12·592	0·707	3·707	16·812	0·834
7	2·365	14·067	0·666	3·499	18·475	0·798
8	2·306	15·507	0·632	3·355	20·090	0·765
9	2·262	16·919	0·602	3·250	21·666	0·735
10	2·228	18·307	0·576	3·169	23·209	0·708
11	2·201	19·675	0·553	3·106	24·725	0·684
12	2·179	21·026	0·532	3·055	26·217	0·661
13	2·160	22·362	0·514	3·012	27·688	0·641
14	2·145	23·685	0·497	2·977	29·141	0·623
15	2·131	24·996	0·482	2·947	30·578	0·606
16	2·120	26·296	0·468	2·921	32·000	0·590
17	2·110	27·587	0·456	2·898	33·409	0·575
18	2·101	28·869	0·444	2·878	34·805	0·561
19	2·093	30·144	0·433	2·861	36·191	0·549
20	2·086	31·410	0·423	2·845	37·566	0·537
21	2·080	32·670	0·413	2·831	38·932	0·526
22	2·074	33·924	0·404	2·819	40·289	0·515
23	2·069	35·172	0·396	2·807	41·638	0·505
24	2·064	36·415	0·388	2·797	42·980	0·496
25	2·060	37·652	0·381	2·787	44·314	0·487

Degrees of freedom	*Probability* = 0·05			*Probability* = 0·01		
	t	χ^2	r	t	χ^2	r
26	2·056	38·885	0·374	2·779	45·642	0·479
27	2·052	40·113	0·367	2·771	46·963	0·471
28	2·048	41·337	0·361	2·763	48·278	0·463
29	2·045	42·557	0·355	2·756	49·588	0·456
30	2·042	43·773	0·349	2·750	50·892	0·449
40	2·021	55·758	0·304	2·704	63·691	0·393
50	2·008	67·505	0·273	2·677	76·154	0·354
60	2·000	79·082	0·250	2·660	88·380	0·323
70	1·995	90·531	0·232	2·648	100·425	0·302
80	1·990	101·879	0·217	2·639	112·329	0·283
90	1·987	113·145	0·205	2·632	124·116	0·267
100	1·984	124·342	0·195	2·626	135·807	0·254
∞	1·960	–	–	2·576	–	–

References

Anderson, M. L. (1950). *The Selection of Tree Species.* Oliver and Boyd, Edinburgh and London.

Andrewartha, H. G. (1961). *Introduction to the Study of Animal Populations.* Methuen, London.

Anon. (1963). *Science Out-of-doors. Report of the Study Group on Education and Field Biology.* Longmans, London.

Anon. (1965). *A-level Biology. Aims and Outline Scheme.* Nuffield Foundation Science Teaching Project.

Anon. (1966). *Nuffield Biology Teacher's Guide I. Introducing Living Things.* Longmans, London. Penguin Books, Middlesex.

Anon. (1967). *Nuffield Junior Science. Teacher's Guide I.* Collins, London and Glasgow.

Anon. (1969). 'The great Rhine fish-kill', *Biological Conservation*, **2**, 35.

Anon. (1970a). *Conservation Corps 1959–1969.* British Trust for Conservation Volunteers Ltd, London.

Anon. (1970b). 'Trumpeter swan no longer rare and endangered', *Biological Conservation*, **2**, 178.

Arthur, D. R. (1969). *Survival. Man and His Environment.* English Universities Press, London.

Arvill, R. (1967). *Man and Environment. Crisis and the Strategy of Choice.* Penguin Books, Middlesex.

Automobile Association (1970). *Schedule of Estimated Running Costs*, Tech 9A (June 1970).

Bailey, N. T. J. (1959). *Statistical Methods in Biology.* English Universities Press, London.

Bainbridge, R., Evans, G. C. and Rackham, O. Ed. (1966). *Light as an Ecological Factor. Symp. Brit. Ecol. Soc.*, **6**. Blackwell, Oxford and Edinburgh.

Banko, W. E. (1960). 'The trumpeter swan: its history, habits, and population in the United States', *U.S. Fish and Wildlife Service, N. American Fauna*, No. 63.

Bawden, M. G. and Tuley, P. (1966). 'The land resources of Southern Sardauna and Southern Adamawa Provinces, Northern Nigeria', *Ministry of Overseas Development, Land Resources Study*, No. 2.

Berkshire, Buckinghamshire and Oxfordshire Naturalists' Trust (1970). *Projects for Environmental Studies.* Berkshire, Buckinghamshire and Oxfordshire Naturalists' Trust, Oxford.

Berry, R. J. (1971). 'Conservation aspects of the genetical constitution of populations', in *The Scientific Management of Animal and Plant Communities for Conservation.* Ed. E. Duffey and A. S. Watt. *Symp. Brit. Ecol. Soc.*, **11,** 177–206. Blackwell, Oxford and Edinburgh.

Beverton, R. J. H. (1952). 'Long-term dynamics of certain North Sea fish populations', in *The Exploitation of Natural Animal Populations.* Ed. E. D. Le Cren and M. W. Holdgate. *Symp. Brit. Ecol. Soc.*, **2,** 242–259. Blackwell, Oxford.

Beverton, R. J. H. and Holt, S. J. (1957). 'On the dynamics of exploited fish populations', *Fishery Invest., Lond*, Ser. II, **19**, 1–533.

Black J. D. (1968). *The Management and Conservation of Biological Resources.* F. A. Davis, Philadelphia.

Black, J. N. (1964). 'An analysis of the potential production of swards of subterranean clover (*Trifolium subterraneum* L.) at Adelaide, South Australia', *J. appl. Ecol.*, **1**, 3–18.

Black, J. N. (1970). *The Dominion of Man: the Search for Ecological Responsibility.* Edinburgh University Press, Edinburgh.

Blackman, G. E. (1968). 'The applications of the concepts of growth analysis to the assessment of productivity', pp 243–259, in *Functioning of Terrestrial Ecosystems at the Primary Production Level.* Ed. F. E. Eckardt. UNESCO, Liège.

Bloomfield, H. E. (1971). *Field and Computer Studies on Succession in Calcicolous Plants.* Unpublished B.A. Thesis, University of York.

Boer, G. de (1963). 'Spurn Point and its predecessors', *Naturalist, Hull,* **1963,** 113–120.

Brasnet, N. V. (1953). *Planned Management of Forests.* Allen and Unwin, London.

Brauer, A. (1957a). 'A new proof of theorems of Perron and Frobenius on non-negative matrices: I. Positive matrices', *Duke math. J.*, **24**, 367–378.

Brauer, A. (1957b). 'A method for the computation of the greatest root of a positive matrix', *J. Soc. ind. appl. Math.*, **5**, 250–253.

Brauer, A. (1961). 'On the characteristic roots of power-positive matrices', *Duke math. J.*, **28**, 439–446.

Brauer, A. (1962). 'On the theorems of Perron and Frobenius on non-negative matrices', pp. 48–55, in *Studies in Mathematical Analysis and Related Topics*, Ed. S. Gilbarg *et al.* Stanford University Press, Stanford.

Braun-Blanquet, J. (1932). *Plant Sociology.* New York.

Braun-Blanquet, J. (1951). *Pflanzensoziologie.* Second Edn. Springer Verlag, Vienna.

Britten, J. (1892). '*Sagina boydii*', *J. Bot., Lond.*, **30,** 226–227.

Broad, S. T. (1969). 'The educational aspects of nature and conservation', *Advmt. Sci., Br. Ass.*, **26**, 91–98.

Brown, P. and Waterston, G. (1962). *The Return of the Osprey.* Collins, London.

Brunelle, R. (1970). 'Polar Bear Provincial Park, Ontario', *Biological Conservation*, **3**, 147–149.

Burden, P. C. (1970). *Field and Computer Studies in the Pattern of some Calcicolous Plants.* Unpublished B.A. Thesis, University of York.

Burhenne, W. E. (1970). 'The African convention for the conservation of nature and natural resources', *Biological Conservation*, **2**, 105–114.

Burton, T. L. (1967). 'The basic elements in planning for recreation', pp 156–161, in *The Biotic Effects of Public Pressures on the Environment.* Ed. E. Duffey. Monks Wood Experimental Station Symposium No. 3, The Nature Conservancy.

Cajander, A. K. (1943). 'Nature et importance des type de forêt', *Intersylva*, **3**, 169–209.

Champion, H. and Brasnett, N. V. (1958). *Choice of Tree Species.* F.A.O. Forestry Development Paper No. 13, Rome.

Christian, C. S. and Stewart, G. A. (1953). 'General report on survey of Katherine–Darwin region, 1946', *C.S.I.R.O. Land Research Series*, No. 1.

Christian, C. S. and Stewart, G. A. (1964). 'Methodology of integrated surveys'. Duplicated, U.N.E.S.C.O. conference on *Principles and Methods of Integrating Aerial Survey Studies of Natural Resources for Potential Development.*

Christiansen, K. (1964) 'Bionomics of Collembola', *A. Rev. Ent.*, **9**, 147–178.

Clapham, A. R., Tutin, T. G. and Warburg, E. F. (1952). *Flora of the British Isles.* University Press, Cambridge.

Clark, L. R., Geier, P. W., Hughes, R. D. and Morris, R. F. (1967). *The Ecology of Insect Populations in Theory and Practice.* Methuen, London.

Collingbourne, R. H. (1966). 'General principles of radiation meteorology', in *Light as an Ecological Factor.* Ed. R. Bainbridge, G. C. Evans and O. Rackham. *Symp. Brit. Ecol. Soc.*, **6**, 1–15. Blackwell, Oxford and Edinburgh.

Cornwallis, R. K. (1969). 'Farming and wildlife conservation in England and Wales', *Biological Conservation*, **1**, 142–147.

Countryside Commission (1969a). *Digest of Countryside Statistics.* Duplicated report.

Countryside Commission (1969b). *Nature Conservation at the Coast. Special study report Vol. 2.* H.M.S.O., London.

Countryside in 1970 (1965). Report of Study Group No. 2, Training and qualifications of professions concerned with land and water.

Cragg, J. B. (1968). 'Biological conservation: the present', *Biological Conservation*, **1**, 13–19.

Crawford, R. M. M. and Wishart, D. (1966). 'A multivariate analysis of the development of dune slack vegetation in relation to coastal accretion at Tentsmuir, Fife', *J. Ecol.*, **54**, 729–743.

Curtis, J. T. (1956). 'The modification of mid-latitude grasslands and forests by man', pp 721–736, in *Man's Role in Changing the Face of the Earth.* Ed. W. L. Thomas. University of Chicago Press, Chicago.

Danilevsky, A. S., Goryshin, N. I. and Tyshchenko, V. P. (1970). 'Biological rhythms in terrestrial arthropods', *A. Rev. Ent.*, **15**, 201–244.

Dasman, R. F. (1968). *Environmental Conservation*, Second Edn. John Wiley, New York, London, Sydney and Toronto.

Davey, A. J. (1961). 'Biological flora of the British Isles, *Epilobium nerterioides* A. Cunn', *J. Ecol.*, **49**, 753–759.

Davidson, J. (1970). *Outdoor Recreation Information. Suggested Standard Classifications For Use in Questionnaire Surveys.* Duplicated report of the Countryside Commission.

Denson, E. P. (1970). 'The trumpeter swan, *Olor buccinator*: a conservation success and its lessons', *Biological Conservation*, **2**, 253–256.

Douglass, R. W. (1969). *Forest Recreation.* Pergamon, Oxford, London, Edinburgh, New York, Toronto, Sydney, Paris and Braunschweig.

Duffey, E. (1967). 'An assessment of dune invertebrate faunas in habitats vulnerable to public disturbance', pp 112–119, in *The Biotic Effects of Public Pressures on the Environment.* Ed. E. Duffey. Monks Wood Experimental Station Symposium No. 3, The Nature Conservancy.

Duffey, E. (1968). 'Ecological studies on the large copper butterfly *Lycaena dispar* Haw. *batavus* Obth. at Woodwalton Fen National Nature Reserve, Huntingdonshire', *J. appl. Ecol.*, **5**, 69–96.

Duffey, E. (1969). 'Wildlife conservation in Europe', *Handbk. a. Rep. Soc. Promot. Nat. Reserves*, 1969, 1–36.

Duffy, P. J. B. (1969). 'Forest land capability in the Queanbeyan–Shoalhaven Area, A.C.T. and New South Wales', *Aust. For.*, **33**, 195–200.

East Lothian County Council (1955). *The Improvement of Gullane Bents.* Duplicated report by the County Planning Officer.

East Lothian County Council (1962). *Coastal Survey 1961 and Recommendations.* Duplicated report by the County Planning Officer.

Edwards, C. A., Reichle, D. E. and Crossley, D. A. (1970). 'The role of soil invertebrates in turnover of organic matter and nutrients', pp 147–172, in *Analysis of Temperate Forest Ecosystems.* Ed. D. E. Reichle. Springer-Verlag, Berlin, Heidelberg and New York.

Eggeling, W. J. (1964). 'A nature reserve management plan for the Island of Rhum, Inner Hebrides', *J. appl. Ecol.*, **1**, 405–419.

Ehrenfeld, D. W. (1970). *Biological conservation.* Holt, Rinehart and Winston, New York, Chicago, San Francisco, Atlanta, Dallas, Montreal, Toronto, London and Sydney.

Elliston Wright, F. R. (1938). 'Notes on two *Saginas*', *J. Bot., Lond.*, **76**, 361–364.

Elton, C. (1942). *Voles, Mice and Lemmings: Problems in Population Dynamics.* Clarendon Press, Oxford.

Elton, C. and Nicholson, M. (1942). 'The ten year cycle in numbers of the lynx in Canada', *J. Anim. Ecol.*, **11**, 215–244.

Errington, J. C. (1970). *A Study of Factors Determining the Pattern of Woodland Bryophytes.* Unpublished M.Phil. Thesis, University of York.

Farrar, J. L. (1962). 'The use of factor gradients in evaluating site', *Proc. 5th World Forestry Congr.*, **1**, 524–529.

Fisher, J. (1954). *Bird Recognition. I. Sea-birds and Waders.* Penguin Books, London, Melbourne and Baltimore.

Fisher, J., Simon, N. and Vincent, J. (1969). *The Red Book: Wildlife in Danger.* Collins, London.

Fisher, J. L. (1969). 'New perspectives on conservation', *Biological Conservation*, **1**, 111–116.

Fisher, R. A. and Yates, F. (1963). *Statistical Tables for Biological, Agricultural and Medical Research*, Sixth Edn. Oliver and Boyd, Edinburgh.

Ford, E. B. (1945). *Butterflies.* Collins, London.

Forestry Commission (1960). 'The grey squirrel. A woodland pest', Second Edn. *For. Comm. Leaflet* No. 31.

Fraser Darling, F. (1969). 'The Reith Lectures', *The Listener*, **82**, various pages in six numbers, 13 November–18 December.

Frankland, J. C. (1966). 'Succession of fungi on decaying petioles of *Pteridium aquilinum*', *J. Ecol.*, **54**, 41–63.

Frohawk, F. W. (1924). *Natural History of British Butterflies. Vol. II.* Hutchinson, London.

Gama, M. M. da (1964). *Colêmbolos de Portugal Continental.* Dr. Biol. Sci. Thesis, University of Coimbra.

Gisin, H. (1943). 'Ökologie und Lebensgemeinschaften der Collembolen im Schweizerischen Exkursionsgebiet Basels', *Revue suisse Zool.*, **50**, 131–224.

Gisin, H. (1955). 'Recherches sur la relation entre la faune endogée de Collemboles et les qualites agrologiques de sols viticoles', *Revue suisse Zool.*, **62**, *601*–648.

Godwin, H. (1956). *The History of the British Flora. A Factual Basis for Phytogeography.* Cambridge University Press, Cambridge.

Goodall, D. W. (1953). 'Objective methods for the classification of vegetation. I. The use of positive inter-specific correlation', *Aust. J. Bot.*, **1**, 39–63.

Goodall, D. W. (1961). 'Objective methods for the classification of vegetation. IV. Pattern and minimal area', *Aust. J. Bot.*, **9**, 162–196.

Goodall, D. W. (1963). 'Pattern analysis and minimal area – some further comments', *J. Ecol.*, **51**, 705–710.

Gordon, N. J. (1963). *Tentsmuir Point National Nature Reserve, Fife: management plan (first revision).* Duplicated by The Nature Conservancy.

Green, R. H. (1966). 'Measurement of non-randomness in spatial distributions', *Researches Popul. Ecol. Kyoto Univ.*, **8**, 1–7.

Greig-Smith, P. (1961a). 'Data on pattern within plant communities. I. The analysis of pattern', *J. Ecol.*, **49**, 695–702.

Greig-Smith, P. (1961b). 'Data on pattern within plant communities. II. *Ammophila arenaria* (L.) Link', *J. Ecol.*, **49**, 703–708.

Greig-Smith, P. (1964). *Quantitative Plant Ecology.* Butterworth, London.

Gulland, J. A. (1962). 'The application of mathematical models to fish populations', in *The Exploitation of Natural Animal Populations.* Ed. E. D. Le Cren and M. W. Holdgate. *Symp. Brit. Ecol. Soc.*, **2,** 204–217. Blackwell, Oxford.

Harley, J. L. (1971). 'Fungi in Ecosystems', *J. Ecol.*, **59,** 653–668, *J. appl. Ecol.*, **8,** 627–642.

Harrisson, T. (1969). 'The tamaraw and Philippine conservation', *Biological Conservation*, **1**, 317–318.

Hemsley, J. H. and Copland, W. O. (1963). *Studland Heath National Nature Reserve, Dorset. Interim Management Plan.* Duplicated by Nature Conservancy.

Hill, I. D. (1969). 'An assessment of the possibilities of oil palm cultivation in Western Division, The Gambia', *Ministry of Overseas Development, Land Resources Study*, No. 6.

Hills, G. A. (1961). 'The ecological basis for land-use planning', *Ontario Dept. Lands Forests Res. Rep.*, No. 46.

Hills, G. A. and Pierpoint, G. (1960). 'Forest site evaluation in Ontario', *Ontario Dept. Lands Forests Res. Rep.*, No. 42.

Hutchinson, G. E. (1953). 'The concept of pattern in ecology', *Proc. Acad. Nat. Sci. Philadelphia*, **105**, 1–12.

Huxley, J. (1961). *The Conservation of Wildlife and Natural Habitats in Central and East Africa.* Report on a mission accomplished for Unesco, July–September 1960. UNESCO.

Huxley, T. (1964). *Wildfowling at Caerlaverock National Nature Reserve: The first seven years*, Second Edn. A duplicated report by the Nature Conservancy, Edinburgh.

Huxley, T. (1967). 'Is wildfowling compatible with conservation?', pp 25–34, in *The Biotic Effects of Public Pressures on the Environment.* Ed. E. Duffey. Monks Wood Experimental Station Symposium No. 3, The Nature Conservancy.

Jeffers, J. N. R. Ed. (1972). *Mathematical Models in Ecology. Symp. Brit. Ecol. Soc.*, **12**. Blackwell, Oxford and Edinburgh.

Jenkins, A. C. (1970). *Wildlife in Danger.* Methuen, London.

Jones, E. L. (1969). 'The decrease of *Pulsatilla vulgaris* in England', *Biological Conservation*, **1**, 327–328.

Jonkel, C. J. (1970). 'Some comments on polar bear management', *Biological Conservation*, **2**, 115–119.

Kendall, M. G. (1951). *The Advanced Theory of Statistics, Vol. II.* Third Edn. Griffin, London.

Kent, D. H. (1956). '*Senecio squalidus* L. in the British Isles – 1, early records (to 1877)', *Proc. Bot. Soc. Brit. Is.*, **2**, 115–118.

Kent, D. H. (1960). '*Senecio squalidus* L. in the British Isles – 2, the spread from Oxford (1879–1939)', *Proc. Bot. Soc. Brit. Is.*, **3**, 375–379.

Kent, D. H. (1966). '*Senecio squalidus* L. in the British Isles – 8, the recent spread in Scotland', *Glasg. Nat.*, **18**, 407–408.

Kershaw, K. A. (1957). 'The use of cover and frequency in the detection of pattern in plant communities', *Ecology*, **38**, 291–299.

Knuchel, H. (1953). *Planning and Control in the Managed Forest.* Oliver and Boyd, Edinburgh and London. [Trans. by M. L. Anderson.]

Kormody, E. J. (1969). *Concepts of Ecology.* Prentice-Hall, New Jersey.

Köstler, J. (1956). *Silviculture.* Oliver and Boyd, Edinburgh and London. [Trans. by M. L. Anderson.]

Kozlovsky, D. G. (1968). 'A critical evaluation of the trophic level concept. I. Ecological efficiencies', *Ecology*, **49**, 48–60.

Lamb, H. H. (1969). 'The new look of climatology', *Nature, Lond.*, **223**, 1209–1215.

Laws, R. M. (1962). 'Some effects of whaling on the southern stocks of baleen whales', in *The Exploitation of Natural Animal Populations.* Ed. E. D. Le Cren and M. W. Holdgate. *Symp. Brit. Ecol. Soc.*, **2**, 137–158. Blackwell, Oxford and Edinburgh.

Lefkovitch, L. P. (1965). 'The study of population growth in organisms grouped by stages', *Biometrics*, **21**, 1–18.

Lefkovitch, L. P. (1967). 'A theoretical evaluation of population growth after removing individuals from some age groups', *Bull. Ent. Res.*, **57**, 437–445.

Leslie, P. H. (1945). 'On the use of matrices in certain population mathematics', *Biometrika*, **33**, 183–212.

Leslie, P. H. (1948). 'Some further notes on the use of matrices in population mathematics', *Biometrika*, **35**, 213–245.

Levins, R. (1968). *Evolution in Changing Environments: Some Theoretical Explorations.* Princeton University Press, Princeton.

Lewis, E. G. (1942). 'On the generation and growth of a population', *Sankhyā*, **6**, 93–96

Lieth, H. (1970). 'Phenology in productivity studies', pp 29–46, in *Analysis of Temperate Forest Ecosystems.* Ed. D. E. Reichle. Springer-Verlag, Berlin, Heidelberg and New York.

Lipsey, R. G. (1966). *An Introduction to Positive Economics*, Second Edn. Weidenfeld and Nicolson, London.

Lloyd, H. G. (1962). 'The distribution of squirrels in England and Wales, 1959', *J. Anim. Ecol.*, **31**, 157–165.

Lockie, J. D. (1966). 'Territory in small carnivores', *Symp. zool. Soc. Lond.*, **18**, 143–165.

Lockie, J. D. (1967). 'A field study of the effects of chlorinated hydrocarbons on golden eagles in West Scotland', *Proc. 7th Congr. Game Biologists, Beograd–Ljubljana*, 275–278.

Lowe, V. P. W. (1961). 'A discussion on the history, present status and future conservation of red deer (*Cervus elaphus* L.) in Scotland', *Terre Vie*, **1**, 9–40.

Lowe, V. P. W. (1969). 'Population dynamics of the red deer (*Cervus elaphus* L.) on Rhum', *J. Anim. Ecol.*, **38**, 425–457.

MacArthur, R. H. and Wilson, E. O. (1967). *Island Biogeography*. Princeton University Press, Princeton.

Macfadyen, A. (1963). *Animal Ecology: Aims and Methods*. Pitman, London.

McKelvie, A. D. (1970). 'A school survey on the flowering time of broom (*Sarothamnus scoparius*) in the north of Scotland', *J. biol. Educ.*, **4**, 227–233.

Mackenzie, J. M. D. (1952). 'Fluctuations in the numbers of British tetraonids', *J. Anim. Ecol.*, **21**, 128–153.

Margaropoulos, P. (1962). 'Forestry and forest grazing in the Mediterranean region', *Proc. 5th World Forestry Congr.*, **1**, 315–322.

Margalef, R. (1968). *Perspectives in Ecological Theory*. University of Chicago Press, Chicago and London.

Maynard Smith, J. (1968). *Mathematical Ideas in Biology*. University Press, Cambridge.

Maynard Smith, J. (1970). 'Population size, polymorphism, and the rate of non-Darwinian evolution', *Am. Nat.*, **104**, 231–237.

Mellanby, K. (1967). *Pesticides and Pollution*. Collins, London.

Meyer, B. S. and Anderson, D. B. (1952). *Plant Physiology*, Second Edn. D. Van Nostrand Co. Inc, Princeton, Toronto, New York and London.

Moore, N. W., Hooper, M. D. and Davis, B. N. K. (1967). 'Hedges. I. Introduction and reconnaissance studies', *J. appl. Ecol.*, **4**, 201–220.

Moran, P. A. P. (1949). 'The statistical analysis of the sunspot and lynx cycles', *J. Anim. Ecol.*, **18**, 115–116.

Moran, P. A. P. (1952). 'The statistical analysis of game-bird records', *J. Anim. Ecol.*, **21**, 154–158.

Moran, P. A. P. (1953). 'The statistical analysis of the Canadian lynx cycle. I. Structure and prediction', *Aust. J. Zool.*, **1**, 163–173.

Moss, W. W. (1968). 'Experiments with various techniques of numerical taxonomy', *Syst. Zool.*, **17**, 31–47.

Mutch, W. E. S. (1968). 'Public recreation in national forests: a factual study', *For. Comm. Booklet*, No. 21.

Nicholson, A. J. (1954). 'Compensatory reactions of populations to stresses, and their evolutionary significance', *Aust. J. Zool.*, **2**, 1–8.

Odum, E. P. (1959). *Fundamentals of Ecology*, Second Edn. W. B. Saunders, Philadelphia.

Odum, E. P. (1963). *Ecology*. Holt, Rinehart and Winston, New York, Chicago, San Francisco, Toronto and London.

Osmaston, F. C. (1968). *The Management of Forests*. Allen and Unwin, London.

Ovington, J. D. (1962). 'Quantitative ecology and the woodland ecosystem concept', *Adv. ecol. Res.*, **1**, 103–192.

Ovington, J. D. (1965). *Woodlands.* English Universities Press, London.

Peakall, D. B. (1970). 'Pesticides and the reproduction of birds', *Scient. Am.*, **222**, 72–78.

Pennycuick, C. J., Campton, R. M. and Beckingham, L. (1968). 'A computer model for simulating the growth of a population, or of two interacting populations', *J. Theoret. Biol.*, **18**, 316–329.

Perring, F. H. (1967). 'Changes in chalk grassland caused by galloping', pp 134–142, in *The Biotic Effects of Public Pressures on the Environment.* Ed. E. Duffey. Monks Wood Experimental Station Symposium No. 3, The Nature Conservancy.

Peterle, T. J. (1969). 'DDT in Antarctic snow', *Nature, Lond.*, **224**, 620.

Phillipson, J. (1966). *Ecological Energetics.* Arnold, London.

Pielou, E. C. (1969). *An Introduction to Mathematical Ecology.* Wiley, New York, London, Sydney and Toronto.

Pierpoint, G. (1962). 'The sites of the Kirkwood Management Unit', *Ontario Dept. Lands Forests Res. Rep.*, No. 47.

Pigott, M. E. and Pigott, C. D. (1959). 'Stratigraphy and pollen analysis of Malham Tarn and Tarn Moss', *Fld. Stud.*, **1**, 1–18.

Pitt, M. (1971). *Multiple Land Use of Spurn Nature Reserve.* Unpublished B.A. Thesis, University of York.

Polunin, N. (1969). 'Conservation significance of botanical gardens', *Biological Conservation*, **1**, 104–105.

Poore, M. E. D. (1955a). 'The use of phytosociological methods in ecological investigations. I. The Braun–Blanquet system', *J. Ecol.*, **43**, 226–244.

Poore, M. E. D. (1955b). 'The use of phytosociological methods in ecological investigations. II. Practical issues involved in an attempt to apply the Braun–Blanquet system', *J. Ecol.*, **43**, 245–269.

Poore, M. E. D. (1955c). 'The use of phytosociological methods in ecological investigations. III. Practical application', *J. Ecol.*, **43**, 606–651.

Poore, M. E. D. (1956). 'The use of phytosociological methods in ecological investigations. IV. General discussion of phytosociological problems', *J. Ecol.*, **44**, 28–50.

Poore, M. E. D. (1962). 'The method of successive approximation in descriptive ecology', *Adv. ecol. Res.*, **1**, 35–68.

Porter, L. K. (1969). 'Nitrogen in grassland ecosystems', pp 377–402, in *I.B.P., The Grassland Ecosystem: A Preliminary Synthesis.* Range Sci. Dept., Sci. Series 2, Colorado State Univ.

Pritchard, T. (1968). 'Environmental education', *Biological Conservation*, **1**, 27–31.

Radley, J. and Sims, C. (1970). *Yorkshire Flooding – Some Effects on Man and Nature.* Sessions Book Trust, York.

Radovich, J. (1962). 'Effects of sardine spawning stock size and environment on year-class production', *Calif. Fish Game*, **48**, 123–140.

Raven, J. and Walters, M. (1956). *Mountain Flowers*. Collins, London.

Richards, P. W. (1957). *The Tropical Rain Forest: An Ecological Study*. University Press, Cambridge.

Rohlf, F. J. and Sokal, R. R. (1969). *Statistical Tables*. W. H. Freeman, San Francisco.

Rowe, J. S. (1962). 'Soil, site and land classification', *For. Chron.*, **38**, 420–432.

Russell-Hunter, W. D. (1970). *Aquatic Productivity: An Introduction to Some Basic Aspects of Biological Oceanography and Limnology*. Collier-Macmillan, London.

Schofield, J. M. (1967). 'Human impact on the fauna, flora and natural features of Gibraltar Point', pp. 106–111, in *The Biotic Effects of Public Pressures on the Environment*. Ed. E. Duffey. Monks Wood Experimental Station Symposium No. 3, The Nature Conservancy.

Scott, G. A. M. (1965). 'The shingle succession at Dungeness', *J. Ecol.*, **53**, 21–31.

Searle, S. R. (1966). *Matrix Algebra for the Biological Sciences, Including Applications in Statistics*. Wiley, New York, London and Sydney.

Shorten, M. (1953). 'Notes on the distribution of the grey squirrel (*Sciurus carolinensis*) and the red squirrel (*Sciurus vulgaris leucourus*) in England and Wales from 1945 to 1952,' *J. Anim. Ecol.*, **22**, 134–140.

Shorten, M. (1957). 'Squirrels in England, Wales and Scotland, 1955', *J. Anim. Ecol.*, **26**, 287–294.

Silliman, R. P. and Gutsell, J. S. (1958). 'Experimental exploitation of fish populations', *Fishery Bull. Fish Wildl. Serv. U.S.*, **58**, 215–241.

Simberloff, D. S. (1969). 'Experimental zoogeography of islands. A model for insular colonization', *Ecology*, **50**, 296–314.

Simberloff, D. S. and Wilson, E. O. (1969). 'Experimental zoogeography of islands. The colonization of empty islands', *Ecology*, **50**, 278–296.

Slobodkin, L. B. (1961). *Growth and Regulation of Animal Populations*. Holt, Rinehart and Winston, New York, Chicago, San Francisco, Toronto and London.

Slobodkin, L. B. and Richman, S. (1956). 'The effect of removal of fixed percentages of the newborn on size and variability in populations of *Daphnia pulicaria* (Forbes)', *Limnol. Oceanogr.*, **1**, 209–237.

Smith, F. E. (1970). 'Analysis of ecosystems', pp 7–18, in *Analysis of Temperate Forest Ecosystems*. Ed. D. E. Reichle. Springer-Verlag, Berlin, Heidelberg and New York.

Smith, J. E. Ed. (1968). *Torrey Canyon Pollution and Marine Life. A Report by the Plymouth Laboratory of the Marine Biological Association of the United Kingdom*. Cambridge University Press, London and New York.

Smith, R. J. (1968). Unpublished paper, University of Birmingham.

Smith, R. J. and Kavanagh, N. J. (1969). 'The measurement of benefits of trout fishing: preliminary results of a study at Grafham Water, Great Ouse Water Authority, Huntingdonshire', *J. Leisure Res.* **1**, 316–332.

Snaydon, R. W. (1962). 'Micro-distribution of *Trifolium repens* L. and its relation to soil factors', *J. Ecol.*, **50**, 133–143.

Sokal, R. R. and Rohlf, F. J. (1969). *Biometry.* Freeman, San Francisco.

Sokal, R. R. and Sneath, P. H. A. (1963). *Principles of Numerical Taxonomy.* Freeman, San Francisco and London.

Solomon, M. E. (1969). *Population Dynamics.* Arnold, London.

Southern, H. N. (1964). *The Handbook of British Mammals.* Blackwell, Oxford.

Spiegel, M. R. (1961). *Theory and Problems of Statistics.* McGraw-Hill, New York, St. Louis, San Francisco, Toronto and Sydney.

Stamp, D. (1969). *Nature Conservation in Britain.* Collins, London.

Stephens, G. R. and Waggoner, P. E. (1970). 'The forests anticipated from 40 years of natural transitions in mixed hardwoods', *Bull. Conn. agric. Exp. Stn.*, No. 707.

Street, H. E. and Öpik, H. (1970). *The Physiology of Flowering Plants: Their Growth and Development.* Arnold, London.

'Student' (1907). 'On the error of counting with a haemocytometer', *Biometrika*, **5**, 351–360.

Tarrant, K. R. and Tatton, J. O'G. (1968). 'Organochlorine pesticides in rain-water in the British Isles', *Nature, Lond.*, **213**, 725–729.

Taylor, A. E. (1969). *The Role of Bishop Wood and Strensall Common as Natural Resources with Special Reference to Education.* Unpublished B.A. Thesis, University of York.

Teagle, W. G. (1966). *Public Pressure on South Haven Peninsula and Its Effect on Studland Heath National Nature Reserve.* Duplicated by the Nature Conservancy.

Tilly, L. J. (1968). 'The structure and dynamics of Cone Spring', *Ecol. Monogr.*, **38**, 169–197.

Trice, A. H. and Wood, S. E. (1958). 'Measurement of recreation benefits', *Land Econ.*, **34**, 195–207.

Uchida, H. and Fujita, K. (1968). 'Mass occurrence and diurnal activity of *Dicyrtomina rufescens* (Collembola: Dicyrtomidae) in winter', *Sci. Rep. Fac. Lit. Sci. Hirosaki Univ.*, **15**, 36–48.

Usher, M. B. (1966). 'A matrix approach to the management of renewable resources, with special reference to selection forests', *J. appl. Ecol.*, **3**, 355–367.

Usher, M. B. (1967). *Aberlady Bay Local Nature Reserve: Description and Management Plan.* East Lothian County Council, County Planning Department, Haddington.

Usher, M. B. (1969a). 'Some properties of the aggregations of soil arthropods: Collembola', *J. Anim. Ecol.*, **38**, 607–622.

Usher, M. B. (1969b). 'The relation between mean square and block size in the analysis of similar patterns', *J. Ecol.*, **57**, 505–514.

Usher, M. B. (1969c). 'A matrix model for forest management', *Biometrics*, **25**, 309–315.

Usher, M. B. (1970a). 'An algorithm for estimating the length and direction of shadows with reference to the shadows of shelter belts', *J. appl. Ecol.*, **7**, 141–145.

Usher, M. B. (1970b). 'Properties of the aggregations of soil arthropods, particularly Mesostigmata (Acarina)', *Oikos*, **21**, 43–49.

Usher, M. B. (1972). 'Developments in the Leslie matrix model', in *Mathematical Models in Ecology*. Ed. J. N. R. Jeffers. *Symp. Brit. Ecol. Soc.*, **12**, 29–60. Blackwell, Oxford and Edinburgh.

Usher, M. B., Longstaff, B. C. and Southall, D. R. (1971). 'Studies on populations of *Folsomia candida* (Insecta: Collembola): the productivity of populations in relation to food and exploitation', *Oecologia*, **7**, 68–79.

Usher, M. B., Taylor, A. E. and Darlington, D. (1970). 'A survey of visitors' reactions on two Naturalists' Trust nature reserves in Yorkshire, England', *Biological Conservation*, **2**, 285–291.

Van Dyne, G. M. (1969a). 'Some mathematical models of grassland ecosystems', pp 3–26, in *I.B.P., The Grassland Ecosystem: A Preliminary Synthesis.* Range Sci. Dept., Sci. Series 2, Colorado State Univ.

Van Dyne, G. M. (1969b). *The Ecosystem Concept in Natural Resource Management.* Academic Press, New York and London.

Waggoner, P. E. and Stephens, G. R. (1970). 'Transition probabilities for a forest', *Nature, Lond.*, **225**, 1160–1161.

Watson, A. (1967). 'Public pressures on soils, plants and animals near ski lifts in the Cairngorms', pp 38–45, in *The Biotic Effects of Public Pressures on the Environment.* Ed. E. Duffey. Monks Wood Experimental Station Symposium No. 3, The Nature Conservancy.

Watt, A. S. (1957). 'The effect of excluding rabbits from grassland B (Mesobrometum) in Breckland', *J. Ecol.*, **45**, 861–878.

Watt, A. S. (1960a). 'The effect of excluding rabbits from acidiphilous grassland in Breckland', *J. Ecol.*, **48**, 601–604.

Watt, A. S. (1960b). 'Population changes in acidiphilous grass-heath in Breckland, 1936–57', *J. Ecol.*, **48**, 605–629.

Watt, K. E. F. (1955). 'Studies on population productivity. I. Three approaches to the optimum yield problem in populations of *Tribolium confusum*', *Ecol. Monogr.*, **35**, 269–290.

Watt, K. E. F. (1960). 'The effect of population density on fecundity in insects', *Can. Ent.*, **92**, 674–695.

Watt, K. E. F. (1962). 'The conceptual formulation and mathematical solution of practical problems in population input–output dynamics', in *The Exploitation of Natural Animal Populations.* Ed. E. D. Le Cren and M. W. Holdgate. *Symp. Brit. Ecol. Soc.*, **2**, 191–203. Blackwell, Oxford and Edinburgh.

Watt, K. E. F. Ed. (1966). *Systems Analysis in Ecology.* Academic Press, New York and London.

Watt, K. E. F. (1968). *Ecology and Resource Management: A Quantitative Approach.* McGraw-Hill, New York, San Francisco, St. Louis, Toronto, London and Sydney.

Wayre, P. (1969a). 'Swinhoe's pheasant in Taiwan', *Biological Conservation*, **1**, 184–186.

Wayre, P. (1969b). 'The role of zoos in breeding threatened species of mammals and birds in captivity', *Biological Conservation*, **2**, 47–49.

Wells, T. C. E. (1967). '*Pulsatilla vulgaris* and changing land use', pp 143–150, in *The Biotic Effects of Public Pressures on the Environment.* Ed. E. Duffey. Monks Wood Experimental Station, Symposium No. 3, The Nature Conservancy.

Wells, T. C. E. (1968). 'Land-use changes affecting *Pulsatilla vulgaris* in England', *Biological Conservation*, **1**, 37–43.

Wells, T. C. E. (1969a). 'The decrease of *Pulsatilla vulgaris* in England – a reply to Dr. E. L. Jones', *Biological Conservation*, **1**, 328–329.

Wells, T. C. E. (1969b). 'Botanical aspects of conservation management of chalk grasslands', *Biological Conservation*, **2**, 36–44.

Wells, T. C. E. and Barling, D. M. (1971). 'Biological Flora of the British Isles. *Pulsatilla vulgaris* Mill. (*Anemone pulsatilla* L.)', *J. Ecol.*, **59**, 275–292.

Whittaker, R. H. (1970). *Communities and Ecosystems.* Collier-Macmillan, London.

Wilks, H. M. (1960). 'The re-discovery of *Orchis simia* Lam. in Kent. 2. *Orchis simia* in East Kent', *Trans. Kent Fld. Soc.*, **1**, 50–55.

Wilks, H. M. (1966). 'The monkey orchid in East Kent', *Handbk. a. Rep. Soc. Promot. Nat. Reserves*, 1966, 2 pp.

Williams, W. T. and Dale, M. B. (1965). 'Fundamental problems in numerical taxonomy', *Adv. bot. Res.*, **2**, 35–68.

Williams, W. T. and Lambert, J. M. (1959). 'Multivariate methods in plant ecology. I. Association analysis in plant communities', *J. Ecol.*, **47**, 83–101.

Williams, W. T. and Lambert, J. M. (1960). 'Multivariate methods in plant ecology. II. The use of an electronic digital computer for association-analysis', *J. Ecol.*, **48**, 689–710.

Williams, W. T., Lance, G. N., Webb, L. J., Tracey, J. G. and Dale, M. B. (1969). 'Studies in the numerical analysis of complex rain-forest communities. III. The analysis of successional data', *J. Ecol.*, **57**, 515–535.

Williamson, M. H. (1959). 'Some extensions of the use of matrices in population theory', *Bull. math. Biophys.*, **21**, 13–17.

Williamson, M. H. (1967). 'Introducing students to the concepts of population dynamics', in *The Teaching of Ecology.* Ed. J. M. Lambert. *Symp. Brit. Ecol. Soc.* **7**, 169–175. Blackwell, Oxford and Edinburgh.

Williamson, M. H. (1972). *The Analysis of Biological Populations*. Arnold, London.

Wilson, E. O. and Simberloff, D. S. (1969). 'Experimental zoogeography of islands. Defaunation and monitoring techniques', *Ecology*, **50**, 267–278.

Wurster, C. F. (1969). 'Chlorinated hydrocarbon insecticides and the world ecosystem', *Biological Conservation*, **1**, 123–129.

Yarranton, G. A. (1969). 'Pattern analysis by regression', *Ecology*, **50**, 390–395.

Index

This index does not include material that is in the Management Plan of Aberlady Bay Local Nature Reserve in Chapter 10. Table numbers are given in italics, and illustration numbers are in bold type.

Aberlady Bay, 87, 277, 314–15, **10.1**, **10.2**, **10.3**, **10.4**, **10.5**, Plates 9–13, 17–19
Aberlady Bay management plan, 316–61
Acer campestre, *4.9*, *4.10*
Achillea millefolium, 238, 242–3, *7.6*, *7.7*, *7.8*, *7.9*, *7.10*, *7.11*, **7.6**, **7.8**
Acrocladium, 36
African Convention, 15–16
Agropyron junceiforme, 87, 282, 299
Agrostis, 38, 48
Aira praecox, 279
Alder, 8, 88, 141, **I.3**
Allium ursinum, 78
Alnus glutinosa, 88, 141
Alopecurus pratensis, **2.8**
Alpine flora, 13–14, 29–30, 95
Alpine sorrel, 95
Ammophila arenaria, 48–9, 87, 282, 299
Anagallis tenella, 261
Anderson, D. B., 70, 72, 379, **2.6**
Anderson, M. L., 129–31, 372, **4.3**
Andrewartha, H. G., 148, 372
Anoa mindorensis, 247
Anthoxanthum odoratum, 75, **2.8**
Ants,
  associated with large blue, 11–12
  succession on Florida Keys, 25–7, *1.1*
Aquila chrysaetos, 211
Arctic fox, 79
Arenaria, 30
Arrhenatherum elatius, 75, 233–4, 238, *7.6*, *7.7*, *7.8*, **2.8**, **7.6**
Arthur, D. R., 2–4, 372, *I.2*, **I.1**
Arvill, R., 3, 15, 372, *I.1*
Ash, 140, 141
Association-analysis, 135, 235–7, *7.7*, **4.2**, **4.3**
Australia, land classification system, 146
Australopithecus, 1
Automobile Association, 292, 372
Azotobacter, 111
Baboon, 15
Bacopa rotundifolia, 101–3, **3.1**, **3.3**
Bailey, N. T. J., 285, 372
Bainbridge, R., 70, 372
Balaenoptera musculus, 177
Balaenoptera physalus, 177
Banko, W. E., 223, 372
Barbary ape, 15
Barling, D. M., 215–18, 384, *7.2*
Barns Ness, 260
Bawden, M. G., 146, 372
Bean goose, *9.3*
Beckingham, L., 153, 380
Bee orchid, 232, 239
Bellis perennis, 280
Berkshire, Buckinghamshire and Oxfordshire Naturalists' Trust, 270, 372
Berry, R. J., 31–2, 373
Betula nana, 8
Beverton, R. J. H., 172–5, 373, *5.12*, *5.13*
Bindweed, 280
Biogeography, 20–30
Birch, 8, 89–91, 299, *2.8*, **I.3**
Birth-rate, 4–5, *I.3*
Bishop Wood, 265–8, 269, *8.2*, *8.3*, *8.4*, **8.1**
Biston betularia, 31
Black, J. D., 206, 373
Black, J. N., xii–xiii, 195, 203–4, 373, **5.19**, **5.21**
Blackgame, 80–2, *2.4*, *2.5*
Blackman, G. E., 193, 373, **5.20**
Blackthorn, 86
Bloomfield, H. E., 235, 373
Bluegill, 226
Boer, G. de, 285, 373
Bog pimpernel, 261
Bog rush, 130
Botanical gardens, 244–6
Brachypodium pinnatum, 233
Bracken, 85, 130
Bramble, 86

Brasnet, N. V., 143, 308, 373, 374
Brauer, A., 91, 369, 373
Braun-Blanquet, J., 131–3, 373
Britain,
climate, 9, **I.4**
waste land classification, 129–31, *4.3*
woodland classification, 140–2, *4.7*, *4.8*, *4.9*, *4.10*
British Ecological Society, xi
Britten, J., 245, 373
Broad, S. T., xiii, 256–7, 262, 265, 273, 373
Broom, 85–7, 271–2, **2.11**, **8.2**
Brown, P., 14, 254–5, 374
Brunelle, R., 220, 374
Bunce, R., xiii, 140–2, *4.7*
Burden, P. C., 241–2, 374
Burhenne, W. E., 15, 374
Burton, T. L., 274–5, 374, *9.1*, *9.2*

Cabbage white, 71–2
Caerlaverock National Nature Reserve, 277
Cairngorms, 254, 279
Cajander, A. K., 128–9, 374, *4.1*
California deer, 183, 191
Calluna vulgaris, 88, 130, 136–40, 299, 301, *4.3*, *4.4*, *4.5*, *4.6*, **4.2**, **4.3**
Callunetum, 136
Calcium cycle, 113, **3.7**
Campanula glomerata, 233
Campanula rotundifolia, *7.6*, *7.7*
Camponotus, 25–7, *1.1*, **1.6**
Campton, R. M., 153, 380
Canada, land classification system, 143–6
Capercailzie, 80–2, *2.4*, *2.5*
Cardiocladius, 101, **3.3**
Carlina vulgaris, 238, *7.6*, *7.7*
Centaurea nigra, *7.6*, *7.7*
Cervus elaphus, 184
Chamaenerion angustifolium, 238, 239, *7.6*, *7.7*
Champion, H., 143
Chauliodes, 106
Chen hyperborea, 223
Cherleria sedoides, 29, **1.8**
Chickweed, 71
Chimpanzee, 15
Chi-squared (χ^2), 44–5, 136–40, 235, 285, 370–1
Christian, C. S., 146, 374
Christiansen, K., 49, 374
Circadian rhythms, 67–73
environmental modification, 67–70
photoperiodic responses, 70–3
Cladonia, 88
Clapham, A. R., 238, 374, *7.6*
Clark, L. R., 148, 374
Classification,
Braun–Blanquet system, 132–3
for British forestry, 129–31, *4.3*
for land-use planning, 143–6, *4.11*, **4.4**
numerical, 133–42, **4.1**
of a Callunetum, 136–40
of British woodlands, 140–2, *4.7*, *4.8*, *4.9*, *4.10*
of ecosystems, 127–46
of forests in Finland, 128–9, *4.1*, *4.2*
of land, *4.11*
Clawson analysis, 293, *9.7*, **9.6**
Clostridium, 111
Clover, 48
Clustered bellflower, 233, 239
Cocklebur, 72
Collared dove, 33
Collingbourne, R. H., 95–6, 374
Colobopsis, 27, **1.6**
Colobus monkey, 15
Colonization curves, 27–8, **1.7**
Committee for Environmental Conservation, 263
Cone Spring, 101–9, *3.1*, *3.2*, *3.3*, *3.4*, **3.3**, **3.4**, **3.5**
Conservation,
botanical gardens, 244–6
definition, 204
Countryside in 1970, 202–3
Cragg's, 203
Fisher's, 203–4
Huxley's, 208
Macfadyen's, 202
Margalef's, 201–2
Oxford English Dictionary, 201
East African aims, 208–9
of animals, 219–24
of large copper butterfly, 227–31, *7.4*, *7.5*
of monkey orchid, 250–4
of osprey, 254–5
of Pere David's deer, 249
of plants, 214–18
of Swinhoe's pheasant, 248
of tamaraw, 247–8
relation to biological resources, 205
supplementation of populations, 224–31
zoological gardens, 31, 246–50
Conservation Corps, 270
Contingency table, 135–6, 285
Convolvulus arvensis, 280
Copland, W. O., 299, 377
Cornish Naturalists' Trust, 12
Cornwallis, R. K., 205, 374
Correlation coefficient, 370–1
Corylus avellana, 140
Council for Environmental Education, 263
Countryside Act 1968, 15

Countryside Commission, xiii, 285, 289, 297, 374, *I.4*
Countryside in 1970, 202, 374
Countryside (Scotland) Act 1967, 15
County Conservation Trusts, 263
Cragg, J. B., xiii, 203, 374
Crataegus monogyna, 234, 239, 243, *7.6*, *7.7*, Plate 14
Crawford, R. M. M., 88, 374
Crematogaster ashmeadi, 25–7, *1.1*, **1.6**
Crematogaster atkinsoni, *1.1*
Crossley, D. A., 100, 375
Curtis, J. T., 20, 375, **1.1**
Cypripedium calceolus, 246

Dactylis glomerata, 78, 233–4, 238, *7.6*, *7.7*, *7.8*, **2.8**
Daisy, 280
Dale, M. B., 91, 238, 384
Dandelion, 71
Danilevsky, A. S., 67, 71, 375
Daphnia pulicaria, 159–60, 162–3, 167, *5.2*, *5.3*, **5.5**, **5.7**
Darlington, D., 257, 288, 383
Dasman, R. F., 206, 375
Davey, A. J., 34–8, 375
Davidson, J., 285–9, 375, *9.5*
Davis, B. N. K., 21, 379, **1.2**
Dawn cypress, 245–6
Death-rate, 5, *I.3*
Denson, E. P., 223, 375, *7.3*
Diapensia lapponica, 245
Dicranum, 260
Dicyrtomina rufescens, 67–9, **2.5**
Difference equations, 120
Diurnal rhythms, 67–73
Douglass, R. W., 276, 375
Drosera rotundifolia, 261
Dryas octopetala, 7
Duffey, E., xii, 207, 209, 214, 227–30, 282, 375, *7.4*, *7.5*, **7.3**, **7.4**, **9.2**
Duffy, P. J. B., 146, 375
Dwarf birch, 8

East Lothian County Council, xiii, 260, 283, 314–15, 375
Economic analysis of survey data, 289–96
Ecosystem,
  attributes of models, 122
  classification of, 127–46
  concept of, 93–126
  definition of, 93–4
  effects of recreation on, 276–84
  energy flow models, 116–26, *3.5*, *3.6*, **3.9**, **3.10**
  modelling, 116–26
  modelling errors, 125–6
  nutrient cycle models, 116–26, *3.5*, *3.6*, **3.9**, **3.11**
  size of, 93–4
Education, 209, 256–73, 288, 302
  adult, 258–63
  competing form of land use, 273
  extensive surveys, 271–3
  facilities for school use, 265–8
  Nuffield science syllabi, 263–5
  planning for school use, 268–70
  school projects, 270–3
  schools, 263–73
  survey of school use, 265–8, *8.2*, *8.3*, *8.4*
Edwards, C. A., 100, 375
Eggeling, W. J., 309, 375
Ehrenfeld, D. W., 177–8, 182, 206, 375
Elaphurus davidianus, 249
Elderberry, 86
Elliston Wright, F. R., 245, 375
Elm, 8, **I.3**
Elton, C., 78–9, 375–6
Elymus arenarius, 87
Empetrum nigrum, 88
Energy flow, 94–109
  diagram of, 106, **3.5**
  in Cone Spring, 101–9
  input of energy, 95–7, *3.1*
  utilization of energy, 98–100, *3.2*
Epilobium nerterioides, 34–8, **1.10**, Plate 2
Erica cinerea, 88, 136–40, *4.3*, *4.4*, *4.5*, *4.6*
Erica tetralix, 88, 136–8, *4.3*, *4.4*, *4.5*, *4.6*, **4.2**, **4.3**
Errington, J. C., 47, 49, 55–9, 376, **1.13**, **1.16**, **1.17**
Euonymus europaeus, *4.9*
Euphrasia nemorosa, *7.6*, *7.7*
Euphrasia officinalis, 260
Euproctis similis, 72
Eurhynchium striatum, 49–51, *1.4*, **1.13**
European Conservation Year 1970, xi
Evans, G. C., 70, 372
Exponential growth curve, 148
Eyebright, 260

Falco peregrinus, 211
False oatgrass, 75
Farrar, J. L., 143, 376
Festuca ovina, 36, 48, 197, **5.22**
Festuca rubra, 75, 87, 280, **2.8**, **9.1**
Field Studies Council, 265
Filipendula ulmaria, 88, *4.10*
Finland, forest classification, 128–9, *4.1*, *4.2*
Fire, 208, 301
Fisher, J., 73, 214, 219, 222, 245, 247, 376, **2.7**
Fisher, J. L., xiii, 203, 376
Fisher, R. A., 370, 376

Floods, 92
Florida Keys, 23–9, 32, 33, **1.5**
Flour beetles, 158, 167
Folsomia candida, 159, 161–2, 164–5, 307, *5.2*, **5.6**, **5.8**
Folsomia quadrioculata, 45
Food chain,
  decomposer, 97–8
  food chain efficiency, 99–100, 108–9, *3.4*
  grazing, 97
  gross ecological efficiency, 99–100, 108–9, *3.4*
  production efficiency, 109, *3.4*
Food web, 104–6, **3.4**
Ford, E. B., 11–12, 227, 376
Forestry Commission, xii, 15, 38, 260, 265, 376
Fraser Darling, F., 204, 376
Frankland, J. C., 85, 376
Fraxinus excelsior, 140, 141
Free recreational benefit, 290–2
Frenesia missa, 101–6, **3.3**
Frohawk, F. W., 11–12, 227, 376
Fujita, K., 68, 382, **2.5**
Fulmar, 33, 73, **2.7**
Fulmarus glacialis, 33, 73

Gale, sweet, 299
Galium saxatile, 51, 78, 93, *4.10*, **1.14**, **1.15**
Galium verum, 238, *7.6*, *7.7*, **7.6**
Gama, M. M. da, 131, 376
Game-bird cycles, 80–2, *2.4*, *2.5*
Gammarus pseudolimneus, 101–6, **3.3**
Gannet, 73, **2.7**
Geier, P. W., 148, 374
Gentianella amarella, 238, *7.6*, *7.7*, *7.8*
Giant hogweed, Plate 16
Gibraltar Point, 284
Gisin, H., 131, 376
Godwin, H., 7, 376, **I.2**
Golden eagle, 211
Goodall, D. W., 48, 238, 241, 376
Gordon, N. J., 87, 376
Gorilla, 15
Gorse, 12, 86
Goryshin, N. I., 67, 71, 375
Grazing,
  by goats, 196
  by rabbits, 197–8, 234
  by sheep, 12, 196
Great water dock, 227
Green, R. H., 44, 376
Greig-Smith, P., 44, 47, 48, 53, 243, 377
Grey squirrel, 33, 38–41, **1.11**, Plate 3
Greylag goose, *9.3*
Grouse, 80–2, *2.4*, *2.5*
Growing season, 271

Guppy, 165–6, *5.2*, **5.9**
Gulland, J. A., 175–6, 377
Gullane, 283
Gutsell, J. S., 165–6, 381, **5.9**
Gwydyr Forest, 260, Plate 8

Haddock, 172
Harley, J. L., 85, 377
Harrisson, T., 247, 377
Hawthorn, 234, Plate 14
Hazel, 8, 140, **I.3**
Heather (see also *Calluna* and *Erica*), 299
Hedgerows, 21, 205, **1.2**
Helianthemum chamaecistus, 280
Hemsley, J. H., 299, 377
Heracleum mantegazzianum, Plate 16
Heracleum sphondylium, 239, *7.6*, *7.7*
Hieracium pilosella, 88, 238, 243, *7.6*, *7.7*, *7.8*, *7.9*, *7.10*, **7.6**, **7.8**
Hill, I. D., 146, 377
Hills, G. A., 143–6, 377, *4.11*, **4.4**
Hippophaë rhamnoides, 8, Plate 15
Holcus lanatus, *7.6*, *7.7*
Holly, 86
Holt, S. J., 175, 373
Homo erectus, 1–2
Homo habilis, 1
Homo sapiens, 2–5
Honkenya peploides, 88
Hooper, M. D., 21, 379, **1.2**
Horses, trampling by, 279–80, **9.1**
House sparrow, 33
Hudson Bay, 220–1, **7.2**
Hughes, R. D., 148, 374
Hutchinson, G. E., 48, 377
Huxley, J., 208–9, 377
Huxley, T., 277–8, 377, *9.3*
Hylocomium splendens, 88, 260
Hypnum, 260

Ice-age, 7–8
Ilex aquifolium, 86–7, **2.11**
Impatiens capensis, 101, **3.1**, **3.3**
Intrinsic rate of natural increase, 148, 153
Island biogeography, 21–9
Isotoma sensibilis, 49

Jeffers, J. N. R., xi, 377
Jenkins, A. C., 247, 377
Jones, E. L., 215, 377
Jonkel, C. J., 220, 377
Juncus balticus, 88
Juncus effusus, 88
Juncus gerardii, 88

k-factor analysis, 229–30, *7.4*
Kale, **5.20**

Kavanagh, N. J., 293
Kendall, M. G., 61, 63, 377
Kent, D. H., xiii, 33–4, 378
Kershaw, K. A., 47, 378
Knuchel, H., 308, 378
Kormondy, E. J., 95, 96, 100, 103, 106, 110, 378, **3.2**
Köstler, J., 129, 378
Kozlovsky, D. G., 100, 378

Lady's slipper orchid, 246
Lakenheath Warren, 51, 93
Lamb, H. H., 9, 378, **I.4**
Lambert, J. M., 135–40, 235–7, 384, *4.4*, *4.5*, *4.6*, **4.2**, **4.3**
Lance, G. N., 91, 384
Land-use planning, 143–6, *4.11*, **4.4**
Large blue, 11–13
Large copper, 31, 213, 227–31, *7.4*, *7.5*, **7.3**, **7.4**, Plate 6
Laws, R. M., 177–82, 378, **5.14**, **5.15**
Leaf area index, 193–6, **5.19**, **5.20**, **5.21**
Lefkovitch, L. F., 155, 156, 378
Legislation, 13, 211, 246, 248, 256
Lemming cycle, 78–9
Lemna minor, 101–3, *3.1*, **3.3**
Lemuroids, 15
Lepidoptera,
  cabbage white, 71–2
  circadian rhythms, 70, 71–2
  large blue, 11–13
  large copper, 31, 213, 227–31
  photoperiodic response, 71–2
Leontodon, 238, 243, *7.6*, *7.7*, **7.6**
Leslie, P. H., 150, 378
Lettuce, 71, **2.6**
Levins, R., 101, 378
Lewis, E. G., 150, 378
Lieth, H., 75, 378, **2.8**, **2.9**
Light,
  intensity, 95
  wavelength, 95
Lipsey, R. G., 285, 378
Lloyd, H. G., 39, 378, **1.11**
Lockie, J. D., 10, 211, 378
Logarithmic growth, 148
Logistic growth, 149
Lolium perenne, 280, **9.1**
Longstaff, B. C., xiii, 159, 161–2, 164–5, 383
Lonicera periclymenum, 87, *4.10*
Lophura swinhoei, 248
Lotus corniculatus, 87, 242–3, *7.6*, *7.7*, *7.9*, *7.10*, *7.11*, **7.8**
Lowe, V. P. W., 184–9, 379, *5.4*, *5.5*, **5.17**, **5.18**
Lucilia cuprina, 158, 160–1, 162, *5.1*, *5.2*
Lycaena dispar, 31, 213, 227–31, *7.4*, *7.5*, **7.3**, **7.4**, Plate 6
Lynx cycle, 79–80, 82–3, 207, **2.10**

MacArthur, R. H., 21, 100, 379, **1.3**
Macfadyen, A., xiii, 202, 379
McKelvie, A. D., 271–2, 379, **8.2**
Mackenzie, J. M. D., 80, 379
Maculinea arion, 11–13
Malham Tarn, 8, **I.3**
Mallard, *9.3*
Mammals, small, in bottles, 302
Man: Nature relation, 203, 264
Management plan, 299–302, 307–14
  Aberlady Bay Local Nature Reserve, 314–15, 316–61
  pro forma for, 309–14
Maple, 89–91, *2.8*
Margaropoulos, P., 196, 379
Margalef, R., xiii, 202, 379
Marram grass, 48–9, 87, 282, 284, 299
Matrix algebra, 365–9
  addition, 366
  definition, 365–6
  eigenvalue, 367
  eigenvector, 367
  inversion, 367
  latent roots, 91, 152, 179, 191, 367–9
  latent vectors, 91, 152, 179, 191, 367–9
  multiplication, 366–7
  non-negative matrices, 91, 153, 369
  subtraction, 366
Matrix models of population growth, 149–58, **5.2**, **5.3**, **5.4**
  exponential growth, 149–53
  forest succession, 90–1
  harvesting, 156–7
  insect life stages, 155
  red deer, 189–91
  sigmoid growth, 155
  trees, 155–6
  two sexes, 189
  whales, 178–9
Maynard Smith, J., 32, 149, 379
Medicago lupulina, 238, 239, 243, *7.6*, *7.7*, *7.9*, *7.10*, **7.8**
Melanogrammus aeglefinus, 172
Mellanby, K., 210, 379
Metasequoia glyptostroboides, 245–6
Meyer, B. S., 70, 72, 379, **2.6**
Mindoro, 247
Mnium undulatum, 50–1, 260, *1.4*, **1.13**
Molinea caerulea, 130, 136–40, 299, 301, *4.3*, *4.4*, *4.5*, *4.6*, **4.2**, **4.3**
Monkey orchid, 250–4, *7.12*, **7.9**, **7.10**
Monkeys, 15
Monomorium flavicola, 27, *1.1*, **1.6**

Moore, N. W., 21, 379, **1.2**
Moran, P. A. P., 65, 79–82, 379, *2.4*, *2.5*, **2.10**
Morris, R. F., 148, 374
Moss, W. W., 134, 379
Mount Iglit Game Refuge, 247
Musanga cecropioides, 84
Mutch, W. E. S., 257, 379
Mute swan, *9.3*
Myrica gale, 299, *4.3*
Myrmica, 11–12
Myxomatosis, 197

National Parks and Access to the Countryside Act 1949, 14
National Trust, 14
Nature Conservancy, xii, xiii, 14, 140, 215, 228, 299–303, *4.8*
  Caerlaverock National Nature Reserve, 277
  Studland Heath National Nature Reserve, 297–304
  Study Group for Field Education, 268–9
  Tentsmuir National Nature Reserve, 87–8, 282
  Woodwalton Fen National Nature Reserve, 227
Nature trail, 258–62, 288, 302
  advantages and disadvantages, 261–2
  content, 258–9
  definition, 257–8
  formal, 258, 260–1, 269, *8.1*, Plate 7
  informal, 258, 260
  school use, 269
  semi-formal, 258, 260, 269, *8.1*, Plate 8
Neodiprion sertifer, 72
New Forest, 136, **4.3**
New Zealand willow-herb, 34–8, **1.10**, Plate 2
Newmarket Heath, 279–80
Nicholson, A. J., 158, 160–1, 379, *5.1*
Nicholson, M., 79, 376
Nitrobacter, 112
Nitrogen cycle, 110–12, **3.6**
Nitrosomonas, 112
North Sea fishery, 172, 196, **5.12**
Nuffield Foundation, xiii, 263
Nutrient cycles, 109–15
  calcium, 113, **3.7**
  closed, 109–10
  nitrogen, 110–12, **3.6**
  open, 109–10
  potassium, 114, **3.8**
  the ecosystem concept, 115
Nutrient reservoir, 114–15

Oak, 8, 89–91, 114, 140, 141, *2.8*, **I.3**, **3.8**
Odum, E. P., 95, 106, 110, 379
Oikopleura, 174
Olor buccinator, 219, 222–3, *7.3*
Ononis repens, *7.6*, *7.7*
Ophrys apifera, 232
Öpik, H., 70, 382
Orchidaceae, 246
Orchis simia, 250, *7.12*, **7.9**, **7.10**
Organisation of African Unity, 15
Organochlorine insecticides, 210
Osmaston, F. C., 308, 380
Osprey, 14, 254–5
Ovington, J. D., 113–14, 380, **3.7**, **3.8**
Oxalis acetosella, *4.9*
Oxford ragwort, 33–4, **1.9**, Plate 1
Oxyria digyna, 95

Pandion haliaetus, 254
Paracryptocerus varians, 25–7, *1.1*, **1.6**
Parathecina bourbonica, *1.1*
Pasque flower, 215–18, *7.1*, *7.2*, **7.1**, Plate 5
Passer domesticus, 33
Pattern, 41–53, 241–3
  analysis of, 45–8, 241, *1.3*, *1.5*, *7.9*, *7.10*, *7.11*, **7.8**
  computer programme, 54–5
  correction for trend, 53
  correlation between species, 51
  detection of, 41–5
  implications of, 48–53
  simulation of, 53–9, **1.16**, **1.17**
  types of, 41–3, 48, **1.12**
  woodland bryophytes, 49–51
Peakall, D. B., 211, 380
Pearlwort, 38
Pennycuick, C. J., 153, 380
Pentaneura, 101, **3.3**
Peppered moth, 31
Pere David's deer, 249
Peregrine, 211
Perring, F. H., xiii, 279–80, 380, **9.1**
Peterle, T. J., 210, 380
Phagocata vetata, 101–6, **3.3**
Phenology, 73–8, **2.8**, **2.9**
Phillipson, J., 98–100, 380
Phleum pratense, 75
Photoperiod,
  in animals, 71–2
  in plants, 70–1, 72
Phryxe vulgaris, 231
Physa integra, 101, **3.3**
Pielou, E. C., 53, 380
Pieris brassicae, 71–2
Pierpoint, G., 146, 377, 380
Pigott, C. D., 8, 380, **I.3**
Pigott, M. E., 8, 380, **I.3**
Pine, 8, 45, 92, 114, 128, **I.3**, **3.7**, **3.8**
Pinguicula vulgaris, 36
Pinus contorta, 92
Pinus sylvestris, 114, 128

Pitt, M., 285, 380
Plagiothecium, 260
Plaice, 172–5, **5.12**, **5.13**
Plantago, 8, **I.3**
Plantago lanceolata, 87, *7.6*, *7.7*
Pleuronectes platessa, 172–5
Poa annua, 279
Poa pratensis, **2.8**
Poa trivialis, **2.8**
Pochard, *9.3*
Poisson distribution, 43–4
Polar bear, 219–22, **7.2**
Pollen record, 7–8, **I.2**, **I.3**
Pollution, 210
Polytrichum, 260
Polytrichum juniperinum, 279
Polytrichum piliferum, 279
Polunin, N., 244, 380
Poore, M. E. D., 131, 132–3, 143, 380
Population,
  characteristics of over-exploitation, 166–9, 176–83, 198
  characteristics of under-exploitation, 166–9, 183–92, 198
  exponential growth equation, 148–9
  fish stocking, 226
  logistic growth equation, 149
  matrix algebra models, 149–58
  models of fish populations, 175–6
  of *Daphnia pulicaris*, 159–60, 162–3
  of *Folsomia candida*, 161–2, 164–5
  of guppy, 165–6
  of large copper butterflies, 227–31, *7.4*, *7.5*
  of *Lucilia cuprina*, 160–1, *5.1*
  of man, 2–6, *I.1*, *I.2*, **I.1**
  of plaice, 172–5
  of plants, 192–8
  of polar bear, 219–22
  of *Pulsatilla vulgaris*, 215–18, *7.2*
  of red deer, 184–91, *5.4*, *5.5*, *5.6*, *5.7*, *5.8*
  of sardine, 169–72
  of trumpeter swan, 222–3, *7.3*
  of whale, 176–82
  over-exploitation of, 176–83, 221
  response to harvesting, 158–69
  simple models of population increase, 148–53
  supplementation of, 224–31
  under-exploitation of, 183–92
Porter, L. K., 115, 380
Potassium cycle, 114, **3.8**
Potato, **5.20**
Potentilla erecta, *4.9*
Poterium sanguisorba, 280, **9.1**
Prickly saltwort, 284
Primula vulgaris, *4.9*
Pritchard, T., 273, 380
Protection of Birds Act 1954, 14
Prunella vulgaris, 238, *7.6*, *7.7*
Prunus spinosa, 86–7, **2.11**
Pseudomonas, 111, 112
Pseudomyrmex elongatus, 25–7, *1.1*, **1.6**
Pseudomyrmex 'flavidula', 27, **1.6**
Ptarmigan, 80–2, *2.4*, *2.5*
Pterydium aquilinum, 85, 130, 136, *4.4*, *4.5*, *4.6*
Pulsatilla vulgaris, 215–18, *7.1*, *7.2*, **7.1**, Plate 5
Purple moor grass (see also *Molinea*), 299

Quercus, 140, 141
Quercus robur, 114
Questionnaires, design of, 285–9, *9.4*

Rabbit, 11, 197–8, 234
Rackham, O., 70, 372
Radley, J., 92, 381
Radovich, J., 170, 381, **5.11**
Random distribution, 43–4
Raven, J., 13, 29, 381, **1.8**
Recreation, 209, 274–304
  and forestry, 275–6
  aspects of economic analysis, 289–96
  assessment of demand, 284–96
  classification of, 274–5, *9.1*, *9.2*
  effects on ecosystem of, 276–84
  on sand dunes, 281–4
  planning for, 296–304
  planning on Studland Heath National Nature Reserve, 297–304, **9.7**, **9.8**
  questionnaires, 285–9
  trampling of vegetation, 279–81, **9.1**
  wildfowling, 277–8
  zoning of users, 303–4
Red Book, The, 147, 214
Red deer, 184–91, *5.4*, *5.5*, *5.6*, *5.7*, *5.8*, **5.17**, **5.18**
Red fescue, 280
Red mangrove, 24
Red Rock Lakes National Wildlife Refuge, 222, **7.3**
Red squirrel, 38–41, **1.11**, Plate 4
Reichle, D. E., 100, 375
Rhacomitrium, 260
Rhine fish disaster, 211
Rhizobium, 111
Rhizophora mangle, 24
Rhum, 130, 184–5, 309, *5.4*, *5.5*, **5.17**, **5.18**
Rhytidiadelphus triquetrus, 50–1, 88, *1.4*, **1.13**
Richards, P. W., 83, 381
Richman, S., 158–60, 162–3, 381, *5.3*, **5.5**, **5.7**

Rockrose, 280
Rohlf, F. J., 285, 370, 381, 382
Rowe, J. S., 146, 381
Royal Society for the Protection of Birds, 14, 254, 262
Rubus fruticosus, 86–7
Rumex acetosa, 38, 78
Rumex acetosella, 279
Rumex hydrolapathum, 227
Rush, 12, 88, *4.3*
Russell-Hunter, W. D., 177, 381
Rye-grass, 280

Sagina, 38, 245
Sagina boydii, 245
Salad burnet, 280
Sallow, 299
Salix, 86
Salix repens, 88
Salsola kali, 284
Salvia, 71, **2.6**
Sambucus nigra, 86–7, **2.11**
Sand dunes, 87–8, 281–4, 297–9, **9.2**
Sandwort, 30
Sanicula europaea, *4.10*
Sardine, 169–72, **5.10**, **5.11**
Sardinops caerulea, 169–72
Sarothamnus scoparius, 85, 271
Saxifraga caespitosa, 29, **1.8**
Saxifraga stellaris, 36
Schoenus nigricans, 130
Schofield, J. M., 284, 381
Sciurus carolinensis, 33, 38–41, Plate 3
Sciurus vulgaris, 38–41, Plate 4
Scott, G. A. M., 85, 381, **2.11**
Sea buckthorn, 8, **1.3**, Plate 15
Sea couch, 282, 284, 299
Searle, S. R., 365, 381
Sedum villosum, 38
Selection, 31
Senecio squalidus, 33–4, **1.9**, Plate 1
Serial correlation, 65, 80, *2.5*, **2.4**, **2.10**
Shelduck, *9.3*
Shorten, M., 39–41, 381, **1.11**
Silene maritima, 87, **2.11**
Silliman, R. P., 165–6, 381, **5.9**
Silver Springs, 99
Simberloff, D. S., 23–9, 381, 385, *1.1*, **1.5**, **1.6**, **1.7**
Simon, N., 214, 219, 222, 245, 247, 376
Sims, C., 92, 381
Simulation, 53–9
Skipwith Common Nature Reserve, 260–1, Plate 7
Skokholm house mouse, 31
Slobodkin, L. B., 148, 158–60, 162–3, 381, *5.3*, **5.5**, **5.7**
Small mouth bass, 225
Smith, F. E., 120–3, 381, *3.7*, **3.11**
Smith, J. E., 211, 381
Smith, R. J., 293, 382
Snaydon, R. W., 48, 382
Sneath, P. H. A., 134, 382
Snow goose, 223
Sokal, R. R., 134, 285, 370, 381, 382
Solar radiation, 95–7, 195, **3.1**, **5.21**
Sole, 172
Solea vulgaris, 172
Solomon, M. E., 148, 382
Southall, D. R., 159, 161–2, 164–5, 383
Southern, H. N., 39, 382
Spatial distribution, 19–59
  genetical implications, 31–2
Species, number of, 27–8, 100–1
Species-area curve, 21–3, **1.3**, **1.4**
Sphagnum, 260, *4.3*
Spiegel, M. R., 63, 382
Spurn Peninsula Nature Reserve, 285, *9.4*, *9.5*, *9.6*, *9.7*, *9.8*, **9.4**, **9.5**, **9.6**
Stamp, D., 13, 15, 209, 247, 270, 382
Statistical tables, 370–1
Stephens, G. R., 88–92, 382, 383, *2.8*
Stewart, G. A., 146, 374
Stoat, 11
Strawberry, 70
Street, H. E., 70, 382
Streptopelia decaocto, 33
'Student', 43, 382
Student's *t*, 370–1
Studland Heath National Nature Reserve, 297–304, **9.7**, **9.8**
Subterranean clover, 195, **5.19**
Succession, 24, 83–92, 299, 307
  in chalk grassland, 233–4, *7.8*
  in fungi, 85
  in sand dunes, 87–8, 282
  in temperate forests, 88–91
  on a shingle beach, 85–7, **2.11**
  to tropical forest, 83–4
  transition probabilities, 89–90, *2.7*, *2.8*
Succisa pratensis, 38
Sugar beet, **5.20**
Sula bassana, 73
Sundew, 117, 261
Survival curve, 180–2, 222, **5.16**
Swinhoe's pheasant, 248

Tamaraw, 247–8
Tapinoma littorale, 25–7, *1.1*, **1.6**
Tarrant, K. R., 210, 382
Tatton, J. O'G., 210, 382
Taylor, A. E., 257, 265–7, 269, 288, 382, 383, *8.2*, *8.3*, *8.4*
Teagle, W. G., 297–303, 382, **9.8**

Teal, *9.3*
Tentsmuir National Nature Reserve, 87–8, 282
Tetracanthella wahlgreni, 43–4, *1.2*
Teucrium scorodonia, *4.9*
Thalictrum alpinum, 36
Therfield Heath, 218
Thuidium, 260
Thyme, 11–12, 280
Thymus drucei, 11, 238, 243, 280, *7.6*, *7.7*, *7.9*, *7.10*, **7.8**
Tilly, L. J., 101–7, 382, *3.1*, *3.2*, *3.3*, *3.4*, **3.3**, **3.4**, **3.5**
Time series, 60–6
  analysis of, 60–6
  components of, 61–2, **2.2**
  cyclical movements, 62, **2.2**
  random movements, 62
  seasonal movements, 62, **2.2**
  trend, 61, 82, 83, **2.1**, **2.2**
Tobacco, 70, 71
Tomato, 71
Torrey Canyon, 211, 263
Tracey, J. G., 91, 384
Trampling of vegetation, 279–81, **9.1**
Tribolium confusum, 158, 167, *5.2*, *5.3*
Trice, A. H., 290–2, 382, *9.6*
Trichophorum caespitosum, 136, 140
Trifolium repens, 48, 87
Trifolium subterraneum, 195
Trophic level, 98–101, 107–8, *3.3*, *3.4*, **3.2**, **3.3**
  number of species, 100–1
Trumpeter swan, 219, 222–3, *7.3*
Tuley, P., 146, 372
Turkey, 32
Tussilago farfara, 238, *7.6*, *7.7*
Tutin, T. G., 238, 374, *7.6*
Tyshchenko, V. P., 67, 71, 375

Uchida, H., 68, 382, **2.5**
Ulex europaeus, 86, **2.11**
Ursus maritimus, 219–22
Urtica dioica, 87
Usher, M. B., 45, 49, 58, 87, 96, 125, 155–6, 159, 161–2, 164–5, 241, 257, 277, 288, 309, 382–3, **5.6**, **5.8**

Van Dyne, G. M., 125, 383
Veronica fruticans, 29, **1.8**
Vetch, 280
Viburnum opulus, *4.9*
Vincent, J., 214, 219, 222, 245, 247, 376
Violet, 70

Waggoner, P. E., 88–92, 382, 383, *2.8*
Walters, M., 13, 29, 381, **1.8**
Warburg, E. F., 238, 374, *7.6*
Water flea, 159–60, 162–3
Waterston, G., 14, 254–5, 374
Watson, A., 279, 383
Watt, A. S., 51, 78, 93, 197, 241, 383, **1.14**, **1.15**, **5.22**
Watt, K. E. F., 109, 122, 125, 126, 158, 166, 167–9, 172, 177, 182, 183, 206, 225–6, 383–4, *5.3*, **5.10**, **5.11**
Wayre, P., 248, 384
Weasel, 11
Webb, L. J., 91, 384
Wells, T. C. E., 215–18, 384, *7.1*, *7.2*, **7.1**
Whales,
  blue, 177–82, **5.14**
  fin, 177–82, **5.14**, **5.15**
Wharram Quarry Nature Reserve, 232–44, *7.6*, *7.7*, *7.8*, **7.5**, **7.6**, **7.7**, **7.8**
Whittaker, R. H., 110, 112, 384
Wigeon, *9.3*
Wild garlic, 78
Wildfowling, 277–8, *9.3*
Wilks, H. M., 250, 384
Williams, W. T., 91, 135–40, 235–8, 384, *4.4*, *4.5*, *4.6*, **4.2**, **4.3**
Williamson, M. H., xii, 78, 79, 148, 151, 153, 156–7, 189, 202, 229, 384–5, **5.2**
Willow, 8, 86, *4.3*, **I.3**
Wilson, E. O., 21, 23–9, 100, 379, 381, 385, *1.1*, **1.3**, **1.5**, **1.6**, **1.7**
Wishart, D., 88, 374
Wolds, 49, 233, *1.4*, **1.13**
Wood, S. E., 290–2, 382, *9.6*
Woodland fragmentation, 20–1, **1.1**
Woodwalton Fen National Nature Reserve, 227, **7.4**, Plate 6
Wurster, C. F., 210, 385

Xenomyrmex floridanus, 27, *1.1*, **1.6**

Yarranton, G. A., 47, 385
Yates, F., 370, 376
Yorkshire Naturalists' Trust, xiii, 232, 261, 263, 287, *9.4*

Zerna erecta, 251
Zoological gardens, 31, 246–50